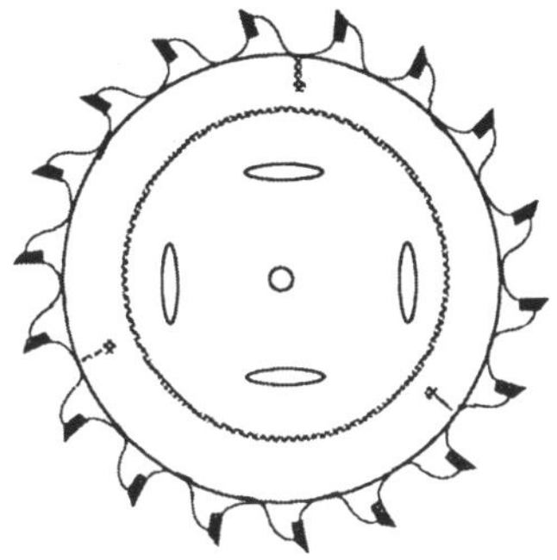

Chisels on a Wheel:

A Comprehensive Reference to Modern Woodworking Tools and Materials

by

Jim Effner

P.O. Box 8623
Ann Arbor, MI 48107

Library of Congress Card Catalog No. 92-64012

ISBN: 0-911168-82-6

Acknowledgments

Many people I have encountered over the past several years have agreed on the need for a text that explains the basic wood-cutting tool processes. A number have contributed to my writing of this book, and I wish to thank the following for their support and enthusiasm and for sharing their wisdom. I owe each of them a considerable debt of appreciation: Dr. Henry Huber, Dr. Otto Suchland, Dr. Alan Sliker, Dr. George Ferns, and Dr. Cass Gentry at Michigan State University; Rich Peters and Paul W. Jones at Steelcase Corp.; Dr. Gary Schajer and Norm Orlander at Weyerhaeuser Corp.; Dan Mote at the University of California at Berkeley; Dr. Carl Eckelman at Purdue University; Orville Wood at Woodcutters Inc.; Sandy Mullin at Barr-Mullin; Dr. Ed Clark and Dr. Bob Gilmore at North Carolina State University; Dr. Ryszard Szymani; Herbert Moitzi, Eugene Trittler, Herman Engert, Heinz Kisslinger, Wolfgang Grundler, and Roland Fuerst at Leitz; Carsten Jespersen at K-Scandia; Edgar Sauder at Sauder Woodworking; Martin Oertli at Oertli Inc.; Eb Kirbach and Tom Bonac at Forintek-Canada; Harold "Sandy" Stewart at Mississippi State University; Karin Deutschler at GUHDO—USA; Holger Nagel at HoWeMa—USA; Scott Erickson and Hans-Peter Kunz at Leuco—USA; Charles Milbury at Stehle—USA; Hugh Reynolds at USDA-U.S. Forest Service; and the many other individuals who have contributed to this book in various small and large ways.

Individuals at the following professional trade magazines and organizations have contributed significantly with both information and enthusiastic support: Richard Rea at *Furniture Wood Digest*; Sandy Berliner and Bruce Plantz at *Furniture Design and Manufacturing*; James Powell at *Modern Woodworking;* Harry Urban at *Wood and Wood Products*; B. Leslie Hart at *Kitchen and Bath Business*; Ray Helmers at Helmers' Communications Expositions and Trade Shows; John Zinn at International Woodworking Fair; High Point Woodworking and Furniture Supply Show; Ligna Machinery Show in Hannover, Germany; InterZum Furniture Supply Fair in Cologne, Germany; Inter-Bimal/Acimall Woodworking Supply Fair & Show in Milan, Italy; Institute of Wood Technology in Rosenheim, Germany.

Preface

In the several years of preparation of this text, a major concern has been the intended audience. Of course, I believe that the information is important to everyone in woodworking: student, hobbyist, artisan, designer, and industrial worker. As I have attempted to compose and structure this work, however, an interesting word returned to me: *epistemology*, the study of the origin, nature, and limits of knowledge. What must be said now is that after the 20 or so years I have studied cutting tools, I surely do not know all there is to know, especially given their diversity and complexity. This is neither an apology nor an admission of any shortcoming. Cutting tools, despite the fact that many of them are common to all of us, are very complex when we consider the metals the bodies are made of, the alloyed knife materials, the heat treatment, the geometry of all of the cutting and relieved surfaces and structural features, and the hundreds of materials that we cut in the woodworking industry. Domestic tree species alone number over a hundred.

What I hope this text will do, then, is explain what the tools are and how they function, and condense and clarify much of this complex information. Throughout, every attempt will be made to address the needs of most tool users.

I also hope that every reader can in some way profit from what is provided here. Anyone who has the occasion to purchase or use cutting tools should make the effort to learn as much as possible about them regarding safety, cost, performance, life, limitations, quality, machine characteristics, and more. Cutting tools are at the same time costly, dangerous, complex, and a necessity. Are you getting what you paid for? Will the tool do what you want?

As regards epistemology (at least the nature of it), I suspect that knowledge is more a process than a collection of facts and figures. Knowledge is also an understanding that there are relationships among all areas of study: physics as it relates to chemistry, mechanics as it relates to physics, and so on. Today, advances of technology in all areas of study are accelerating. By way of explaining this technological advance, it is interesting to note that there are now more scientists and engineers involved in active research than in all of the rest of recorded history.

In woodworking as elsewhere, we have to avail ourselves of the technology as it is brought forward. Too often, industrial advances seem to be disseminated too slowly. For instance, many of the cutting tools illustrated in this text have been common for several years, and still many woodworkers are not aware of the advantages that they will provide. What I would hope for every reader is to be able to better comprehend the function of the cutting tools that you need in the pursuit of your work. The cutting tools that we purchase need to be safe, well made, and capable of performing our work precisely.

In this text, you will not find how to build cabinets, case goods, or furniture, at least not their design and assembly. You will no doubt be better prepared to deal with the stuff they're made of, though.

I wish you each best regards, and great success.

Jim Effner
July 1992

Contents

Introduction

For thousands of years, we have harvested trees for our benefit. Wood has given us shelter, kept us warm, and driven our industrial revolution. Trees have provided us furniture and eating utensils, and in many ways, wood has shown us its great beauty and strength. We have expressed our creativity in wooden art works including numerous unique and still functional machines. Our first printing press, the Gutenberg, was made entirely of wood—even the type. Today, we still cherish wood and are searching for yet more new wood products and uses for this versatile material. A listing of products made in part or entirely of wood would require several texts larger than this one.

Recent years have brought unique demands on the cutting tools we have designed to work wood into desired products. As the availability of clear, solid wood has diminished, we have developed new man-made sheet panels and larger blocks made of wood fibers, chips, and particles. The new (in the last 45 years) panel materials have many of the desired characteristics of wood solids and a few of their own that require unique consideration. The net effect of bringing more materials into woodworking industries has been the creation of a considerable collection of cutting tools not seen before, as well as new cutter knife materials, new tool geometry, and in some instances entirely new machine processes.

Toolmakers worldwide in recent years have had to expand their manufacturing abilities greatly to keep pace with the accelerating change taking place about them; not all have survived. Tool research and design have to be ongoing processes. If toolmakers are not willing to invest the time and cost for such activities, their products will soon be obsolete. Like all technologies, cutting tools are evolving in enormous leaps; beyond basic hand tools, much of what is in use today did not exist 35 years ago.

A number of tool manufacturing companies have developed marketing strategies to extend and protect their products. A very basic premise of these strategies is to examine the needs of tool users and new materials coming into the marketplace. Very few tool manufacturers want or have the capacity to address all market segments. Cutting tools, even the most basic, are designed to perform within a well defined task range. It is not that one is better or worse than another, but rather that the tool user has to understand his own needs and make the choice. There are three broad market areas that cutting tool manufacturers identify:

1. Hobby tool users who do not earn their livelihood from woodworking. Most hobby tools can be well maintained and sharpened, but many are disposable.
2. Craft tool users earn at least part of their livelihoods with cutting tools. Quality, precision, size, and cost are the principal differences between their tools and those for hobbyists. A cutting tool for the hobbyist may cost $30 and have a life expectancy of 20 hours. For the craftsman, a similar tool may cost $90, but the life will be 200 hours. For both, the tool selected probably was correct and appropriate despite the considerable difference in performance characteristics and cost. In tool selection and use, users must understand their own specific needs.
3. Industrial tools are without doubt the largest, most complex, and most costly ones made. The machines and processes are often

similarly large and complex, with, in some instances, more than 30 cutting tools at work in a single machine at the same time. For the industrial tool user, initial cost is probably not the first consideration. Beyond any doubt, the most interesting cutting tools are those developed for industry in recent years.

Technology that evolves in industry gradually finds its way into other areas, and this text will illustrate that repeatedly. The tool manufacturers have, in many instances, reduced costs of very refined cutters through methods of mass production, making them more available to all of us. Also feeding the mainstream of technology is the competition of the worldwide marketplace.

A thorough examination of current technology in wood cutting tools will have to include at least some review of the machines and machine processes for which many of the tools were developed. Like tools, woodworking machines are very quickly undergoing a transformation of their own. In adapting computer numerical controls to every conceivable machine movement, the processes are finding speed and precision beyond the imagination just 20 years ago. Even hand-held routers have solid-state controllers that sense motor load and can seek optimum horsepower and rpm. Hand-held scroll saws have variable speed and inclination angles controlled by a small electronic brain. These are but small examples of the explosion of technology that is just gaining momentum. Overall, the cutting tools in the market today are safer, faster, more precise, and longer lasting than ever. That idea will, perhaps, be met with initial opposition that can be easily dispelled.

We can choose to utilize the technology that surrounds us, or we can choose to do otherwise. In the real world, this is really not a choice, though. I do own a fairly good typewriter and a still serviceable drafting table, and I could do a number of math exercises on my old slide-rule; but I am not likely to give up my Macintosh IIcx. I, for one, would not have believed that 25 years after my college days I would be seeking out advanced classes on topics like CIM, MacDraw II, word processing, ceramics in metal technology, advanced CAD/CAM design theory, and MGM Station CAD. With all that in mind, I sincerely hope that with this text I can bring a better comprehension of the technology brought to woodworking in recent years.

Chapter 1

Basic Concepts of Wood Cutting Tools

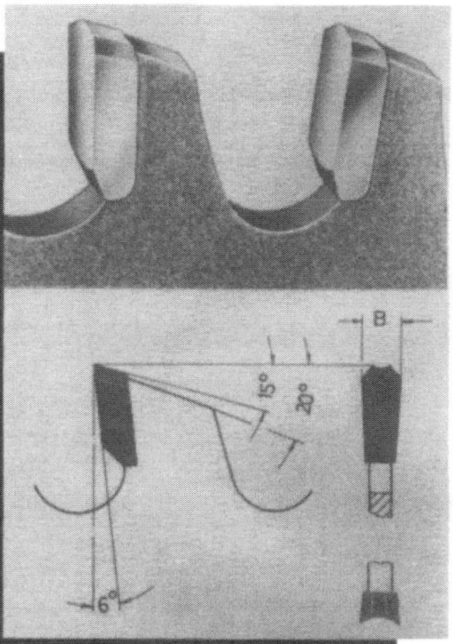

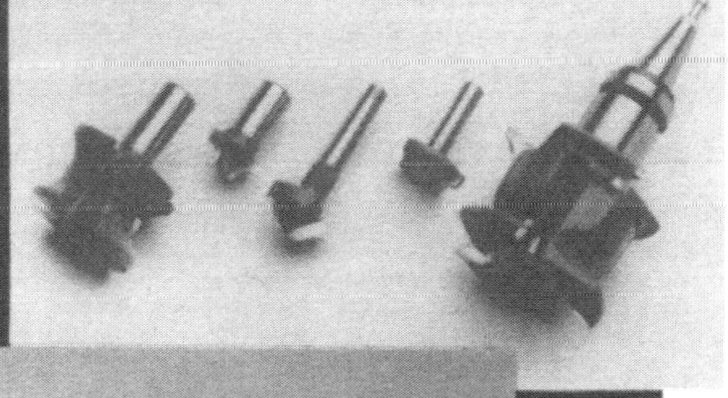

Photos courtesy of Guhdo USA, Inc., Leitz Tooling Systems, Inc., and Oertli Woodworking Tools, Inc.

Basic Concepts of Wood Cutting Tools

Any tool extends what can be accomplished with one's hand. Tools for power equipment similarly extend what can be accomplished with hand-held, hand-driven tools. Hand tools, of course, have a history back beyond what is recorded; people were using tools of all types long before development of any means to document their activities. The cutting tools for power equipment in use today resulted from development and study over the past 175 years. The first cutting tools to extend hand power were for sawing, followed closely by those for drilling. These two kinds of tools still account for most shaping of wood into practical and attractive products. There are, of course, numerous other tools at our disposal, but we still use more saw blades and drills than any others.

All industries have progressed along a time frame aligned to the development of different forms of usable power. The progression of power development moved from animals, to water, to steam, and later to electricity. This development of safe, dependable power that could be easily moved about brought about the Industrial Revolution, which continues today.

The spread and the acceptance of electricity greatly accelerated tool development. Cutting tools have always been developed in unison with particular machine processes, and, as electric motors became commonplace, the numbers and types of machine processes greatly expanded. In the earliest instances, toolmakers were probably also machine designers and builders. Today, however, the research, design, and manufacture of cutting tools is a large, well developed industry within itself, but there remains a close relationship between the manufacturers of cutting tools and the manufacturers of the machines on which the tools are used. Another more recent set of relationships exists among toolmakers, the producers of some of the exotic materials that tools are made of (tungsten carbide, polycrystalline diamond, etc.), and the developers of unusual new man-made materials that require special tools (e.g., wood-based panel materials).

In recent years, the rate of development in the woodworking industry and its consequent need for special cutting tools has been influenced by several factors:

1. Development of more man-made materials;
2. Need for larger, faster machines and processes;
3. Worldwide competition on several levels;
4. Need to make better use of all resources, including:
   a. raw materials,
   b. labor utilization and costs, and
   c. time as it relates to productivity and costs.

Today, toolmakers worldwide identify and direct manufacturing efforts at a particular segment of the market. That is, not all tool manufacturers have the capacity or desire to produce every type of tool. Given the enormous range of tools available for woodworking machines and materials, specialization is needed.

All cutting tools manufactured today attempt to provide geometry and design to satisfy the broadest range of applications and materials possible. For the end user, this can be a mixed blessing. General-purpose cutting tools can be mass produced at a lower relative cost, but if the specific application or material is too far from the design parameters, the tool may not function well at all. Overall, every cutting tool has a fairly well defined range of design and capability. There is not really a "do everything cutter" that will last indefinitely and cost nothing.

Kinds of Cutting Tools

Cutting tool manufacturers, American, European, and Asian, competing for a share of the market, have aimed efforts at specific users—or more correctly, at the particular work load normally placed on tools. They have generally identified and produced cutting tools for three broad and often overlapping market areas:

1. *Hobby work tools* are designed primarily for small portable and stationary machines. They are intended to be in use only intermittently for short periods of time. The tool knife materials are usually designed to cut only natural woods using machines of low horsepower. They have less complex tool geometry and a fairly wide selection of common shapes and sizes for use by woodworkers who do not make their livelihood with these tools.

2. *Craft work tools* are designed for medium-sized stationary and the largest portable machines. These tools are designed for extensive use, such as by cabinet makers, building tradespeople, and small manufacturers. Craft tool users make their livelihoods with their tools, and cost efficiency is a factor in their purchase. At the craft level, the materials cut and shaped are as diverse as in industry. Hence, the tools have to be as well designed and of similar materials.

3. *Industrial work tools* are for medium and large stationary machines. They cut and shape a very diverse range of materials and function in the most demanding applications. The processes are directed at efficiencies not necessary with other wood cutting tools. Cost justifications are a primary consideration, driven by the reality of having to produce a product at the lowest incurred cost. Cutting tools made for industry are geometrically complex, have knives made of the most modern materials, and are efficient as never before.

Each type of cutting tool is designed to perform within a fairly well defined task range, that is, for given processes and for materials with certain physical properties. Hobby tools may differ from craft and industrial types in several ways beyond the obvious—cost. Hobby tools are not, for instance, expected to

1. Run extensively, hours at a time per set up;
2. Be sharpened repeatedly;
3. Withstand high horsepower and torque loads;
4. Be designed to cut specific materials;
5. Be examined for cost effectiveness;
6. Hold stringent dimensional tolerances;
7. Be made of unusual, costly materials;
8. Conform to extreme standards of balance and detail.

Craft and industrial tools are designed to support manufacturing processes not encountered by the hobby worker. None of this implies that hobby tools are in any way inferior; they are just different by design, structure, and intended use.

Most tool manufacturers do not attempt to address all three market areas, though there are obvious overlaps. Where similar tools, such as drills, saws, and router bits, apply to all areas, the craft and industrial versions probably will not look much like anything intended for the hobby market. (As we describe the construction of all of the tool families, the reasons will become apparent.)

The overlap between craft and industrial tools is minimal and most often relates to machine size or horsepower, where most design criteria are the same. A three-man cabinet shop may use tooling nearly identical to that in a cabinet factory with three hundred employees, but the latter will surely have larger, more refined machinery. The development of cutting tool technology has followed the needs of industry. The small craft shop and hobbyist benefit from access to these advances, however costly.

To begin describing cutting tools and how they function, we need a common language. In explaining geometric features of wood cutters, we have to include some of the buzzwords that have evolved, for instance, the "rake angle" of a cutting tool, a commonly accepted term among woodworkers (at least here in the midwest United States) that identifies the "cutting angle." To someone who prefers the physics terminology, the same feature is "angle of incidence."

Geometric Features of Cutting Tools

The most common wood cutting tool is the circular saw blade. (It is also the easiest to visualize.) We can use a saw as a basis for identifying and explaining the most important geometric features of all cutters. The illustrations here use a saw blade with tungsten carbide tips that are shaped to achieve specific characteristics. We will later illustrate the earliest saw blades, made entirely of tool steels and high-speed steels. Unlike current tipped saws, the first saw blades were manufactured quite differently, with the teeth themselves distorted to achieve particular clearance angles. Later, as we examine characteristics of different tools, we can add those features to complete the explanations.

At the periphery of all cutting tools are four primary geometric features. (See Illus. 1-1, 1-1a, 1-2, 1-3, and 1-4.)

1. The *cutting angle* is defined by the leading face surface of the tool knife. The outermost edge of this angle also locates and identifies the cutting circle, which is always the largest diameter of any cutting tool. There may be a need to identify other diameters to describe totally the function of more complex tools, but

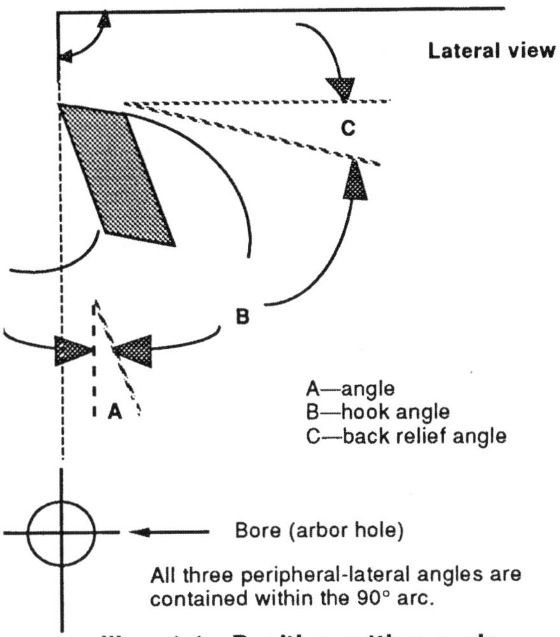

Illus. 1-1—Positive cutting angle

the cutting circle is always the largest diameter. It is important to understand where and how the cutting, body, and clearance angles begin and end. To define the laterally viewed angles (see Illus. 1-1), we draw a line through the center of the tool and intersect that with another line at (tangent to) the periphery. The 90° angle formed by this intersection defines the cutting angle, hook angle, and back relief angle. The sum of these three, then, totals 90°. When the three angles are constrained within the 90° arc, the cutting angle is *positive*. A negative cutting angle does not fall within the 90° arc (see Illus. 1-1a). A tool knife that falls exactly on the vertical line has a cutting angle of 0 (zero).

The cutting angle is a design parameter specified for a particular application, process, or material. Further, the cutting angle is in part only relative, in that we can often manipulate the machine process to alter the true angle of incidence with the material being cut, for example by raising or lowering the height of the saw blade. Cutting angles for most wood and wood-based materials will vary between -6° and +30°. As we more completely describe tool types and functions, we will explain further the effects of cutting angle. There is an interdependent relationship among all geometric features of every tool that must be understood.

2. *Hook angle* is the distance between the cutting angle and the back (relief) clearance angle. The size varies with the design of the other two angles. The hook angle is essentially the tool post that the knife material is attached to and consequently is an important design feature in calculating work load. Hook angle is a distance into which we design the tool knife on the post that supports it. This design feature absorbs all the impact, force, and shock of the work, in most instances an immense task.

3. *Lateral clearance angle* is viewed looking down toward the top of a saw blade tooth with the face of the tooth/knife toward you, as in Illustration 1-2. This

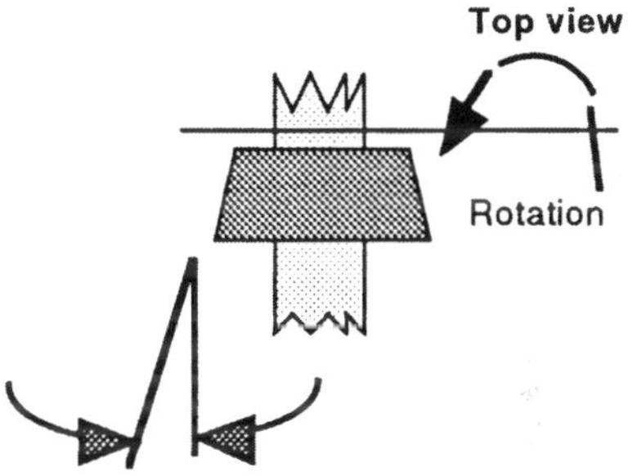

Illus. 1-2—Lateral clearance angle

geometric feature allows the tool knife to pass through the material being cut with as little contact as possible. Only the leading edge is in contact during the cutting process, with the remainder of the knife angled away from the now cut surface. The illustration is exaggerated for clarity; the angle, at least for saw blades, is seldom more than 6°. (We will detail this feature more fully for other tools later.) If no lateral clearance were provided, cutting tools would very quickly overheat. In the case of a saw blade, the tool would probably become unstable and wobble.

Clearance angles are, like other features, design characteristics that may vary considerably for different materials or machine processes. Cutting, for instance, very dense or wet materials requires a larger lateral clearance angle than cutting dry softwood such as white pine. This feature on any tool has to be designed and manufactured to be accurate and completely functional (i.e., with the correct angle, exactly centered).

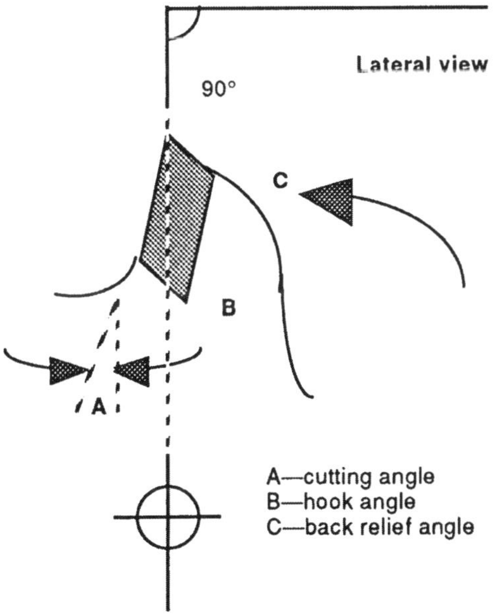

Illus. 1-1a—Negative cutting angle

4. *Lateral relief to the periphery* is viewed looking

directly at the face of a saw blade tooth/knife. (See Illus. 1-3.) This angle, like lateral relief, is provided to allow the tool to pass through the material being cut with as little contact as possible. The nomenclature may be

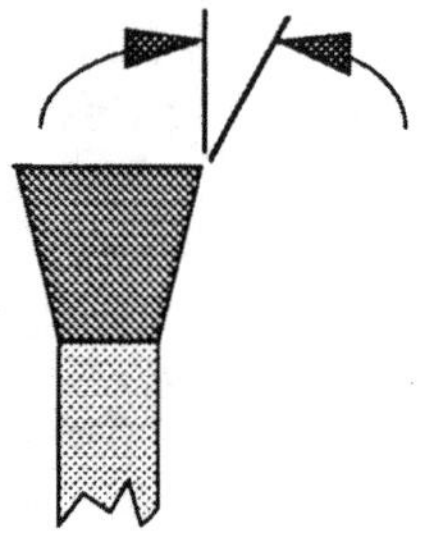

Illus. 1-3—Lateral relief to the periphery

more correctly "from the periphery." The design allows only the leading edge of the tool to contact the material during the cutting process. The tool knife tip is wider at the periphery and narrower toward the tool body.

Primarily, this relief angle prevents friction heat from building in the tool. The cutting tool would overheat in a matter of seconds if the lateral portions (sides) were allowed to be in constant and complete contact with the material being cut. Even with wood chips and sawdust carrying away generated heat, measures must be taken to keep the tool from overheating. Instances when the angle of lateral relief to the periphery can be varied will be explained later. This angle is never large, due mostly to the relatively small distances between the tool body and the tool knife. As will be illustrated often, very small changes to the features of a cutting tool have very dramatic effects.

A fifth angle should be noted here. *Shear angle*, when present, is viewed looking directly down on the tool knife/tooth. (Illus. 1-4 shows the tool, facing toward the bottom of the drawing.) Shear angle has little effect on heat dissipation or reduction of friction, but changing it can greatly affect the performance of the

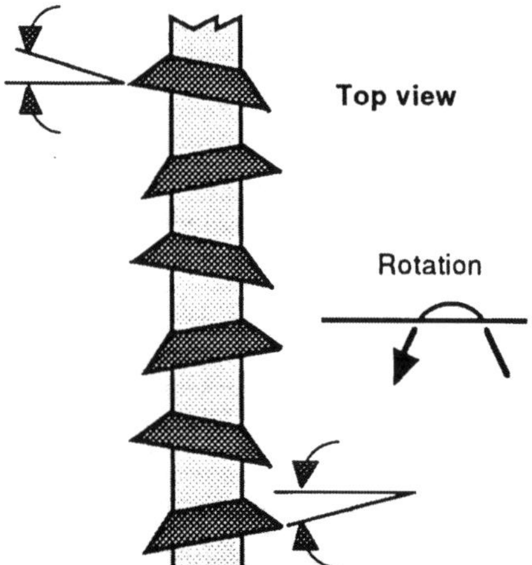

Illus. 1-4—Shear angle (or decline to axle)

cutting tool. This is particularly true in cutting any material that has a grain or layered structure. In cutting natural wood with a tool having an alternate shear (as illustrated), the knife arrangement makes the cutting process more mechanically harmonic (smooth), often requiring less horsepower and improving finish quality.

What is important to remember about shear angle is that as the knife tip makes contact with the material, the leading edge (side to side) enters the work progressively, that is, not all at once, as would be the case with a tool having a squared edge. The knife following repeats the process, but shears from the opposite direction. This tends to make the entire process easier. There are also instances when we want to have every cutting tooth shear angled in the same direction. Summarily, shear is a performance characteristic designed for particular processes or materials.

In Illustration 1-5, the three views indicate the basic geometric features common to all cutting tools.

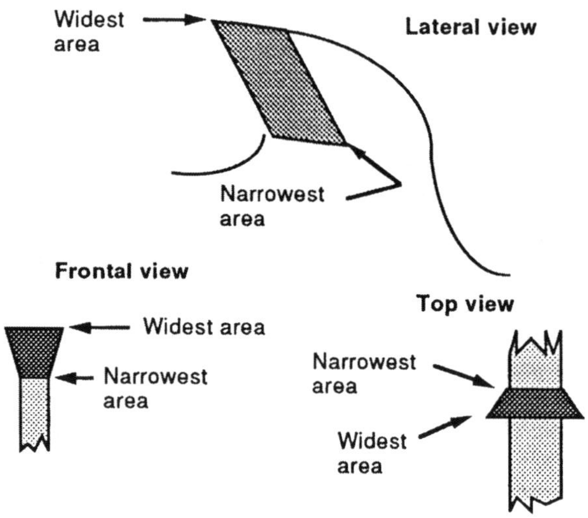

Illus. 1-5—Cutting tool geometry

The leading edges of all tool knives are widest at the periphery and narrowest at the lower end of the tooth segment. This is true of tools tipped with tungsten carbide (or another hard metal), as well as those made of a solid tool steel alloy. The tool knife also narrows in the direction away from the cutting-edge face, and thus is effectively relieved in two directions simultaneously. Illustration 1-5a shows the relative size difference between the width of a saw blade tooth at its widest and the saw blade body that it is attached to. The lateral clearance distance indicated is proportional to that found on all similar saw blades. With such a small distance to provide the correct angularity of relief, precision and attention to detail are required.

For all cutting tools, the relationship between accurate details and performance characteristics is paramount. Even slight and subtle change to any geometric

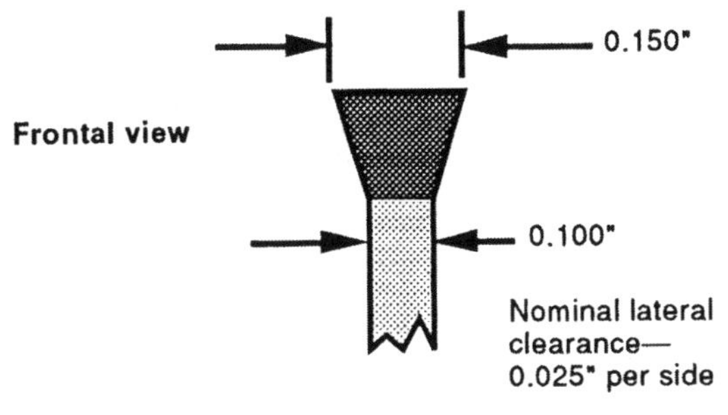

Illus. 1-5a—Saw tooth dimensions

feature can have dramatic effects, as will be illustrated as we examine the performance of cutting tools as they function when in motion.

Cutting Tool Performance Characteristics

To begin an examination of cutting tool performance characteristics, three observations may prove helpful:

- Wood cutting tools travel at enormous speeds, greater than nearly anything in our normal work environment.
- To comprehend the work being done, a large number of knife/tooth marks have to be examined collectively. The work of a saw blade or router bit, for instance, may involve 1,000 marks per inch of travel.
- The work done in the cut is an abstraction (effect) of the cutter itself; that is, if we examine a complex, profiled shape, the cutter is like a photographic negative of it. To design, produce, and control this apparent phenomenon takes up most of the toolmaker's time, especially when two profiles are joined together, as in a complex window, door, or cabinet frame.

The first characteristic of wood cutting-tool performance is the enormous speed, at least as defined in relative terms. A 12" saw blade will normally turn at about 4,000 rpm and has 72 teeth. Let's examine what is taking place with just that basic information. (Note: At the end of this chapter we will provide the tool performance formulas for use with metric dimensions.)

We need to first solve for *rim speed*, that is, the velocity of blade at its outermost diameter (symbol: V, expressed in ft./sec.). The formula is:

$$V = \frac{\Pi D}{12} \times \frac{rpm}{60}$$

(3.14159 × diameter × rpm)

We divide ΠD by 12 to convert the circumference into feet, and the rpm by 60 to place the figure in one second. Consequently,

$$3.14159 \times 12 = \frac{37.699''}{12''/\text{ft.}} = 3.14'$$

$$3.14' \times 4{,}000 \text{ rpm} = \frac{12560'/\text{min.}}{60 \text{ sec./min.}} = 209'/\text{sec.}$$

The periphery of a saw blade traveling at 209'/sec. does not actually seem fast until we compare it to other moving objects in our environment. We most commonly express speed in miles/hour. One mile/hour is 1.466"/sec. So the saw blade is traveling at 142 miles/hour. Here are a couple of other comparisons: an automobile, with 26" tires turning at 4,000 rpm would speed down the road at 309 miles per hour. A semi-tractor trailer with 42" tires turning at the same 4,000 rpm would travel at exactly 500 miles/hour. Actually, when a semi-tractor passes you at 70 miles/hour, the wheels are turning at only 560 rpm. In our examination of cutting tools, we should become comfortable with making comparisons like these to understand completely the complex relationships that define what our cutters are doing at any given moment.

Rim speed, then, is the velocity of the blade at the largest diameter, measured at the periphery. We make use of tools with different diameters that turn at different speeds (rpm) for varied purposes. As redundant as all of this may appear, we often overlook pertinent and obvious facts when making decisions regarding tool performance. If the saw machine has a fixed-speed motor arbor, our selection of saw blades has to be guided within that constraint, so we must approach the number of other variables available to us with some thought.

For instance, most saw machines can effectively use saw blades of a given range of diameters. Most often the machine is identified (model, series, or whatever) by the largest possible diameter tool—a 12" table saw, a 10" table saw, and so on. The point is that we can make choices. Selecting tools that have only minor differences in diameter or number of teeth/knives can make a remarkable difference.

The saw blade mentioned earlier, having 72 teeth and traveling at 4,000 rpm, is remarkable in the rate at which the teeth enter the material being cut:

$$72 \times 4{,}000 = 288{,}000/\text{min.}$$

OR

4,800 times each second a saw tooth is entering the material being cut. Given such speed, we can begin to understand that even minor changes to the smallest detail can and do have dramatic effects. (See Illus. 1-6.)

Saw blades are like other cutting tools in rim speed and the rate at which the teeth/knives are brought to the material being cut. Router tools operate in the rpm range of 16,000 to 25,000, shaper tools from 3,500 to 10,000, and so on. The important point to remember is that rim speed is a relationship of diameter to rpm. The number of teeth/knives on a given cutter (and their rate of pro-

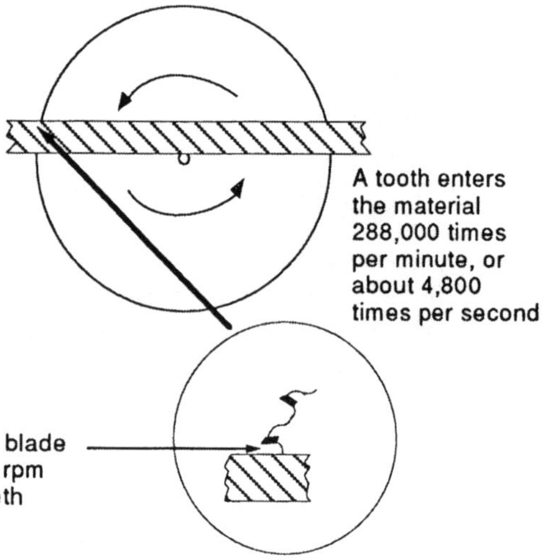

Illus. 1-6—Effect of rim speed

gression) relates more to feed rate and measurements of chip size, which we will address later in this chapter.

There is often a significant advantage in using a larger diameter cutting tool rather than a smaller one. Later calculations will illustrate this advantage in terms of performance characteristics such as feed rate, finish quality, and horsepower consumption. (See Illus. 1-7.)

The next performance characteristic is *tooth/knife progression*, that is, how many teeth/knives make contact with the material being cut over a particular distance—expressed as number of tooth/knife marks per 1/1000" (or length of cutter mark). The formula:

$$Sz = \frac{U \times 12}{N \times Z} = \text{progression per 1/1000"}$$

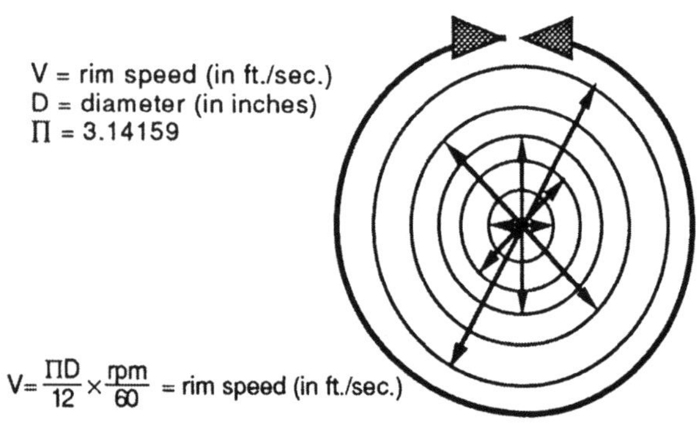

Illus. 1-7—Effect of tool diameter on rim speed

The symbols are:

Sz = tooth progression expressed as 1/1000" (also called length of cutter mark);
U = feed rate expressed as feet/min.;
N = rpm;
Z = number of teeth/knives on a cutting tool.

The calculation:

a. Multiplying the feed rate by 12 changes the number from feet/min. to inches/min.;
b. Multiplying rpm by the number of teeth/knives yields the number that enter the material being cut each minute;
c. Dividing the lower half of the equation into the upper tells us how far the material traveled while in contact with a single cutter tooth/knife and, consequently, how many marks per inch this has created with another simple division.

Staying with the 12" saw with 72 teeth turning at 4,000 rpm, to complete the tooth progression formula, we need to add the feed rate—say 45'/min.:

$$45'/\text{min.} \times 12"/\text{ft.} = 540"/\text{min.}$$

$$4{,}000 \text{ rpm} \times 72 \text{ (teeth)} = 288{,}000$$

$$540/288{,}000 = 0.001875"$$

Each saw tooth removes 0.001875" of material. Dividing 1 by 0.001875 tells us that there are 533 saw tooth marks per inch in this particular cutting process. This tooth progression exercise for a saw blade is only an example and would not be appropriate for every situation. Such a small tooth progression would most often apply to cutting thin, dense material.

By altering the number of teeth on the tool, the feed rate, or rpm, we can achieve a desired outcome. For instance, if we know that a certain number of tooth marks per inch will give the finish quality needed, and we know the rpm of the machine to be used, we can vary the feed rate and number of teeth/knives on the cutting tool to achieve the finish we want. And there is a simple formula to do this very thing.

The formula to select feed rate requires that we already know the tooth progression, expressed as progression per 1/1000", not as cutter marks per inch.

$$U = \frac{Sz \times Z \times N}{12} = \text{ft./min.}$$

U = feed rate, in feet/min.;
Sz = tooth/knife progression in 1/1000";
Z = the number of cutter teeth/knives;
N = rpm.

With the saw blade, as before, having 72 teeth and turning at 4,000 rpm, but now working on material that cuts very well at a tooth progression of 0.0045" (rather than the earlier 0.001875"), we calculate the feed rate to achieve it:

$$\frac{0.0045 \times 72 \times 4000}{12} = 108'/\text{min.}$$

The 108'/min. feed rate is hypothetical for our exercise, but it is not unrealistic. Some materials can be cut with great precision at three times that rate and more. The point is that we woodworkers have variables that we can alter and direct toward what we need and want cutting tools to do.

Another tool performance formula that is useful, though not as often needed, is *cutter mark depth.* It is used when feed rates are fast. In industry, this serves to calculate the amount of stock removal necessary during subsequent sanding or planing.

As a cutting tool turns in a circular motion about a given radius, it leaves the cut surface with shallow scalloped marks. (See Illus. 1-8.) The depth of each mark in most woodworking is too small to cause any great concern. Only when there are too few cutter marks per inch do they even become visible. Altering the feed rate, number of cutting teeth/knives, tool diameter, and rpm will affect the depth of each cutter mark, though the normal range for each element in the relationship is broad.

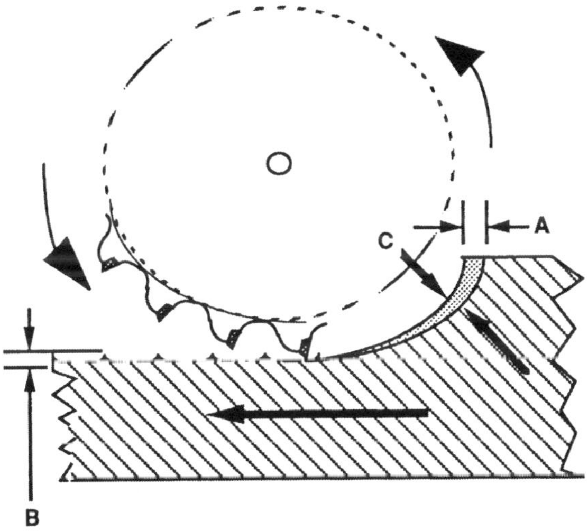

A—(Sz) Tooth/knife progression: chip formation and travel during conventional abrasive cutting.
B—(t) Cutter mark depth: effects of the cutting process as chips are formed and removed.
C—(hm) Mean chip thickness: the average chip formed by each tooth/knife of the cutter.

Illus. 1-8—Tool performance characteristics

The depth of the cutter mark is a chordal section of the tool knife path as it passes through one revolution of travel in relation to the forward movement of the material being cut. With the following formula, it is possible to select the desired cutter mark depth for a particular material or process and design tools and machine functions to satisfy it. This is of particular use for man-made materials:

$$t = \frac{Sz^2}{4 \times D} = \text{depth of radius as } 1/1000"$$

t = cutter mark depth;
Sz = tooth/knife progression;
D = diameter.

The last basic tool performance formula is a means to calculate the average size of the chip mass gathered by individual cutting tool teeth/knives. *Mean chip thickness* is a relationship of tooth/knife progression, depth of cut, and the diameter of the cutting tool used. The product of this calculation, when examined with the other performance formulas will give a complete view of what any cutting tool is doing at any given moment. While chip thickness will not be of first interest to woodworkers, it will give them another way to diagnose tool operation processes. Of greater importance to the toolmaker, perhaps, chip thickness in a given process will direct a part of the design criteria, namely, those characteristics that determine knife progression and tool geometry:

$$hm = Sz\sqrt{\frac{a}{D}} = \text{average thickness as } 1/1000"$$

hm = mean chip thickness;
Sz = tooth/knife progression;
a = depth of cut;
D = diameter.

For the woodworker, chip thickness has more to do with the cutting characteristics of different wood species. For instance, cutting very dense hardwood will require forming smaller chips than will cutting softer, less dense wood. What should be stressed are, again, the interrelationships within the performance formulas and that the woodworker should learn how to control the variables in seeking optimum conditions for specific processes.

As can be seen from examination of the five basic tool performance calculations, there is an interdependent relationship of tool diameter, rpm, feed rate, and the number of cutting teeth/knives. With the careful use of the formulas, we can remove the trial and error of solving problems related to most tool use. Even if, for instance, the machines that we use have single-speed motor spindles, there are still other variables that we can manipulate to find the ideal feed rate, tooth progression, and rim speed.

A summary of these formulas follows:

1. *Tooth progression* is the distance relationship of each cutting tooth/knife and the

"travel" of the material being cut, expressed as length of cutter mark. The finish quality of most cutting tool processes is most often expressed as a given number of tooth/knife marks per inch. (This will be addressed later in the section on profile cutters.)

2. *Feed speed* is the calculated rate of moving the material to be cut to a cutting tool to achieve a desired outcome. It is a relationship of tooth progression, rpm, and the number of teeth/knives on the cutting tool, expressed in time and distance. In seeking the ideal feed speed, the criteria usually are finish quality and time.
3. *Rim speed* is peripheral tool velocity and is a relationship of circumference and rpm. Rim speed is the first characteristic that we examine to determine all other performance criteria. Tool diameter reveals how many teeth/knives are possible, while rpm tells us what our choices are to seek feed speed and tooth progression.
4. *Cutter mark depth*, though not often sought, is an important criterion in a more in-depth tool study. The relationship is between tool diameter and tooth progression, with finish quality and product net size the outcome.
5. *Mean chip thickness* is the relationship among tooth progression, depth of cut, and tool diameter. This, like cutter mark depth, is a criterion that is not often of great importance to the woodworker in that tool design usually provides what is needed. Chip thickness will be of importance when cutting unusual materials that may extend the design range of tools; for instance, increasing feed speeds, changing the rpm (in relation to the number of teeth/knives), and so on.

Later in the text, we will list suggested rim speed, cutting angles, and tooth progression for most woodworking materials. Even with tables and charts, there is no possible way to provide all information for every application. Cutting natural solid woods is a diminishing part of what the woodworker has to deal with today.

How tools form chips during the cutting process is usually of more importance to the toolmaker/designer than to the woodworker. Woodworkers most often simply check sawdust and chip size visually as a gauge of tool performance, and that is usually an accurate estimation based on the history of a particular tool cutting a specific material—with some precautions.

For instance, the same saw blade used to cut several different materials in differing orientations will produce sawdust and chips that appear radically different from one another. The list of conditions that affect chip size would include:

1. The make up of the material—solid natural wood, plywood, particleboard, hardboard, and so on;
2. Orientation of the cutting process—ripping, cross cutting, mitering, jointing, etc.;
3. Relative strength of the material—grain makeup (if present), density, hardness, and the other measures of strength;
4. Moisture content of the material;
5. Condition of the tool—sharpness, balance, and concentricity;
6. Tool design features—cutting and relief angles, gullet size, and shape and number of cutting teeth/knives;
7. Process characteristics—feed speed, rim speed, etc.;
8. Machine characteristics—vibration, tool or stock clamping and holding method;
9. Conditions of the process—adequate methods and means to remove the waste offal, chips, and sawdust;
10. Climatic conditions—humidity, heat, and barometric pressure can affect tool performance.

From this list, we can see that the appearance of sawdust and chips is the result of many factors. Most often, though, tool sharpness is identified as the primary reason why sawdust looks as it does, and that may be only partially true. For instance, if a perfectly sharp, new saw blade is placed in a condition where there is no adequate means to remove the the waste chips during cutting, the chips could stay in the area and be re-cut repeatedly into dust before being pushed away from the tool. Another instance may be when a saw blade with a large number of teeth is fed too quickly or too slowly for the intended design or function. Either condition will have dramatic effects on the performance of the tool.

All chip formation is a stress reaction process, that is, the cutting tool has to overcome the strength of the material being cut and remove the waste offal. This is, again, not as simple as it might appear.

Characteristics of the Cutting Process

At this point we should more fully define a couple of the pertinent characteristics of the cutting process.

The *cutting circle* is the circular area in which the tool functions. It is the largest diameter of all cutting tools and includes areas that need additional explanation. (See Illus. 1-9.)

- In the *contact zone*, where the tool knife/tooth makes contact with the material and the cutting takes place, tremendous forces are at work. The force of the tool movement overpowers the strength of the material,

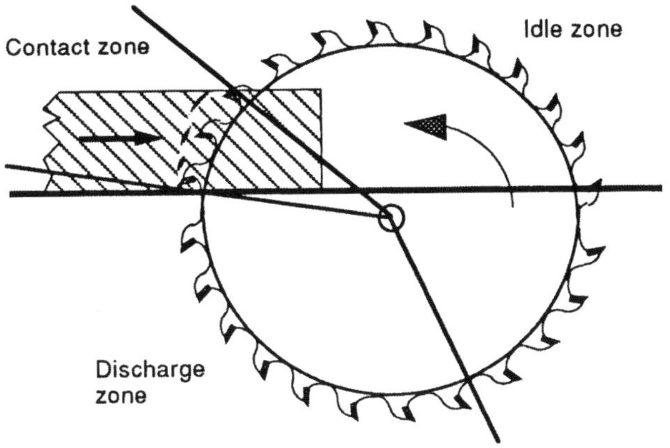

Illus. 1-9—Cutting circle (saw blade)

and the chip is forced against the face of the cutter and into the gullet. As the chip is formed, the angle of the tool knife face (cutting angle) and shape of the gullet determine how the tooth/knife progressively breaks the chips into segments and compresses them. The size and shape of each chip result from several factors in the process—feed rate, geometric features of the tool, and characteristics of the material itself. For each tool tooth/knife, from the point of initial contact until it leaves the material, there are complex forces at work simultaneously—physical, thermal, and chemical. The physical property that is most important is the fact that the material is at the same time very strong and somewhat elastic.

As the tool knife/tooth makes contact, the strength and elasticity create a slight compression ahead of the tool before the leading edge is able to sever the chip(s) away. The elasticity affects how the chips are compressed, broken, and forced onto the tool knife face, into the gullet, and along the walls of the tool path. The combined effects of the strength of the material being cut and its elasticity contribute to the chemical and thermal forces at work in each cutting process.

The thermal properties of the cutting process vary somewhat, even for similar materials. As mentioned, compression takes place in the formation of the chip and in the material surrounding the area where cutting is actually taking place. Close examination of any cut surface provides clear evidence of this. For instance, a wood specie with large, well-defined cell structures will show collapse and crushing of cell walls even when the cells were not in contact with the tool at all. In addition, heat is created by the friction of contact of the tool and wood surfaces and by the chemical processes taking place. The unique chemical properties of water and water vapor also significantly affect the thermal properties of the cutting process.

The chemical processes that take place in forming the chips are dramatic, though short lived. We need not fully examine all the chemical reactions that take place, but it is important to understand what causes them and the effects that they may have on cutting tools and materials being cut.

At the moment a chip is formed at a saw blade tooth, the forces present create conditions that include very high temperature, a mix of different chemical elements, and high pressure—all in the presence of water or water vapor and natural atmospheric gasses (most notably, oxygen and nitrogen). To conclude that the cutting process is simply making sawdust would be an enormous understatement. Recent studies show that most tool wear does not result from abrasion, but rather from the effects of chemical and thermal activity in the process. This is particularly true of the hard metal tool knife materials, tungsten carbide, polycrystalline cubic boron nitride (PCB-Borazon™), and other trade-protected alloys.

The thermal and chemical reactions are as brief as dramatic, in that the length of time for a saw blade tooth to travel through the contact zone may only be 1/4,800 sec. As quickly as the cutter tooth leaves the contact zone, the pressure diminishes to that of the surrounding atmosphere, and the temperature also falls as wood chips are thrown away from the tool and dissipate heat. A portion of the heat building and later dissipation can be explained by the compression of chip formation and the decompression as the compacted chips are thrown away from the tool as it leaves the contact zone. Another way generated heat is carried away and dissipated lies in the molecular structure of water. During the compression of forming the chips, the pressure keeps the water from vaporizing into steam and allows it to be stored, but as soon as the pressure drops, the water flashes into steam and dissipates the heat very efficiently by both evaporation and decompression. Even with relatively little moisture in wood and wood-based materials, evaporation and decompression account for the successful removal of heat.

- In the *discharge zone*, chips and sawdust are evacuated from the tool. For most materials this takes place in the first 45° to 60° of arc travel from the point of leaving the contact zone. This assumes ideal conditions that are not always present—correct tool geometry, feed speeds, etc. During this time, heat that has accumulated in the cutting tool while in the contact zone begins to dissipate. As the chip waste is thrown away from the tool, changes take place in the mechanical behavior of the tool. In the instance of a saw blade, the forward movement of the tool is impeded during a part of each revolution by the friction of contact with the material being cut and the impact of the work being performed. Further, the mass of the tool at its periphery is modified as the gullet spaces fill with accumulated chip waste.

A saw blade is unlike all other cutting tools in one basic aspect: It derives most of its strength from the fact that it is in motion. While in the discharge zone, as the heat dissipates, the tool begins to stabilize both thermally and mechanically. The geometric features of the cutting tool assist in the means by which chips are gathered as well as evacuated. Gullets are usually constructed with large, open-radiused spaces to allow easy entry and exit of waste chips. Small, angular gullet spaces would tend to compact the waste beyond what the centrifugal force could remove. This condition, though uncommon, could result from an overly high feed rate or perhaps from the condition of the material being cut.

● In the *idle zone* of the cutting circle, the tool continues to cool and stabilize. All heat accumulated in the tool during each revolution has to be completely removed. For the cutting process to continue, the tool has to return to a normal temperature for that given process. If even a slight gain in temperature occurs during each revolution, the cutting process will not continue for long. Many factors that must be understood and monitored contribute to the conditions of every cutting process. Some have nothing to do with the tools. The size of each of the zones within the cutting circle varies with tool diameter and cutting depth or material thickness. What is important to comprehend is that cutting tools perform tremendous amounts of work at great, great speed.

Illustration 1-10 shows the same router bit in three different applications. There are dramatic differences

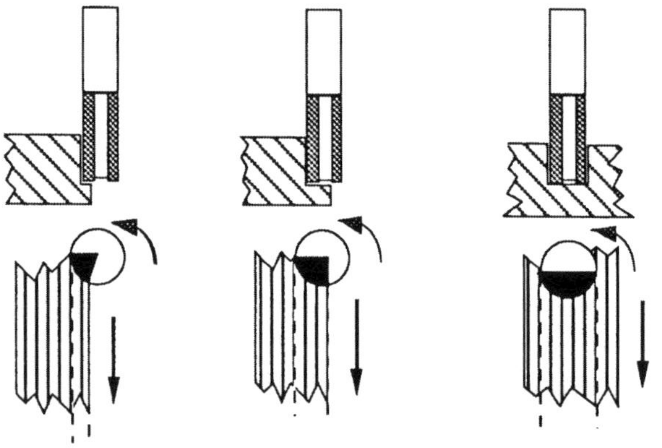

Illus. 1-10—Cutting circle (router)

of performance to be expected, given the amount of work and cutting circle characteristics of each. First, the length of the contact zone varies from less than a quarter of the cutting circle to half of it. Similarly the discharge and idle zones are radically different from one application to the other. What we can expect is obvious: The tool, cutting with its entire diameter, will use more horsepower, have a higher operating temperature, dull more quickly, and perhaps have to be fed into the wood/material more slowly than in the other two situations.

It is important to consider all the conditions and variables in every cutting process when selecting tools. Cutting tools are designed, in geometry and size, to have a range of capabilities. That is, saw blades of a given diameter are designed to cut materials of a certain range of thicknesses. The width of the teeth and body and the number and shape of the teeth further define performance capabilities. Unfortunately, tool manufacturers too often have no means to communicate the specifics of their engineering and design criteria to woodworkers and craftspeople. The remainder of the text will address at least the most common needs of woodworkers.

Manipulating the Cutting Process

Most areas of woodworking present situations that call for manipulating the cutting processes to gain a particular advantage over the material being cut or shaped. Among the reasons for this is the grain orientation of a solid wood or perhaps the need to increase the feed speed of a given process. Most wood cutting is what we have come to call *abrasive cutting*, that is, with the cutting tool working in the opposite direction to the feed of the material being forced against it, as with a table saw. The opposite of that is called *climb cutting*, in which the tool travels in the same direction as the material being cut. (Illus. 1-8 shows abrasive cutting.)

Illustration 1-11 depicts climb cutting. There are advantages to both orientations of stock feeding. The first and most noteworthy difference between the two is chip formation. Abrasive cutting builds the chip gradually through the arc of tooth/knife travel; climb cutting takes its full cut entering the material, then tapers the chip toward the end of travel through the arc. These two chip-forming processes have significantly different results in the appearance of the material cut, the effects on tools, and the energy required.

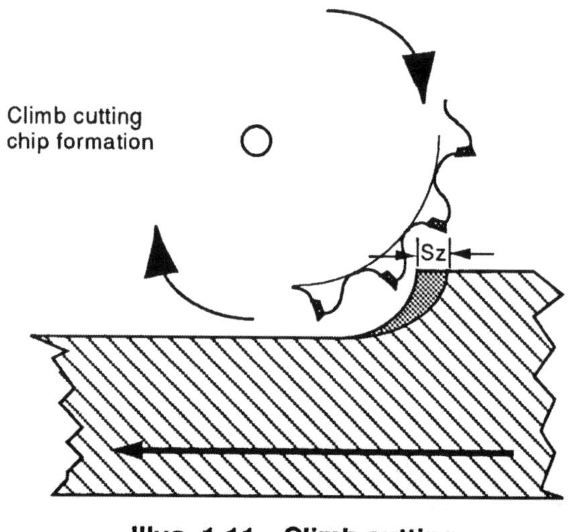

Illus. 1-11—Climb cutting

Climb cutting has the following advantages:

1. Straight grain orientation yields a smoother finish because, as the tool tooth/knife completes its arc, the leading edge rubs against the material and burnishes the surface.
2. When cutting other than straight grain, such as end grain or at an angle, climb cutting can be directed to have the least grain-tearing effect. This is the reason for machines such as tenoners, shapers, and routers with reversible or left- and right-hand rotating motor spindles.
3. With significant improvement of finish quality, greater feed speeds are possible.

The disadvantages of climb cutting are:

1. The material must be mechanically and very securely held and fed to the cutting tools.
2. It usually requires more horsepower to cut in this manner, as building the short, blunt chips takes more energy than building the longer chips of abrasive cutting.
3. There is more stress on machine parts caused by impact vibration and the heavy torsional load of the chip-building process.
4. Cutting tools give less run time between sharpenings due to the rubbing/burnishing at the end of the cutting arc.

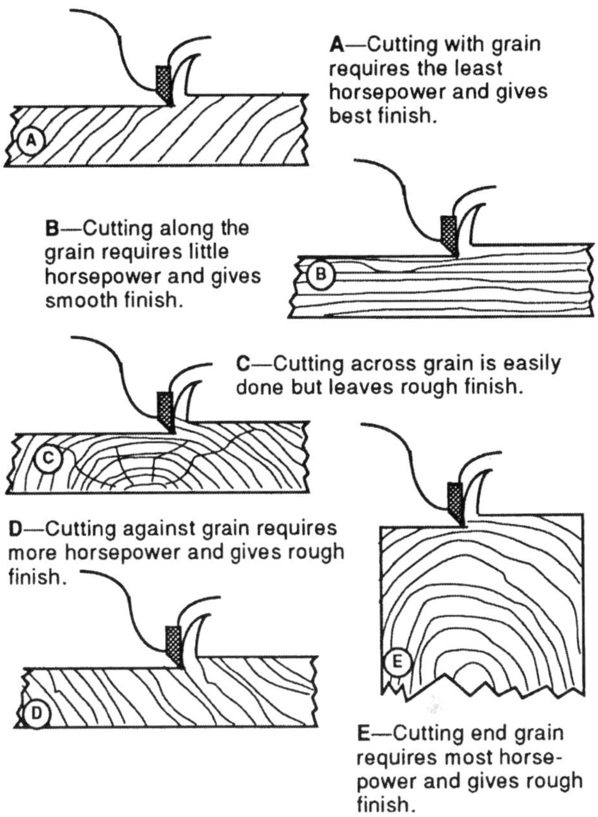

Illus. 1-12—Grain orientation

The disadvantages of climb cutting do not outweigh the advantages by any measure. For many processes, there is no better alternative. In constructing many solid wood panels, such as cabinet doors, climb cutting offers the solution to grain tear-outs that would be difficult or costly to avoid by any other means.

In cutting wood solids, woodworkers have to contend with the natural strength of the fibers and their complex arrangement. Let's consider the organization of wood cell structure as a bundle of very small drinking straws held tightly together. (This is perhaps a dangerous oversimplification, but for the moment will serve our purposes.) The length of wood-tissue cells varies from only a few thousandths of an inch to over a quarter inch. In diameter, individual cells are 50 to 100 times longer than thick. In cross-sectional view they are most often round. Individually, the wood-tissue cells look like a straw with the ends pinched. As abstract as this description of wood tissue may be, it does help to explain the different surface qualities of wood that is cut in different grain orientations. (See Illus. 1-12.)

By examining a wood sample, we can, with experience and careful study, tell where in the tree the piece came from. We can easily identify where the sample came from cross-sectionally; direction of growth is a bit more difficult to discern, and of more importance. When cutting any wood, the direction of growth will determine the success of the process. Feed rate, tool wear, horsepower consumption, and finish quality are all affected dramatically by the orientation of the cutting process in relation to wood grain and growth direction.

The importance of orienting the cutting process to wood grain cannot be overstressed. We often seek improvements in cutting tools and machinery without examining the wood carefully enough. Wood is a difficult and complex material to cut and shape; there is not an instance when this is not true.

Comparatively, cutting and shaping almost any metal is like a walk in the park. Cutting wood is easier only in that less horsepower is required and that it is much faster. Other comparisons of cutting metal to cutting wood include the following:

1. Metal can be cut in any direction with equal success; that is never true of wood.
2. Metal has fairly predictable internal stress; wood is far less predictable.
3. Metals of known alloy content have predictable expandability; wood varies much more, even within a single specie.
4. Metals are engineered and designed to have certain properties; wood is the product of a complex organism.
5. Metals will change, but minimally, with temperature variations; wood can change

dramatically with temperature and humidity differences.

6. Metals are homogeneous in make-up; wood is not.

The list could continue for perhaps several pages, but the point is that we are all so comfortable with and accustomed to the presence of wood in our environment that we often overlook its complexities. None of the listed differences make cutting and shaping wood any more or less difficult, but they should indicate that we must approach wood work more thoughtfully.

Craftsmanship begins to take on different dimensions when we look at, for example, every part of a very complex piece of "art work" furniture and all the processes involved in its construction. Woodworking has been considered both a craft and an art form for thousands of years, and modern machines and tools have not greatly changed basic shaping and cutting processes. The most refined and complex wood cutting tools perform the same basic functions that woodworkers have performed for hundreds of years. What is dramatically different is that modern tools can do the work at amazing speed for long periods of time with precision not possible even 30 years ago.

In later chapters, we will examine all families of cutting tools. This chapter will have to be referred to often in explaining the geometric functions of each. The basic features and performance formulas remain the same for all. Where there are significant differences, they will be explained in the context of the tool involved. For instance, with the use of disposable-knife tools, unique methods have developed to achieve specific clearance angles.

Tool performance formulas for metric dimensions

1. Tool/knife progression (Sz) .... $\frac{U \times 1000}{N \times Z}$ = progression in mm.

2. Feed speed (U) .... $\frac{Sz \times Z \times N}{1000}$ = M/min.

3. Rim speed (V) .... $\frac{D \times \Pi \times N}{1000 \times 60}$ = M/sec.

4. Cutter mark depth (t) .... $\frac{Sz^2}{4 \times D}$ = depth of cut in mm.

5. Mean chip thickness (hm) .... $Sz\sqrt{\frac{a}{D}}$ = average thickness in mm.

D = diameter in mm
pi = 3.14159
N = rpm
Z = number of teeth/knives
a = depth of cut in mm

Sz = tooth progression in mm
U = feed rate in M/min.
V - rim speed in M/sec.
t = cutter mark depth in mm
hm = mean chip thickness in mm

The formulas do not allow for normally present deviations caused by machine tolerance, vibration, or stock clamping and feeding. Normal deviations will not affect the working accuracy of the caluclations.

Chapter 2

Wood and Wood-Based Materials

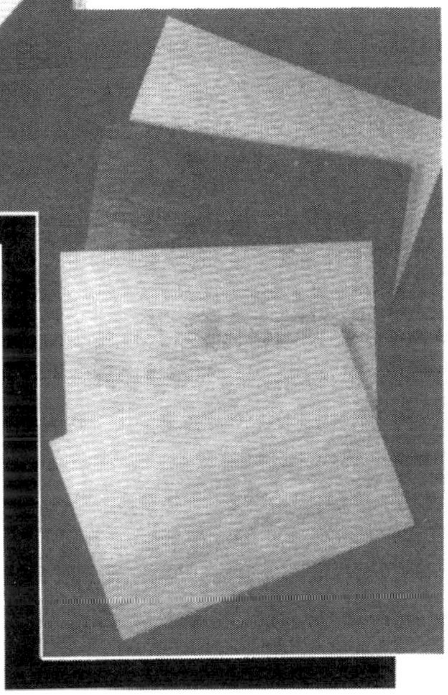

Photos courtesy of Wood Digest and Tech Directions

Wood and Wood-Based Materials

Describing even a simple, homogeneous material such as steel is a broad and complex process. Wood is not a simple material; neither is it homogeneous. Wood has at least the following characteristics:

- Wood is *orthotropic*—in biology, meaning growing vertically.
- Wood is *isotropic*—in physics, meaning having unique and independent mechanical properties in the directions of three mutually perpendicular axes: longitudinal, radial, and tangential.
- Wood is *anisotropic heterogeneous material*—in biology, meaning subject to specie differences, wide biological variability, and a similar array of naturally occurring irregularities and defects.
- Wood inherently reacts to stresses applied to or from different directions.

All wood and wood-based materials are products of a once-living organism—a tree. Whatever qualities the material has, at least in part, can be traced back to that fact. Like all living organisms, trees are enormously complex. The wood developed as tissue of a living plant, quite beyond the control of you or me as woodworkers.

Wood is essentially the skeletal remains of once-living tissue. Plants are unlike animals in perhaps hundreds of ways, but the most basic difference is the composition of the cell walls of the organisms. Animal cell walls are complex proteins; plant cell walls are a complex hydrocarbon that we call *cellulose*, which very closely resembles sugar.

Trees, when viewed one to the next, have only a single commonality: They grow from the ground upward and from the outside. From that point, the description gets more complex.

We broadly categorize trees into two families that describe how the trees propagate, i.e., the form of their seeds:

Angiosperms are plants having seeds in an ovary. Trees in this group also have broad leaves, and we call them hardwoods, a term that has nothing to do with relative strength or density. Since trees in this class have cell structures that interconnect tissue groups—*vessels*—they are also called porous woods.

Gymnosperms are plants having seeds not enclosed in an ovary. All trees in this group are referred to as *softwoods*. Softwoods do not have vessel cell structures and are also called *nonporous woods*. The term softwoood has nothing to do with relative hardness, strength, or density. The leaf structures of all softwoods are significantly different from those of hardwoods: There are no broad-leaf softwoods or needle-leaf hardwoods.

The main body of the tree that we take lumber from is generally called the *stem*, *trunk*, or *bole*. The upper body of the tree having most of the limb structures is called the *crown*. The lower portion of the tree body is called the *stump*, and the below-ground structures collectively are called the *root system*. (See Illus. 2-1.) The crown and trunk areas of some species are not clearly discernible because small limb structures may grow over the entire body length. Within softwood and hardwood groups, there are methods of identifying exact species by the relationship of limb structure to body size and shape, as well as by leaf forms.

Wood, as noted earlier, is made up of once-living cells. Cells of similar type or function are collectively

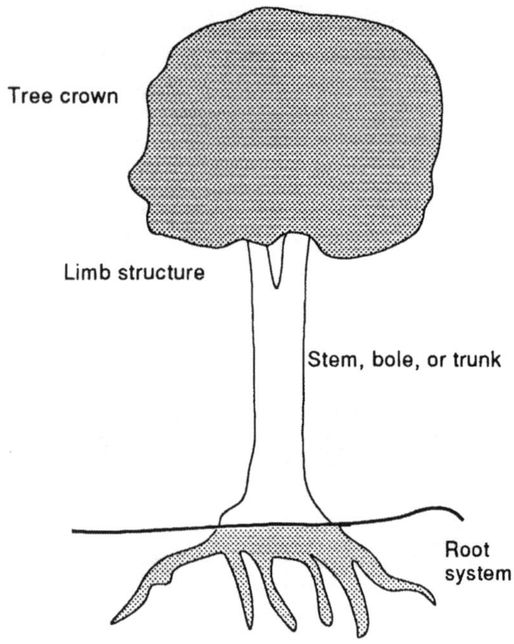

Illus. 2-1—Tree structure

referred to as tissue. Trees contain numerous tissue types: bark, cambium, growth rings, limb stems, leaves, and their supportive tissue groups.

The outermost layer of a tree, the *bark*, is made up of both dead and living tissue in two or more layers. Beneath the bark is the most active tissue level, the *cambium*, and beneath that lies the first *annual growth ring*, that is, the growth added to the tree during this growing season. Annual growth rings appear in what is called *sapwood*. Cells in the growth rings may or may not contain living, functional cells, but most participate in the tree's growth cycle in some way. Beneath the sapwood is the largest area of a tree that we would harvest, the *heartwood*. All tissue in heartwood is nonliving. At the center of the tree is a small area called *pith*, the original stem. (See Illus. 2-2.)

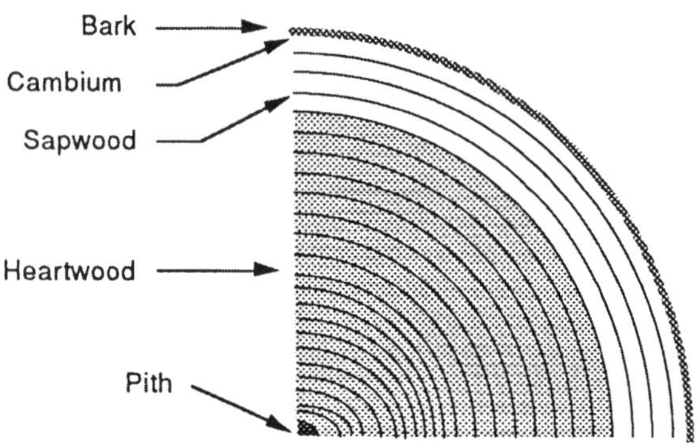

Illus. 2-2—Wood tissue types

Cell division takes place in the cambium. A "mother" cell divides lengthwise, forming two "daughter" cells. One enlarges to become another cambial mother cell, and the smaller cell begins to mature into a bark cell, if located at the outer layer of the cambium, or an annual growth ring cell, if located near the inner cambium layer.

Individual wood cells have several characteristic shapes. (See Illus. 2-3.) The most prevalent look like drinking straws with the ends pinched. These are called *tracheids* and make up from 85 to 90 percent of the tree's volume. Tracheids are arranged vertically within the tree body; cells and tissue groups are arranged about the limb structures in an upward spiral.

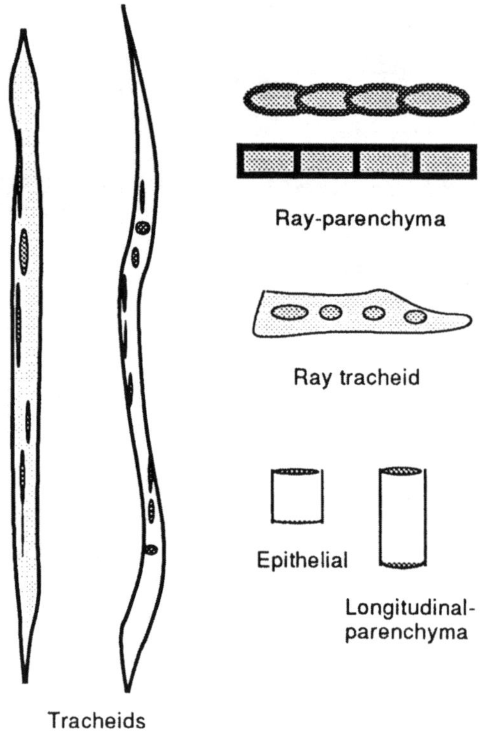

Illus. 2-3—Wood cell types

Most of the remainder of wood cells are *ray-parenchyma* cells, most often called just "rays." Ray cells look like short barrels and are arranged along a horizontal plane, interconnecting groups of other tissues in the annual growth rings and between growth rings in the sapwood. Some softwood species have a small percentage of epithelial cells that line the resin canals. This brief description of the wood cells, tissues, and their arrangements can help us understand the physical and mechanical properties of wood.

As mentioned earlier, the predominant chemical in the structure of wood is *cellulose*, having the general formula $(C_6H_{10}O_5)_n$ and an approximate specific gravity of 1.5. Another family of chemicals in wood is the *hemicelluloses*, which are in most aspects similar to cellulose itself. Their exact chemical makeup varies between hardwoods and softwoods and also significantly between species.

The last major group of chemicals that compose wood is *lignins*. Generally, lignins are the adhesive that

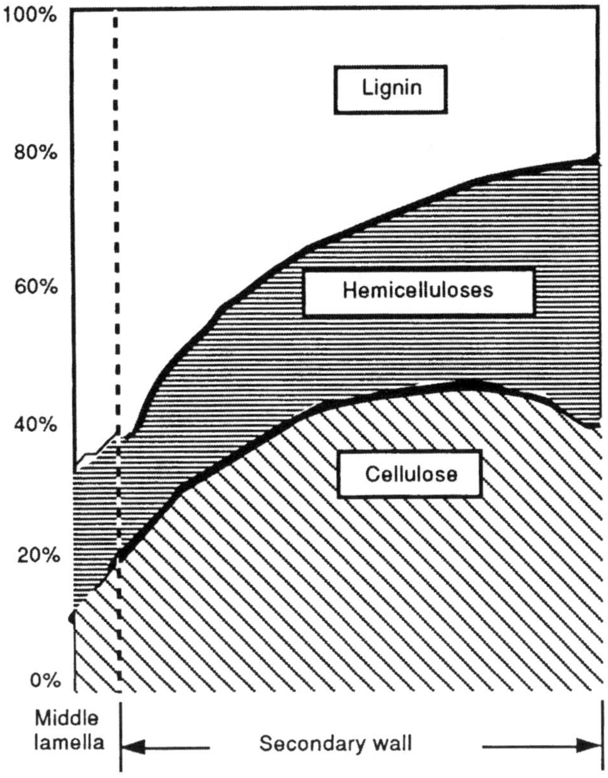

Illus. 2-4—Chemical distribution in wood cells

Component	Percentage	
	Softwoods	Hardwoods
Cellulose	42 ± 2	45 ± 2
Hemicellulose	27 ± 2	30 ± 5
Lignin	28 ± 3	20 ± 4
Extractives	3 ± 2	5 ⊥ 3

Chemical composition of woods

holds the wood cells, tissues, and fibers together. Their exact composition is not yet known, but it is broadly accepted that lignin is a three-dimensional, cross-linked polymer made up of phenylpropane units, for which no single structural scheme can be established. The problem in ascertaining the structure of lignin is that there is no way to remove it intact from wood without altering its makeup. (See Illus. 2-4 and table above.)

A cross-sectional examination of a tree, such as at the stump, allows a view of the entire history of its growth. There are differences in the relative width of the annual growth rings that tell us something of each particular growing season. Details of early and late growth within the annual rings offer more specific information.

While it is not necessary to know all the physical characteristics of the wood that we work with, it is important to understand what created them. A cross-sectional view of the tree that we take our lumber from would give us a glimpse of how the time taken to produce the wood also affected it. Over its entire length and girth, the tree changed and grew through "growing up" problems that would include such things as the following:

1. Climatic conditions over its life span;
2. Nutrition and access to sunlight;
3. Relative health and freedom from parasites;
4. Location—hillside, swamp, mountain top, or open field; thick, crowded forest, or relative open area;
5. Damage from weather—wind, hail, and storms;
6. Soil conditions—ability to hold moisture and nutrients;
7. Relative freedom from damage from insects and animals using it as a source of food.

That a tree can live to maturity, in some instances for over a hundred years, is a considerable achievement. Even trees of the same species differ greatly in response to their surroundings. Additionally, within each tree, we can find evidence of its relative success in adapting to everything in the environment. The results of stresses within each tree, often not apparent until it is cut into lumber, are what each woodworker has to learn to overcome. For instance, trees that grow on unusual hillsides may show stress throughout their entire length. Similar signs of stress may appear in trees that grow in areas unprotected from strong winds. Overall, it must

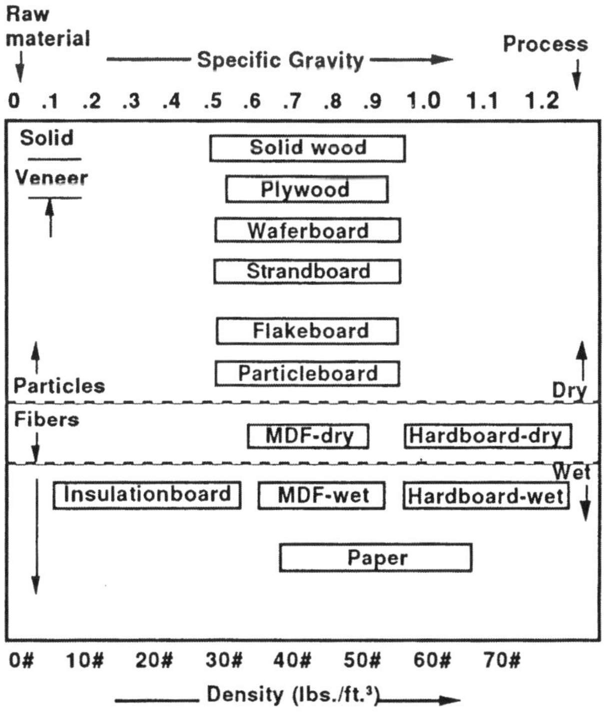

Illus. 2-5—Density/specific gravity of wood

be said that each piece of wood has to be treated with respect and, at least to some extent, studied to understand its particular characteristics.

One such characteristic is density, a measure of wood's relative strength that is usually expressed as *specific gravity.* (See Illus. 2-5.) Density is weight per unit of volume; specific gravity is the same ratio expressed in terms of a standard substance—water. Water has a density of 62.4 lbs./cu. ft. (1 gm./cu. cm.). A wood sample having a density of 37.44 lbs./cu. ft. (0.6 gm./cu. cm.) is thus 6/10 as heavy as water and has a specific gravity of 0.6.

As mentioned earlier, cellulose, the primary chemical in wood tissue, has a specific gravity of 1.5, that is, 1-1/2 times greater than water. So we know that in addition to the complex chemicals involved, something has to account for the fact that wood materials do not sink in water. If wood were nothing but solid cellulose, all species should weigh about 93.6 lbs./cu. ft. and would not float. The seeming paradox here is not that much of a mystery at all because other chemicals and water are also present in woods.

The amount of water present in wood, or its *moisture content*, is perhaps the most important characteristic of wood to understand. Several properties of wood depend on moisture content, most notably strength, weight, and shrinkage.

In living trees, moisture content may range from 30 percent to over 200 percent of the weight of the wood substance. In most softwoods, sapwood has a much higher moisture content than heartwood; in hardwoods, the differences are not often as great. Moisture in living trees can exist inside the cell lumens, or cavities, as water or water vapor. Water can also be present in cell walls, bound chemically to the substances that compose the walls. Free water can also be present between cells, in tissue groups, in ray cells, and throughout the tree body.

Because the entire body of a tree contains enormous amounts of water, we must consider how to remove it. Once the tree is cut off the stump and into logs, billets, or lumber, it loses its stored water. Harvested wood may be dried by any of several methods, depending on specific needs. Most commonly, the wood is stacked and allowed to air dry to the moisture content of the climatic air. This could take from 3 to 12 months. At that point, the wood may be recut into smaller dimensions before being kiln dried. Recutting reduces the stresses, time, and costs of drying. (Moisture content is usually expressed as a percentage of kiln-dried wood.)

Kiln drying may involve any of several processes that use steam to extract water from wood tissue. Precision is necessary in every step of these procedures to ensure success. If the process is not carefully monitored, stresses will occur inside the lumber that are very difficult to compensate for during manufacturing processes such as making furniture or framing members for construction. Trees are complex enough without adding stresses in the drying process.

As moisture is removed from wood, the volume of wood mass reduces through shrinkage. Cells shrink mostly in diameter as they lose moisture, causing change in the tangential dimension. (See Illus. 2-6.) Natural stresses in the tree's growth can be compounded by irregular or incorrect drying methods.

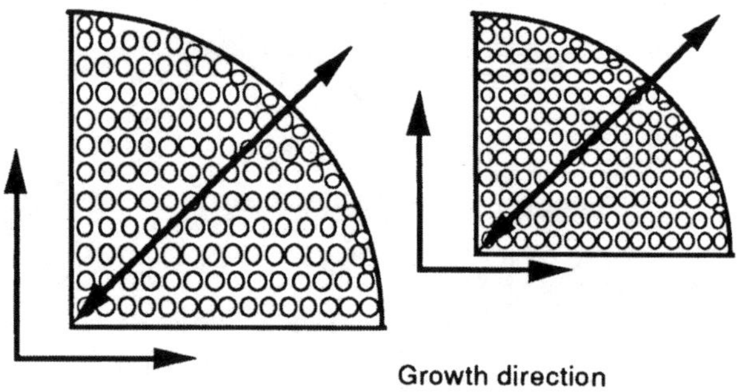

Illus. 2-6—Wood cell shrinkage

The results of stresses are present in every tree. As each tree matures, the exact manner of growth creates internal stresses. With the addition of each annual growth ring, there is evidence along the entire length of the tree how growth was directed to ensure its success. The placement of cells, leaf stems, and limb bodies, as well as reactions to growing conditions, all contribute to the body of harvested wood.

For example, trees that grow on hillsides have eccentric growth rings, with wider ones growing on the uphill side in an apparent attempt to overcome gravity. A tree growing in dense forest will shed most low limbs to direct growth upward toward more sunlight, while a tree of the same specie growing in an open area may have limbs over its entire length and, consequently, will grow faster in height and girth.

Harvesting trees and preparing lumber for manufacturing are but a portion of today's wood industry. Understanding even the basic physical characteristics of wood morphology will greatly improve chances of success in woodworking. This overview is intended only to introduce woodworkers to the fact that each wood piece chosen for a particular product has a life history and characteristics that may require study. Before attempting to work a particular piece of wood, any woodworker who has not harvested the tree and overseen every step of its preparation into dimension lumber should do the following:

1. Identify the exact specie.
2. Know something of the harvesting and drying methods used.
3. Examine the stock to identify how and where it was taken from the tree.

4. Carefully plan the best utilization of the wood with respect to grain direction, yield, and intended product.

Following just these basic steps will greatly improve your chances to avoid most calamities.

Wood-Based Man-Made Panel Materials

The most popular contentions regarding man-made panels are that they are a recent phenomenon and that products made of them are inherently inferior to those made entirely of wood solids. Both of these observations need clarification, however.

All composition boards, i.e., man-made panel materials, were invented or reached commercial importance in the last 100 years. The same can be said of paper made of wood pulp. The development and refinement of panel, sheet, and solid-wood-based materials continues today with new products coming forward constantly. As in all industries, the two driving forces of production are the availability of raw materials and worldwide competition. The introduction of man-made materials into the woodworking industry first came about because of a need to utilize waste by-products from the manufacture of paper.

Insulation board, a fairly loose panel made of relatively large wood fibers, developed from a paper by-product in England in 1898. The product evolved as an effort to utilize large fiber bundles, called "screenings," that were unsuited for use in paper. Ten years later, a plant for manufacturing insulation board was built in the U.S., near Trenton, New Jersey.

In 1914, a plant to produce another type of insulation board was constructed at the Minnesota and Ontario Paper Company in International Falls, Minnesota. This facility used waste by-products of two area industries—paper and mining—to produce a rigid panel made of wood-fiber grindings and the sulfite screenings that were waste from local mining operations. The product, sold as *Insulite*, was used as sheathing, interior finish panels, and roof insulation. The plant is still in production today.

Meanwhile, others were attempting to make similar panel materials using the waste products from agricultural sources such as the fiber offal in wheat straw, corn stalks, and *bagasse*, the waste from the extraction of sugar from sugar cane. As efficient energy sources became available, bagasse became the basis for *Cellotex* insulation board. The first plant to produce this panel, built in 1920 near Marrero, Louisiana, remains the largest producer of insulation board in the U.S.

As utilization of the great quantities of sawmill wastes developed, a number of significant advances took place. In 1924, in Laurel, Mississippi, William H. Mason was operating a plant for processing rosin and turpentine. Mason experimented with a unique device for reducing wood chips to wood fiber without the loss of naturally occurring lignin. The device, a "digester," subjected wood chips to high-pressure steam for a short time, allowing the lignin bonds to soften, and, upon expansion, the chips fiberized completely. The resulting pulp, while not suitable for paper, could be formed into a sheet in a hot press to yield a remarkably strong, hard material. Mason's panel was the first to capture and use lignin as the bonding agent. The process, patented in 1926, became the basis of the *Masonite Corporation*, whose Laurel facility is still the largest of its type in the world.

The Mason process had some problems: It required enormous amounts of water and energy to generate the needed steam, and the sheet produced was smooth on only one side. (The back side was imprinted by the screen that allowed water and steam to escape during pressing.) The latter problem was solved with development of another process Mason patented in 1938, which pressed the wet sheet a second time between two smooth, heated platens.

The attempt to produce a similar panel product without such high cost for water and energy took a unique twist a few years earlier, in 1931. A Swedish engineer, Arne Asplund, investigating the thermoplastic properties of wood, discovered that wood chips ground up under steam pressure would "defiberize." This process consumed significantly less water and energy and later led to the development of *dry-process hardboard*, *particleboard*, *oriented strandboard*, and other chipboards. Today, nearly all chip, particle, and fiber panels come from this process, with varying modifications to suit the needs of the product.

Wet-process fiber panels differ from dry-process ones in a couple of significant ways. In the wet process, the raw material is probably already in a water suspension during the course of its "manufacture" as a waste by-product of the paper industry. Because the pulp is suspended in water, the panel produced is generally freer of "tramp materials" (sand, minerals, metal, etc.) that may be found in other sheet materials. Further, most wet-panel boards do not rely on the addition of adhesives beyond lignin as a binder.

The development of *dry-process panels* in the U.S. did not begin until 1946, in Corvalis, Oregon. This technology advanced with the discovery of a way to include the fibers of hardwood species that had previously been too difficult to break down. Principal to dry-process panel manufacturing is the use of compressed air to transport and sort fibers and, in some instances, to load the press platens. Air is also used to blend, mix, correctly meter, and apply adhesive to the fibers prior to pressing. In fiber panels, density is more consistent cross-sectionally than in other sheet panels with larger particles in their makeup.

A recent development is the manufacture of *oriented strandboard* (OSB), which shows considerable success in orienting individual wood fibers in a particular manner—much like arranging drinking straws lengthwise, side by side, as opposed to in a random pile. This is another step in producing a stronger, less costly panel from more-accessible raw materials.

Between 1935 and 1945, there was little progress in products or processes in woodworking industries. Between 1945 and 1960, however, there was an explosion of activity, in both the U.S. and Europe, with many manufacturers attempting to introduce molded wood products for the sheet and structural steel parts then in short supply.

Particleboard was invented in 1943 in Germany, where there was a need for inexpensive building material that could be made primarily of waste by-products. The wood waste available was the offal of lumber and paper industries; petrochemical industries supplied (and continue to supply) the waste material for making the adhesive. The first particleboards were introduced into the U.S. in 1953, but, due in part to users' failure to understand the intended applications for the panels, they were not immediately successful. Later, though, with significant improvements, the panels won wide acceptance in a variety of furniture and construction uses. With improved quality and the use of cleaner, single-specie raw materials, these panels continue to gain in share of the marketplace.

Another German innovation that furthered the production of man-made panels was the *Mende press*, introduced in the U.S. in 1972. Essentially, the Mende is a drum press in which the flow of the dry-panel product is a truly continuous process. Today nearly all dry panels are manufactured with this much faster method.

Plywood and similar sheet panels have had a historical progression similar to that of hardboards and fiberboards. While we have examples of furniture with decorative veneered coverings that are over 2,000 years old, the manufacture of panels made entirely of thin, cross-banded layers developed in this century. There have been significant changes in technology to accomplish this manufacturing. The machinery to hold, cut, dry, and assemble the panels has improved steadily over the past 75 years.

There are three basic methods to cut trees into the thin sheets that will become either veneer stock, molded plywood sheets, or plywood for sheet-panel production. These involve the rotary, guillotine, and oscillating rocker. With the rotary, the log sections are held and turned as a knife is brought to the wood to peel away a thin layer. With the guillotine method, either the log or the knife can be stationary, depending on machine design. The rocker is a variation of the rotary not commonly used.

The choice of cutting method depends on the intended product. Rotary cutting follows the annual growth rings inward to the center of the tree. The guillotine cut yields a cross-section of the entire tree body—sapwood and heartwood. Generally, decorative veneer stock is guillotine cut, while veneers used in construction-panel plywoods are rotary cut. The reason is twofold. First, guillotine-cut veneers can be steam treated and dried in a process that will make the sapwood take on the desired color of heartwood, increasing overall yield; and, second, the cross-cutting method makes it easier to "fault" the panel (i.e., remove imperfections mechanically). Further refinement of drying systems and conveyors has made the production of plywood and veneer stock one of the most automated segments of the woodworking industry.

The original intent for plywood panels was to provide furniture manufacturers with a stable, less costly panel for case goods than was possible with ripsawn and glued wood-solid panels that were often later veneered with a decorative hardwood. But plywood in its many forms has been shaped into more products than any wood-based man-made material—from numerous aircraft and marine parts to enormous structural beams that span hundreds of feet without other support. Molded plywood is used in almost countless applications. Recent research with unusual adhesives and wood preservatives has provided structural panels for use underground and even under water.

Reconstituted wood, the newest man-made wood-based material, is available as a veneer stock or produced in solid blocks from which decorative trim is made. The material is essentially a plywood product that has been cut apart and put back together perhaps several times. The process of cutting the material is essentially the same as for either veneer or plywood manufacturing. After the first cutting, however, the process uses machinery systems that are very refined.

Reconstituted wood developed to meet a need for a clean, fault-free decorative wood with a particular color or grain effect. In this product, thin veneer strips are cut, faulted, and glued back together. The block is then turned 90°, and the process is repeated. In manufacturing, different species are often mixed together and colored for the desired effect. The result is a wood material, with none of the stress effects of solid wood, that is 100 percent usable, though costly. An end view of reconstituted wood looks like square toothpicks glued together; the cross-section of each little stick, however, is often smaller than 0.003", not easily seen without magnification.

When veneers or solids of reconstituted wood are seen in a finished product, it is often not possible to determine that it is not a "natural" wood, and solid blocks of this material are available in nearly any size. An interesting offshoot of this developing technology is the very thin veneers that are now available. Thicknesses of less than 0.010", down to 0.004", are possible.

Reconstituted wood products were developed in Europe and Japan, where the supply of clean, clear wood solids has already diminished. The U.S. still has a fairly abundant supply of all of our native woods; that condition will not exist for long, however, especially for decorative hardwoods. For this reason, there has always been considerable study in industry directed at how we use wood. In a given product, is wood the gross structure or decorative? Is the surface meant for contact or simply for visual effect? As cost rises and availability of "real wood" diminishes, these questions will have a more pronounced effect on how and of what we build our homes, furniture, and other products.

In all areas of woodworking, materials have changed dramatically, even in the past 10 years. Illustration 2-7 shows the most recent distribution of materials in use, excluding the construction industry (in which the distribution is similar, though). As manufacturers have progressed in making man-made materials, there has been increasing acceptance and use of them.

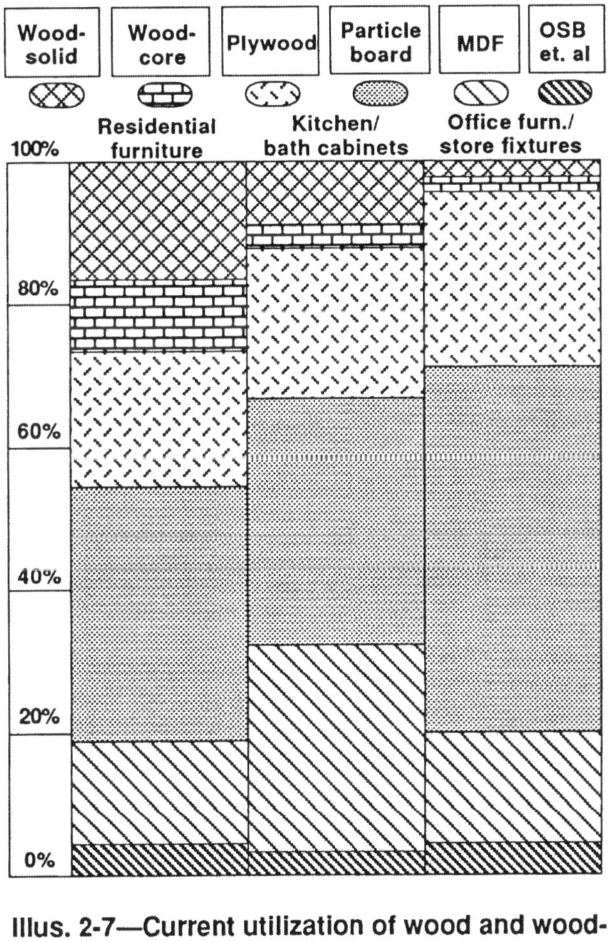

Illus. 2-7—Current utilization of wood and wood-based materials

As mentioned earlier in this chapter, natural woods and man-made panels share many physical characteristics. Similar measurements of strength apply in some manner to all materials used in the woodworking industry. For wood solids:

1. *Modulus of rupture*—the load-carrying capacity;
2. *Maximum load in bending*—combined strength and toughness;
3. *Maximum crushing strength*—compression parallel to grain;
4. *Compression perpendicular to grain*;
5. *Shear strength parallel to grain;*
6. *Impact bending*—shock/stress absorption;
7. *Tensile strength perpendicular to grain*;
8. *Hardness.*

Less common properties not often measured include torsion, toughness, creep, creep rupture, fatigue, and rolling shear strength.

We could not, in all honesty, afford to design, develop, and produce a man-made material that has *all of the properties* of even the simplest natural wood. The measurements that we make of natural wood samples make up only a quality grading system.

The array of man-made panel, sheet, and solid-block materials is easier to describe than are natural wood solids. Man-made materials are "engineered" to have a certain range of physical properties—not that all are completely understood or controlled. Measurements of strength of man-made materials closely parallel those of natural woods. The measured and controlled properties of commercial panel stock are:

1. *Internal bond*;
2. *Tensile strength*;
3. *Modulus of rupture*;
4. *Modulus of elasticity*;
5. *Linear expansion*;
6. *Water absorption*—as a percentage of both weight and volume;
7. *Thickness swelling*;
8. *Specific gravity*;
9. *Moisture content.*

As mentioned earlier, man-made materials are engineered to have a predictable range of properties; that does not, however, say that any such materials approach the strength of natural wood. The strength of solid wood used in constructing furniture, houses, bridge beams, and even aircraft parts has been studied and appreciated for years. Not everything woodworkers build, however, requires the enormous strength of, for instance, shagbark hickory, which has a modulus of elasticity of 2,160,000 psi and modulus of rupture of 20,200 psi, or even of eastern white pine, which has elasticity of 1,460,000 psi and rupture of 9,400 psi. Even though not as strong, man-made panels, sheets, and blocks, if used for the correct applications, will in all likelihood perform very well.

The single thing that woodworkers should keep in mind, whether working with natural woods or man-made wood-based materials, is the importance *monitoring and controlling moisture content.* Wood, in all its forms, is more dramatically affected by moisture than by anything else.

The most telling property of wood and wood-based materials is density, expressed as pounds per cubic foot or as specific gravity. Different wood species and wood-based materials behave most like each other when they have similar density, and they have similar strength and other features when they are alike in that important characteristic.

Effects of Wood Materials on Tools

An examination of the effects that wood and wood-based materials have on cutting tools is a study of chemistry, physics, biology, mechanics, economics, and more.

Chemically, wood tissue is a complex of hydrocarbons, water, and trace minerals. Through their excellent hydraulic system, trees are capable of lifting water and soil nutrients over their entire height. Along with the nutrients and water, a few trees, a couple of maple species for instance, can also pump unsuspended mineral solids into their tissues, causing what is called "mineral streaking."

Different soil nutrients and conditions add to the mixture of substances that trees can be made of. The physical differences between species and their unique chemical makeup do not radically affect short-term performance of cutting tools, but they can cause definite variations in tool wear characteristics. The chemicals that compose wood and man-made products do not pose any insurmountable problems. Natural solid wood products without glue lines can be cut without problems using tool-knife materials that have been in use for hundreds of years.

The addition of adhesives to wood, even the animal-hide types first seen several centuries ago, bring another set of circumstances for toolmakers to deal with. The wide range of chemicals in adhesives introduces conditions into the cutting process that are not easily seen or understood. The great speed of modern cutting tools generates enormous heat, and, with the presence of moisture, air, and a mix of chemicals, the effects can be dramatic. While you and I shape wood into a product, the processes we employ are by far more complex than is visible.

Natural solid woods, plywoods, fiberboards, and particleboards all have different cutting properties that affect how well tools perform on them. Their effect on cutting tools is measured by some of the following:

1. Abrasive wear,
2. Thermal corrosion and mechanics,
3. Chemical wear (oxidation),
4. Stress mechanics,
5. Overall tool costs,
6. Tool-knife material,
7. Power consumption,
8. Production efficiencies and costs.

These characteristics are the elements that are most often examined in design, material selection, and manufacture of cutting tools. All of the materials found today in various segments of the woodworking industry have physical properties and contents that are different from one to the other and require at least a basic understanding. For all wood-based materials, there are common basic cutting characteristics:

1. The cutting process generates large quantities of heat.
2. The heat generated is not sustained but rather is carried away and dissipated by saw dust and chips.
3. Management and control of heat is the most important facet of all cutting processes.
4. Moisture content is the most important variable that the woodworker can control (and often the only one of consequence).
5. All wood cutting or machining is a stress-failure process.
6. All wood and wood-based materials being cut display the results of stress.
7. Cutting any wood material with any tool-knife material will eventually dull the cutting edge by abrasive wear, thermal corrosive wear, or chemical abrasive/corrosive wear.
8. Given the complexity and range of materials cut, there is not ever a perfect nor permanent method for any cutting application.

Cutting *natural solid woods* does not offer any problem that has not been encountered over the past several hundred years. Tools made of all types of steel alloys are as successful today as in the past. Today, there are several patented alloys that are much better at withstanding the great range of temperatures of current production processes and speeds.

The best illustration of heat as it relates to cutting tools is a saw blade. A saw blade, while cutting, has constantly changing zones of temperature across its diameter. The coolest area is near the center, which is affected only by the normal operating temperature of the motor driving the blade and the friction inherent in the bearings that support it. Most electric motors operate in a range of 145° to 180° F. At the periphery, the leading tip of the cutting edge may reach 900° F and more.

Moving from the outer rim toward the center, the temperature decreases, though not often in a smooth, even progression. There are wave-like thermal zones that are constantly in motion because the blade is not always in contact with the workpiece. (A complete explanation of this would require a text as large as this one.) Moreover, the amount of heat varies according to the interruption of the work and the effects of the chips that dissipate it.

As mentioned repeatedly, moisture and the unique chemical structure of water have a dramatic effect on many aspects of cutting any wood-based material. A 1 percent change in the moisture content of nearly any wood substance can change the cutting temperature by as much as 70° F.

An important point here is the difference between *heat* and *sustained heat*. The cutting edges of tools can often reach elevated temperatures, but this is a normal condition of the process and is not sustained. If, for instance, the 900° F. were sustained, wood subjected to any cutting process would literally burn. Most wood species and products have ignition points well below that temperature.

The chemical makeup of natural woods—cellulose hydrocarbons, trace minerals, and water—does not constitute any serious problem for most cutting tools (assuming appropriate design and application). The biological characteristics of natural woods that are most difficult to deal with are beyond the control of the woodworker. The effects of stress in natural wood are a part of each tree's normal growth history and of the processes used to dry it after harvesting, and they relate directly to precautions that must be taken later. At best, the woodworker must understand the various stress-related conditions that the wood may have and be prepared to deal with them.

The addition of even a single glue line to a natural wood product alters the choice of cutting tool-knife material. While steel alloy tool-knife material will cut wood well enough, it probably will not cut well for very long. Newer alloys of tungsten carbide or similar material would be more practical. Most adhesives have negative effects on steel alloy cutting tools because they are at the same time abrasive and thermally and chemically corrosive.

Cutting *plywood* requires consideration of cutting tool-knife materials. Plywood, measured by any means, contains a fairly large amount of different adhesives. It does not cut with any more difficulty than most wood solids, nor at a much higher temperature. Moreover, the moisture content does not often vary greatly. The chip formation and chip size allow heat to be captured and dissipated. For cutting some forms of plywood, such as reconstituted woods, the tool-knife material would be the same as for sheet or molded plywood. Sheet plywood does have unusual strength, as it is cross-banded, having every other sheet arranged 90° from the next. This effect stabilizes most cutting tools by creating a harmonic balance not possible in similar wood solids. (See Illus. 2-8.)

Illus. 2-8—Plywood

Cutting *wet-process hardboards* and *fiberboards* requires several cutting tool considerations. Of greatest importance, these man-made materials, when cut by any means, redefine the term *saw dust*. The offal is truly like dust—very small particles that are difficult to capture and dissipate. The panel stock is largely free of tramp materials such as sand, clay, or metals; there is most often no adhesive added to it; and it is more consistent cross-sectionally than most other man-made panels. The remarkable heat generated in cutting these panels is easily overcome with correct tool geometry and tool-knife materials and adequate dust collection systems. (See Illus. 2-9.)

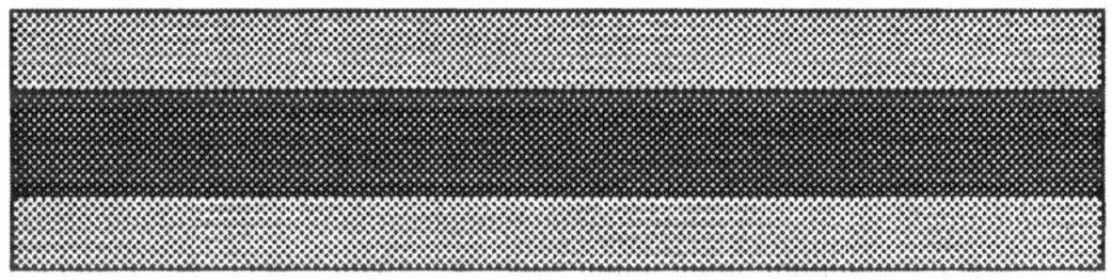

Wet-process hardboard

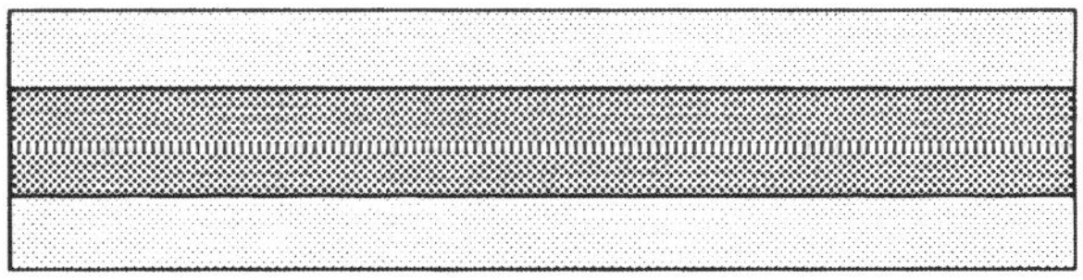

Wet-process fiberboard

Illus. 2-9—Wet process panels

Cutting *dry-process hardboards* and *fiberboards* is similar to cutting wet-process panels, with a couple of exceptions. First, these man-made panels most often contain a variety of adhesives, making them more abrasive to tools. Additionally, in their manufacture, they may gather tramp materials. The waste is similar to that of wet-process panels and contributes to elevated tool temperatures during cutting. (See Illus. 2-10.)

Cutting particleboards, chipboards, flakeboards, and oriented strandboards adds another set of conditions for the choice and use of cutting tools. When these panels were first developed 35 years ago, they were the most abrasive wood-based materials ever introduced. This was particularly a problem since tungsten carbide cutting tools were not in common use then. The earliest

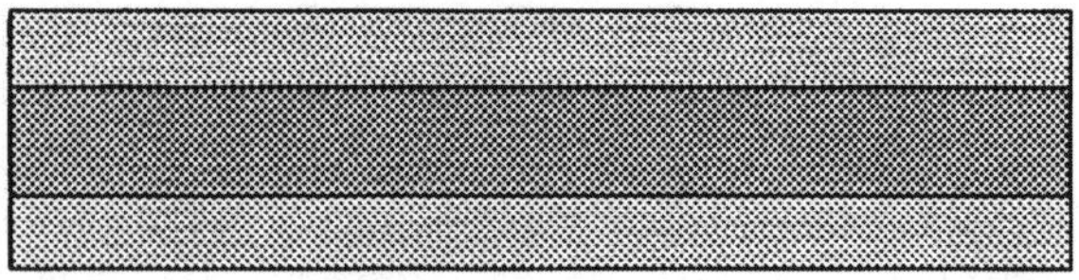

Dry-process hardboard

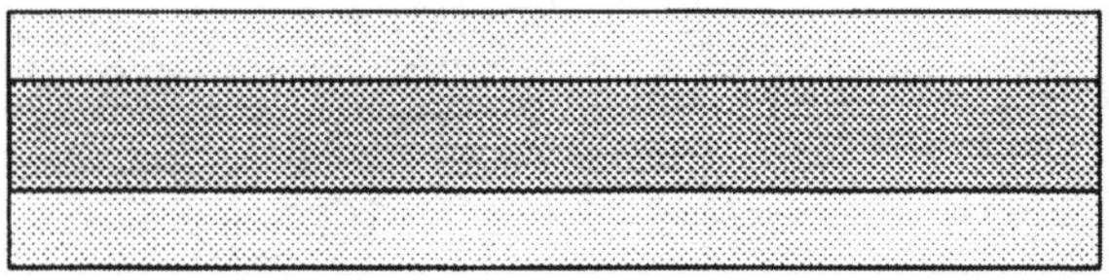

Dry-process fiberboard

Illus. 2-10—Dry-process panels

particleboard panels were constructed of a wide mix of wood-waste materials, several adhesive types, and, too often, large amounts of tramp materials such as sand, clay, and parts of the machinery that produced them. Today the same type of panels are remarkably clean, compared to those of even 10 years ago, and are widely accepted.

The characteristics that make all such panels difficult to cut are their chemical contents, abrasive makeup, and relatively hot cutting properties. As with other man-made panels, the chips produced in cutting are too small to capture and carry heat away. Current studies indicate that the most significant improvements in chipboard and similar panels are in those that are made of single-specie raw materials. Other refinements in manufacturing have produced panels with ever-increasing strength that are also easier to cut and better to work with. Particleboard and similar panels will continue to be difficult to cut in terms of abrasive tool wear. With correct tool geometry, newer cutter-knife materials, and quality improvements in the panels, however, most of the earlier problems have been overcome.

In today's woodworking industry, there is an enormous range of materials to choose from. Each has unique characteristics to understand when searching for a functional machining process or proper tool-knife material. Following is a summary description of the most prevalent materials:

Natural solid wood. Regardless of specie or where taken from in the harvested tree, natural woods cut easily in comparison to most man-made materials. Moisture content is the most important controllable variable to observe. Chips from cutting are dense and well formed, and control of cutting-generated heat is not often a problem. Excluding unusual mineral deposits, tool wear is consistent and predictable within specie groups. Cutting wet woods is accomplished with alloy-metal-tipped tools capable of withstanding elevated temperatures. (See Illus. 2-11.)

Glue-line solid woods, plywoods, and *reconstituted woods.* Solid woods with glue lines differ from those without in a couple of significant ways. First is the presence of adhesive(s) that makes the material much more abrasive to tools. Also, the arrangement of the individual parts in the now larger block adds to the complexity of natural stress conditions already present.

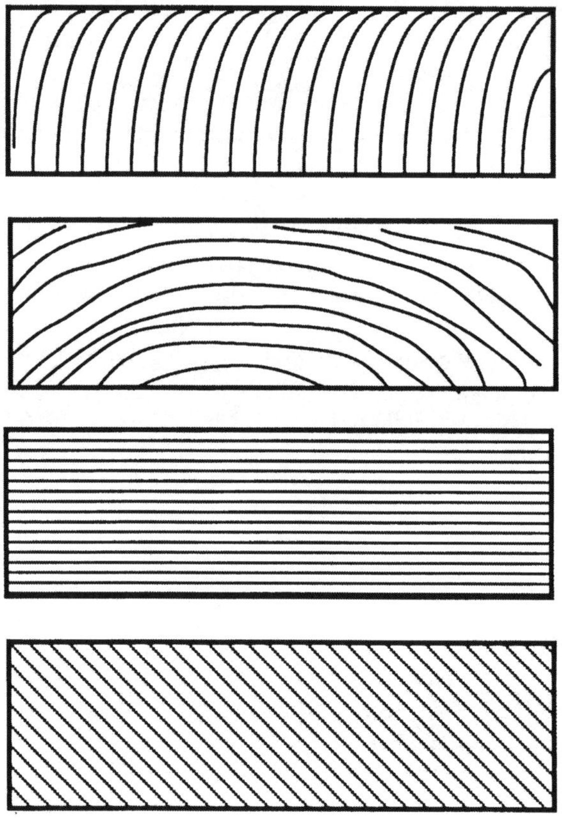

Illus. 2-11 Wood solids

Plywoods, while not difficult to cut, are very abrasive due to glue(s) present and to the cross-patterned grain arrangement. Cutting temperature is not a problem except as it may relate to tool wear. Plywoods are tough, strong, dense, and resilient and, when cut, produce chips that easily capture and dissipate the heat generated.

Reconstituted woods are nearly identical to plywoods but may actually contain a greater percentage (by weight) of glue. Conversely, these materials are produced under more stringent conditions than plywoods and are less subject to the inclusion of foreign substances. Overall, the structure of these products is such that they often cut at higher temperatures than plywoods.

Wet-process and *dry-process hardboards* and *fiberboards.* The wet-process suspends wood-fiber material in a water slurry. Later, during the pressing operation(s), moisture is extracted as the sheets and panels are formed. The sheets are free of any foreign matter and contain only wood fibers, fiber bundles, and the naturally occurring wood lignins and wood chemicals. These panels are not plied into distinct layers, nor do they contain different size particles, though they are less dense at the center. The small particle size of the cut waste does produce very high cutting temperatures with these panels.

Dry-process panels and sheets differ from wet-process panels in several ways. They contain adhesive; they may contain foreign material; and they may have a wider range of fiber(s) and fiber-bundle sizes in their makeup. These panels cut hot and generally are more abrasive to tools. While not distinctly multilayered, dry-process panels display a greater range of density between the center and outer surface and overall are less dense than wet-process panels.

Particleboards, chipboards, flakeboards, and *oriented strandboards (OSB).* Particleboard and chipboard are largely identical: They are both classified as three-ply panels; both have fairly large amounts of adhesive; and they have similar density (only chip/particle size may vary). Both are much denser at the outer surfaces than at the center. Both have an unfortunate history of containing large and varying amounts of foreign materials; that problem, however, has been largely solved. Both panels are abrasive and fairly hot cutting, though both of these problems are also being corrected. (See Illus. 2-12.)

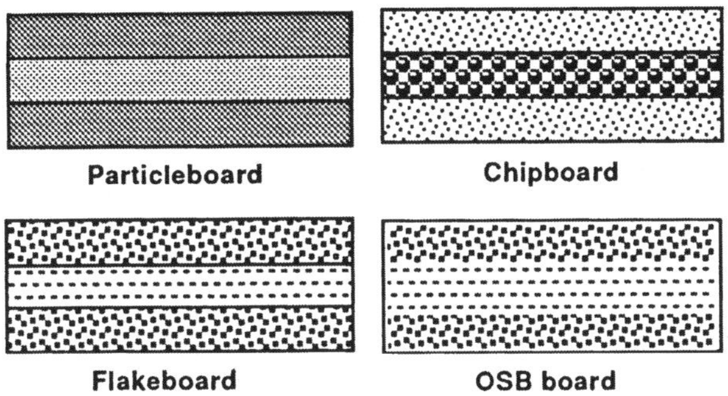

Illus. 2-12—Panels

Flakeboard and OSB are alike, both being constructed of large, flat, thin wood shavings that are steam processed and pressed into panels. Both contain fairly large amounts of adhesive and are more like plywood in many strength qualities. OSB is much like plywood in that the shavings are arranged lengthwise, yielding some unusual properties. Both are abrasive, though not as hot cutting as other man-made panels, and both are three-ply panels.

In the ever-widening selection of man-made wood-based panels, there are two other elements that have altered the effectiveness, selection, and use of cutting tools: the types of adhesives used in such panels and the wide range of coverings (laminations) created for them. Many of these materials increase the number of variables in the makeup of the products and add chemical conditions not encountered before in the cutting process.

Woodworkers have to be more knowledgeable than ever before, and this will require more than reading packaging labels. It is not that the same rules don't apply; it is that there are more materials and machine processes to make new rules for. As technology advances, we have to make every effort to learn as much about it as possible. Many technological advances do not reach a very large audience for relatively long periods of time—even those that could be very profitable for the developers of the particular technology. Suffice it to say that we are seeing an explosion of changes that will be interesting to participate in and observe for many years to come.

Chapter 3
Cutting Tool Materials

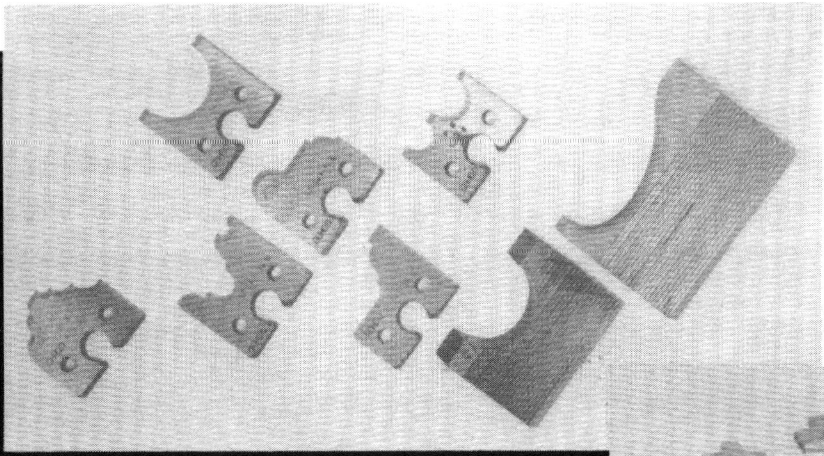

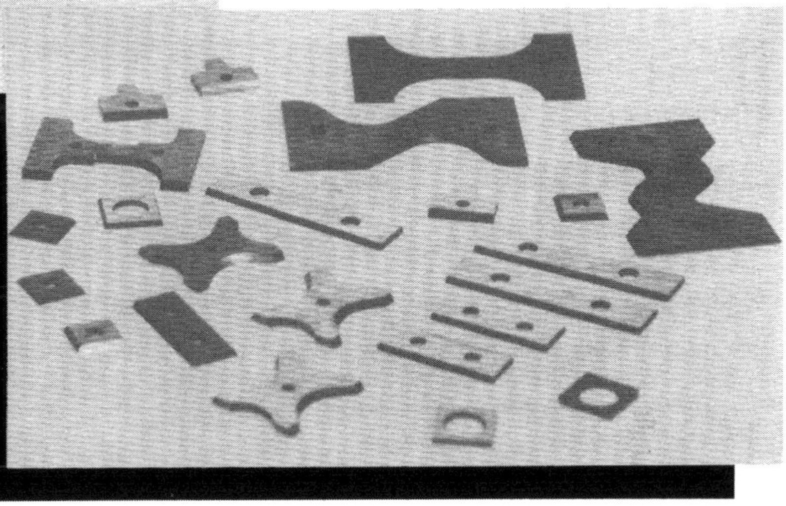

Photos courtesy of Leitz Tooling Systems, Inc., Guhdo USA, Inc., and Leuco Tool Corp.

Cutting Tool Materials

The primary metals that make up cutting tools are steels and alloys of steel. Cutting knives are made of or tipped with a wide range of materials, the most common being *high-speed steel* and *tungsten carbide*. Other rare-earth alloys may in the future become important for knife-tip materials—polycrystalline cubic boron nitride and others—but their advantages do not yet compensate for their considerable cost.

We will not address in depth all the materials that our tools are made of, but a basic examination can provide enough information to guide woodworkers to make better decisions when selecting cutting tools.

A Short History of Iron and Steel

The earliest evidence of iron being shaped into useful objects dates back to 3700 B.C. The discovery of molds used to form sickle blades and other agricultural tools gives us evidence of successful production of iron in China before 1000 B.C., and some of the earliest records of iron making are from China and the area now known as Afghanistan. A technology that has been around for almost 5,700 years can't be called a recent development, but in fairness to more recent technology, the development of metallurgy has flourished significantly only in the last 250 years.

First in our study of cutting tool metallurgy, we must clarify one point: When does iron become steel? This happens when we add something to iron to alter it so that the resulting steel has particular desired properties. We may want a metal that is harder or softer or perhaps less subject to rusting (oxidation). We may want a metal that is more easily formed, has a lower or higher melting temperature, and so on. We can, with the addition of other materials and with various methods of heating and cooling the resulting alloys, dramatically change their properties to suit specific needs. This technology has been advancing quickly for the past 100 years as our ability to control all of the processes involved improves.

Iron is a naturally occurring element. When taken from the earth, it is a reddish-brown substance that crumbles into powder as easily as lump of hard clay. In its natural state, iron ore is iron oxide and is usually mixed with other substances, impurities present in at least trace amounts. An interesting aspect of the history of iron making is that iron ore, unlike many other metals, is plentiful and found throughout the world. Also unlike metals such as copper, nickel, and others, iron is found only in the form of iron oxide ore—not a recognizable pure metal. The quality and makeup of iron ores found in various places in the world differ in trace elements and impurities, which affected many of the earliest metals that were produced.

Discovering how to convert iron ore into a useful material was probably an accident—at least that's the most popular theory. Scholars tell us that ancient hunters first converted iron ore into a mass of metal in a fire at a hunting camp where several ideal conditions were at the same time present.

The earliest iron alloy was probably crude *wrought iron*, a soft metal of uneven properties that was difficult to work and the product of partially melting a mass of ore in an open fire. The semi-molten metal was shaped by pounding it into a useful shape while still white hot. Pounding not only the shaped the metal but also removed most of the ash, slag, and trapped gasses. The product was not iron by modern definition because it was not heated to the point of melting, even though the pounding, or forging, probably removed most of the oxygen, a critical impurity in any metal alloy.

In that ancient camp fire mentioned earlier, the fire was perhaps built over a mass of iron ore. The burning wood produced charcoal, and the glowing charcoal united chemically with the oxygen in the iron ore to form a common gas, carbon dioxide. This would create a spongy mass of nearly melted metal from which the earliest wrought iron was produced. As mentioned before, the heat of an open fire was not sufficient to melt the iron ore into a liquid form.

The next advance in metallurgy was the movement of more air to the fire, first by shaping the fire place to capture the wind, later by forcing air into the fire to increase the rate at which the fuel burned, thus elevating temperatures dramatically. Reduction of iron ore, i.e., the conversion from ore to metal, begins at 2,200° F. The melting point of iron is 2,800°F.

In the earliest production of iron, temperature was elevated by use of a variety of furnaces that used many ingenious methods to move air into the process. A wide assortment of bellows and other means were employed to advance metal production. In many parts of the world, alloys of iron and steel are produced today just as they were 2,500 years ago. Furnaces made entirely of ceramic brick are still used, with only primitive means to elevate temperature sufficiently to melt the metals.

Making Steel

The fuel for the melting process is *charcoal* or *coke*. Charcoal is produced by heating wood in an atmosphere containing as little oxygen as possible to remove all the moisture, gums, tars, and other materials, leaving only a crude form of carbon. Coke is made in a similar process that begins with coal instead of wood. Once ignited, coke or charcoal can be burned in controlled ovens, furnaces, and cupolas to produce temperatures of over 3,000° F.

Carbon, and not just in random amounts, is the primary element that, when added to iron, makes iron into steel. The addition of controlled amounts of carbon affects the properties of iron more than adding any other element. Before describing the effects of carbon on iron, we should make it clear that the melting process removes any carbon that may have been present in the ore or the fuel burned in the process, even though the fuel used, charcoal or coke, is nearly pure carbon. The addition of oxygen to the melting process elevates the temperature, accelerates the burning of the fuel, and effectively burns all carbon away from the molten metal, removing it as carbon dioxide. Pure iron contains no carbon at all, even though other trace elements may be present.

Carbon steel is a general term given to steel whose principal alloying material is only carbon, as opposed to any of the other metals (e.g., copper, manganese, or chromium) often added to iron to create various physical properties. The amount of carbon added to iron ranges from 0.05 percent to 1.7 percent. Pure iron is a relatively soft, weak, easily formed metal that cannot by any method be hardened, but the addition of even 0.05 percent carbon creates steel of remarkable strength. The addition of 1.4 percent carbon yields what we have come to call *carbon tool steel*, from which we make drills, saw blades, and other metal objects that are resistant to abrasive wear and require great strength. The upper limit of 1.7 percent carbon content results in a metal that is too brittle for any practical use. Where produced, this alloy is usually processed further to remove a portion of the carbon.

All metals are classified as crystals, that is, as a metal changes from a (molten) liquid state into a solid, it crystallizes. During the solidification process, the atoms organize themselves into an orderly cubic arrangement. There are several types of cubic patterns that metals take, and the complete chemical content of a mixture alters the patterns further, as does the exact process involved in cooling the metal. Addition of even very small amounts of carbon completely changes the cubic pattern of iron into the remarkably different metal, steel. The property that carbon brings to the mixture is that the atoms of both elements arrange themselves into a cubic pattern completely unlike the arrangement possible without the presence of the other element.

Heat Treating

After adding carbon we *heat treat* metals and alloys of metals to affect their properties to make them more suitable for tools and machinery. Generally speaking, heat treatment is any of the several methods of conditioning a metal or metal alloy and altering its properties. For instance, if a steel alloy having a carbon content of 1.4 percent is heated to white-hot temperature then quenched (cooled rapidly) in water, the metal will be both hard and brittle—easily broken by even moderate impact shock. To eliminate the brittleness, the same metal is reheated to a lower temperature and allowed to cool slowly. The resulting metal will be as hard as that originally created but now able to perform the intended work without failing.

Heat treatment processes may involve many steps requiring several heating and cooling cycles over a period of hours or even days. Many of these processes were no doubt discovered by accident, but today, careful scientific procedures examine all the details and properties of the metals and other additives involved. Alloys of iron respond differently to heat treatment than do alloys of other metals, creating a completely different set of procedures for each.

In carbon tool steel, there are at least four elements

present in addition to iron and carbon, two that are undesirable and two that are added to counteract the effects of the undesirable ones:

1. An undesirable element in any iron alloy is *phosphorus*, which has the effect of segregating the makeup of the alloy and severely weakening it.
2. Sulfur is undesirable in any iron alloy because it makes the resulting metal unpredictable, weak, and more brittle.
3. Silicon added to an iron alloy still molten has several benefits. Silicon easily forms silicon dioxide, removing the excess oxygen and other suspended gasses. Silicon also disperses other contaminates such as phosphorus in the iron mixture, lessening their effects. Silicon that is not dissolved in the molten metal is easily removed as it is lighter and floats. Small percentages of silicon in the steel alloy has little other effect on its properties.
4. To counteract the effects of sulfur, manganese is added to the mixture while it is still molten. Manganese has a considerable affinity for sulfur and easily forms manganese sulfide, in addition to removing oxygen and other suspended gases. Manganese increases the strength and hardness of steel but reduces its plasticity.

Strength and Physical Properties of Metals

There are several metals that are added to steel alloys and as many heat treatment processes to improve or alter them to suit a particular need. For the manufacturers and users of wood cutting tools, there are specific needs to be addressed in successful selection and use of the metals we employ. These include four strength characteristics and five physical properties. The strength characteristics are the following:

1. *Tensile strength* is the maximum load in tension which a material will withstand prior to fracture. This is the most commonly given value for the strength of a material and is expressed as pounds per square inch or kiloNewtons per square meter.
2. *Compressive strength* is the maximum load in compression that a material can withstand prior to a predetermined amount of deformation. An example of compressive strength is exhibited best by cast iron or concrete. Due to their brittle nature, both have greater compressive strength than tensile strength. Steel and most other metals have greater tensile strength than compressive strength.
3. *Fatigue strength* is the maximum load that a material can withstand without failure during a large number of load reversals. Fatigue strength is the normal wear life of any object. For instance, a shaft in a machine has predictable life expectancy. Before any part of any machine is produced, every detail has to be known and predicted with accuracy, for example, the life expectancy of moving parts in an automobile or even the normal flex in an airplane wing
4. *Yield strength* is the maximum load at which a material exhibits a specific deformation. Most engineering calculations for most metals are based on yield strength rather than tensile strength. For instance, the yield strength of copper is only about 50 percent of its tensile strength. For mild carbon steel, the figure is nearly 85 percent. Yield strength is a safer figure to work with in any engineering calculation that deals with the expected function of a material.

These four strength indices yield true measures of what a metal or any other material is capable of withstanding. Each gives a different perspective on how we make use of not just cutting tools, but also all the machines that we employ in our work and their constituent parts. The physical properties we use to describe metals (and other materials) we work with are:

1. *Elasticity*—Since placing a load on a material causes a change in its form, elasticity is the ability of a material to return to its original form after removal of the load. Theoretically, if loaded beyond its elastic limit, the material will be permanently deformed. Though metals are not entirely elastic, there are accurate, workable values to express the extent to which metals exhibit elasticity.
2. *Ductility*—The plasticity exhibited by a material under tension loading is known as ductility, measured by the amount that it can be permanently elongated. Ductility is the property of many metals that permits them to be drawn into wire or other similar forms.
3. *Malleability*—This is the ability of a metal (or other material) to be deformed permanently under compression without rupture. It is this property that permits hammering (forging) and rolling metals into sheets and other commercial shapes—angle iron, beams, forgings, etc.

4. *Toughness*—The description of a material such as a metal that has a high tensile strength and the ability to deform permanently without rupturing is toughness. There is no direct, accurate method of measuring toughness of metals, as it involves both ductility and strength. Toughness is also taken as the ability of a material to absorb energy, impact shock, and stress without failure.
5. *Brittleness*—This physical property is essentially the opposite of plasticity. A brittle metal cannot be visibly deformed without failure. Certain cast iron and fully hardened tool steel may exhibit very little plasticity and could be classified as brittle. This is not, however, a measure of plasticity. Brittle metals exhibit little strength upon tensile loading but may be safely used in compression and, further, show very little shock or impact strength. There are no values for brittleness in metals or other materials, and they may fail in this regard without any warning.

Because the structure of metals is crystalline, there are unique properties that direct how they can be safely used for constructing things. The *notch effect* refers to the great extension of the strength of any metal object by avoiding sharp corners, both inside and outside, in its use. (See Illus. 3-1.) A sharp corner, abrupt change in cross-sectional dimension, or a notch will greatly reduce any metal object's ability to withstand stress or shock. *There is no exception to this rule.* The presence of even small radiuses at inside or outside corners will increase an object's ability to withstand shock and improve all measurements of its strength.

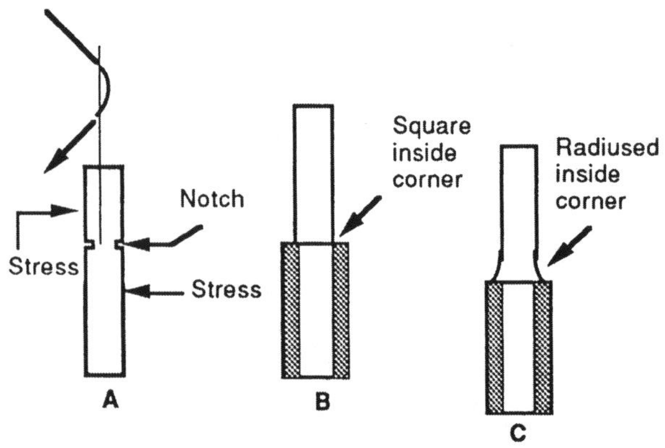

Illus. 3-1—Notch/sharp corner effect

Hardness is not a fundamental physical property of a material, but is related to its elastic and plastic properties. Generally, hardness is defined as resistance to indentation. The greater the hardness, the greater the resistance to penetration. Hardness tests are widely used because of their simplicity, and the results can be correlated to tensile and yield strength. That is, we can define the properties of a given metal more closely if we can gather even general information about it. We know, for instance, that steel without a certain carbon content cannot be hardened, and that hardness, depending on the test method used, varies in direct proportion to the alloy content of the metal in question.There are many different ways to describe hardness in metals, and still others to describe hardness in other materials.

Applying Metallurgy

From the list of strengths and physical properties of metals, it is apparent that they are very complex. As toolmakers and tool users, we need to know, first, what will work safely, and, then, what will work well.

The decisions toolmakers come to when selecting a metal alloy to make a given tool are based in part on information that woodworkers bring to them—the machine process involved, wood specie cut, available horsepower, rpm of the tool, expected feed rate, and such details. To make the correct metallurgical choices, toolmakers have to understand all the cutting processes involved, as well as the strengths and characteristics of the materials that will be used to make the cutters.

The principles that guide toolmakers' choices are essentially twofold—cost and applicability. For the manufacturer who mass produces cutting tools, the choices will be examined carefully. The tools have to perform the task well, operate safely, and be competitive in the marketplace. An incorrect decision affecting any of the factors could be very costly, perhaps in many ways. Some of a toolmaker's decisions are guided by concerns about product liability, though this is usually a minor concern.

As the tool maker begins to design and manufacture a particular cutting tool, there are established criteria that have been researched, documented, and followed for many years, so it is not like reinventing the wheel for each tool. There is, though, considerable research going on every day all over the world to deal with the ever expanding demands of manufacturing that involves woodworking—new materials to be cut, newer and larger machines and machine processes, even larger horsepower motors used to drive the cutting processes faster.

A popular misconception is that cast metal is in some way inferior; it is not. The term *cast* simply means that the metal was melted and poured into a form. Most metals are produced in this manner—iron, steel, stainless steel, brass, bronze, and others. In fact, all metals that are melted in the course of their manufacture at some time are poured into a form before subsequent processing such as rolling, forging, or some other shaping operation.

The mistrust of cast metals may have originated with cast iron, but even this particular metal, though inexpensive to produce, fulfills a wide range of needs in objects that are common to us all, most notably engine blocks of automobiles and farm tractors, bath tubs, bodies of pumps that supply homes with water, and pipe fittings that carry waste water from homes and industries. The list could continue for pages just for cast iron objects.

The point is that for the hundreds of metal alloys we use every day and for every application of them, considerable study has been involved. It would not be reasonable, for instance, to fabricate the engine block for a lawn mower of cast stainless steel because there is no mechanical advantage of that alloy in that application, and the cost would drive the product out of the market. For engine blocks, cast iron is the best available material when weight is not a problem. For small automobiles though, the engine block is probably made of aluminum with cast iron sleeves for the pistons. This has been true only for the past 30 years, as aluminum has become more available and less costly. In summary, our choice and use of metal alloys is directed by their mechanical features, cost, and availability.

In the past 200 years, as metallurgy has become increasingly more scientific, many advances in both quality and quantity of metal alloys have taken place. This rapid progress, interestingly, has brought about a number of unusual metal alloys and processes that are legally protected by patents. There is tremendous commercial value for a unique alloy that will offer any advantage over other metals now in use.

Research into new metal alloys, processes, and heat treatments is pursued today by manufacturing companies, universities, governmental agencies, and independent companies that do only such research. Worldwide, there is an enormous amount of study in pursuit of advances in all areas of metallurgy. The most recent advances have come about because of the tremendous capacity of computers to handle large amounts of information with great speed. Computers are also involved in the complex functions in all areas of the manufacturing and subsequent heat treatment of metals. Today, a wide range of computer systems control the melting of ores, the correct mixing of elements, and the machinery to move, shape, and cool the metal as it is produced.

Metallurgical Demands of Cutting Tools

As we examine the metallurgical needs of the manufacture of wood cutting tool bodies, we start with consideration of the facts and figures that describe the work that cutting tools do. For instance, we know the following:

1. The operating characteristics of the machine and process where the cutting tool will be used;
2. How much of the work will be done within a particular time, at what rate, and whether or not this will be an ongoing process;
3. The nature and characteristics of the material to be cut by the tool, i.e., hard, soft, abrasive, hot cutting, etc.;
4. What the tool and its total costs should be in relation to the other processes involved in the manufacture of a given product.

These and other bits of information begin to direct decisions to help us make the best possible selection of the cutting tool for the particular process.

Exact description of the cutting process leads us to more accurate selection of metals for the tools. For instance, cutting tools are already being turned at maximum velocity before they are brought into contact with the material to be cut. This very basic observation helps describe the torque load that will be placed on the cutting tools. A worst possible case may be a router tool turning at 20,000 rpm being brought very abruptly into the material to be cut. If the motor driving the tool has five horsepower, the tool probably has to withstand the maximum load possible with that motor. The same would be true of a saw blade turning at 4,500 rpm, driven also by five horsepower and brought abruptly into the cutting process.

The point is that there are a number of performance characteristics for the toolmaker to carefully define and measure to proceed in manufacturing a cutting tool. Within the cutting process, there are

- High torque loads created by the work processes and more so by the fact that the work is interrupted and the tools are brought to the process abruptly over many cycles;
- Dramatic temperature changes as the tool goes from room temperature to operating temperature over hundreds and thousands of cycles;
- Great centrifugal and centripetal forces acting on the tool bodies;
- The effects of vibration, work stresses, and other forces.

These and other forces define what is needed in terms of the strength and physical properties of cutting tool bodies. Metallurgically, cutting tool bodies can be of three basic forms:

1. Made entirely of a material from which all features can be shaped—body details, cutting and relief angles, and knives. Such materials include tool steel, high-speed steel, or tungsten carbide. (A solid tungsten carbide tool will probably be small, given the

cost of the material and processes to produce it.)

2. Made of a material that has all necessary strength and performance characteristics, from which all details are shaped, but to which cutting knives are attached by fusion or chemically, as by welding or brazing: a carbide-tipped drill, saw blade, shaper tool, etc.
3. Made of a material similar to #2, except that knives are attached to the body mechanically, with a wide variety of wedges, clamps, and threaded screws. This type of cutting tool will have higher strength needs in that all forces acting on it are greater and its operating life will most likely be much longer.

We cannot suppose to know precisely what metal alloys every cutting tool manufacturer uses, but, of the cutters prevalent in industry today, most are well made and of excellent materials. (See UNS numbers and List of Tool Steel Alloys below and on next page.) The metallurgical makeup of a particular cutting tool is directed by:

- Work-load stresses that it will encounter. The alloy content is a design criterion of known factors relating directly to characteristics of strength, hardness, resistance to abrasive wear and elevated temperatures, etc.;
- Cost of materials and processes involved in its manufacture. The selection of a particular alloy is carefully examined for cost justification in each application. The cost of manufacture of alloyed metals will relate directly to the cost of the tool produced, and must meet the demands of the marketplace.

Alloy Steel

All cutting tools are made of alloys—that is, there are no pure metals used at all, iron or any other. As mentioned earlier, even basic iron contains at least four other elements—sulfur, silicon, manganese, and phosphorus—if only in trace amounts. The element that has the greatest effect on any alloy of iron is carbon, added in small controlled amounts, after which we call the much stronger and radically different metal steel. It is

A joint effort to provide a comprehensive means to identify metals and metal alloys offered by the American Society for Testing and Materials and the Society of Automotive Engineers is called the Unified Numbering System (UNS). The UNS is not a specification number, just a means of correlating the identification for a metal or metal alloy for which the specifications could be found elsewhere. The UNS attempts to avoid the confusion caused where two different numbering systems identify the same metal or alloy.

Metal Classification and Numbering Systems

	UNS Series	Metal
Nonferrous metals and alloys	A00001 to A99999	Aluminum and aluminum alloys
	C00001 to C99999	Copper and copper alloys
	E00001 to E99999	Rare earth and rare earth-like metals and alloys
	L00001 to L99999	Low melting metals and alloys
	M00001 to M99999	Miscellaneous nonferrous metals and alloys
	P0001 to P99999	Precious metals and alloys
	R00001 to R99999	Reactive and refractory metals and alloys
	Z00001 to Z99999	Zinc and zinc alloys
Ferrous metals and alloys	D00001 to D99999	Specified mechanical property steels
	F00001 to F99999	Cast irons
	G00001 to G99999	AISI and SAE carbon and alloy steel (except tool steels)
	H00001 to H99999	AISI H steels (hot-work steels)
	J00001 to J99999	Cast steels (except tool steels)
	K00001 to K99999	Misc. steels and ferrous alloys
	S00001 to S99999	Heat and corrosion resistant steels (stainless steels)
	T00001 to T99999	Tool steels

Tool steels are classified by their contents, heat-treatment process, features, machining, or intended special purpose. There are far too many steel alloys to list here, though there are many industrial texts that provide extensive tables to detail the contents of steel alloys commonly in use.

Classification of Tool Steel Alloys	
Classification	**Letter Symbol**
Water hardening	W
Shock resisting	S O (oil hardening)
Cold work	A (medium alloy D (high carbon and chromium)
Hot work	H (H1-H19 = chromium base) (H20-H39 = tungsten base) (H40-H59 = molybdenum base)
High speed	T (tungsten base) M (molybdenum base)
Special purpose	F (carbon/tungsten)
Mold steels	P

carbon that affects the hardness of any steel alloy, regardless of other alloying elements present (to a maximum of about 68-C on the Rockwell scale, a standard for measuring hardness in metals).

Of all iron alloys produced worldwide, the greatest tonnage is plain carbon steel. An alloy steel is any carbon steel to which other metallic elements have been added in sufficient quantities to alter its characteristics. The addition of other elements to steel is accomplished in any of four ways:

1. Dissolving the element into the steel as a liquid or solid state, thus retaining it at room temperature in solid solution;
2. Combining the element chemically with carbon in the steel to form carbides or, in some instances, with nitrogen as nitrides;
3. Forming an oxide with oxygen trapped in the steel while it is still molten;
4. Remaining uncombined, though suspended, as lead does when added to improve the machinability of several steel alloys.

Several metal elements are commonly added to steel to alter its characteristics, thus creating alloy steels such as the following:

Manganese Steel—Manganese is always added to steel, having direct and indirect benefits. In larger amounts, manganese changes how the alloy is heat treated. In smaller amounts, manganese overcomes the negative effects of sulfur and oxygen.

Tungsten Steel—Tungsten dissolves in steel to form a carbide and imparts both strength and hardness. Tungsten creates a fine, dense grain structure; increases overall toughness at high temperatures; and greatly increases retention of hardness at elevated temperatures.

Molybdenum Steel—Molybdenum is considered the most useful element to be alloyed in steel. Like tungsten, molybdenum forms a carbide (i.e., joins with carbon), but in correct proportions is nearly twice as effective. On the negative side, however, molybdenum added to steel by itself is very difficult to heat treat in any way and for this reason is most commonly used with other metals such as nickel, chromium, or tungsten.

Nickel steels—Nickel dissolves easily in steel and greatly affects it's microstructure. With certain percentages of nickel, the alloy is magnetic, while with other percentages, it is nonmagnetic. Nickel adds corrosion resistance and greatly affects the coefficient of expansion of steel. Joining nickel with another metal such as chromium results in an alloy that remains nonelastic over a great range of temperatures, making its use in thin-section cutting tools, such as saw blades, a great benefit. Nickel does not dramatically affect heat treatment processes. In recent years, cobalt has been substituted for nickel with nearly identical benefits.

Chromium steels—Of all special elements added to alloy steel, chromium has the widest range of application. It dissolves easily in iron and in the presence of carbon combines into a very hard carbide. Chromium imparts hardness, wear resistance, and useful magnetic properties to steel and increases its resistance to corrosion. It also increases steel's ability to be completely hardened without extraordinary measures. In addition to carbon, chromium is added to steel alloys that are to be used as wood and metal cutting tool knives. *Stainless steels* are a large family of chromium steel alloys in which numerous other elements are present. Originally intended for high-grade cutlery, this group of metals has found widespread use in many areas of industry because of its resistance to a wide range of corroding materials.

Vanadium steels—Although not a major constituent in the manufacture of steel, the role that vanadium plays is very important. The addition of even small amounts of vanadium stabilizes the alloy content throughout the heat treatment processes by keeping its makeup homogeneous. Vanadium joins easily with any carbon present, forming a stable carbide.

Silicon steels—While silicon is present and soluble in all steel and iron alloys, when added in precise amounts, it improves the tensile and yield strengths of the metal, especially for products such as springs. Silicon in other alloying mixtures improves certain corrosion and oxidation resistance, especially at elevated temperatures.

High-speed steels—These are so named because when they are used for cutting tool knives they remove the material being cut much faster and at higher temperatures. Their superiority over other steels is the ability to maintain hardness even at a dull red heat—about 1,100° F. Due to this property, such alloys find widespread use over others that would lose hardness. The most common makeup of high-speed steel is called *18-4-1*, meaning 18 percent tungsten, 4 percent chromium, and 1 percent vanadium (carbon content varies from 0.75 percent to as much as 1.30 percent) Recently, molybdenum has been successfully substituted for tungsten, though most often the two are used together in identical percentages (about 5 percent each). Numerous alloy mixtures may be called high-speed steel, though most are identified by the content of their primary alloying elements, tungsten or molybdenum.

Tool steels—This large family of metal alloys has been developed for unique characteristics or for specific cutting functions. Many are similar to high-speed steels, though they are much less expensive to produce, or are considered more suited for specific purposes. Often, they are trade name protected or patented and marketed vigorously as having special properties needed in various segments of metal and woodworking industries. Tool steels, as all alloyed metals, are generally categorized by the percentage of alloying element present, though for this group of metals the greatest concerns are hardness, strength, and abrasive wear. Hardness depends, first, on the amount of carbon present, followed by other elements such as chromium or tungsten.

Plain carbon steel, for instance, can be considered a tool steel, as it is the most common metal used to manufacture blades for band saws, jig saws, and certain circular saws. For most alloys in the category of tool steels, elements other than carbon are present in varying quantities. Knowing the specific amounts of various elements in a particular metal alloy, however, does not mean we can accurately predict how it will behave; heat treating processes can dramatically change its performance. Heat treatment is that important and may vary in a minor way from one sample of metal to another.

Cutting tool manufacturers are not required to divulge the exact metallurgical contents of their products beyond a very generalized description, if at all. There is more than one reason for this seeming paradox. First, cutting tool design features can be trade protected, as can the materials that compose them. For many cutters, such disclosure would perhaps jeopardize any marketing advantage a tool may have. Further, beyond ensuring that tool performance characteristics are stated in any advertising or product description, it is not often necessary to describe in such complete detail as to include their metallurgical content.

For these reasons, tool manufacturers identify cutter materials only very generally and usually include only five categories of metal alloys:

1. *Plain carbon tool steels*, which contain from 0.45 to 0.8 percent carbon, along with other common trace elements;
2. *Mild-alloy tool steels*, which have similar carbon content and are alloyed with not more than 5 percent of other elements;
3. *High-alloy tool steels*, which have similar carbon content and are alloyed with more than 5 percent of other elements;
4. *Special-alloy tool steels*, which have similar carbon content and are alloyed to a maximum of 12 percent with unique elements such as tungsten, molybdenum, or chromium. In this classification, only one or two alloying elements are involved;
5. *High speed steels*, which have similar carbon content and are alloyed to more than 12 percent with a number of metallic elements, e.g., tungsten, cobalt, vanadium, or chromium, in some relationship to each other. Most often, these alloys contain at least three unique materials beyond carbon, iron, and common trace elements, depending on specific use of the metal.

Tool Manufacturing and Marketing

Given the number of tool manufacturers, foreign and domestic, that supply the U.S. market, there is no way to state with confidence what metal alloy is used in any given tool. We can, however, reasonably address what the contents should be. It must be said initially that quality of materials, workmanship, and performance characteristics differ more between segments of the market, i.e., hobby, craft, and industrial, than from one manufacturer to another. Saw blades for the hobby market are more like each other, regardless of who produced them, than they are like nearly any saw blade produced for industrial use.

This is because most manufacturers do not attempt to produce tools for all three market segments. There are tool producers that provide only hobby tools and others that manufacture only more expensive and precise craft and industrial tools. This may seem judgmental, but it isn't. It is simply an observation that each kind of tool has a purpose, and manufacturers address individual market needs. The manufacturing and marketing of cutting tools has already been segmented into specific areas. For instance, a hobby worker or wood artisan could not justify the cost of a saw blade capable of cutting tens of thousands of feet of natural wood in each sharpening cycle, any more than an industrial worker could operate with a saw blade that would last only a few minutes when cutting very abrasive man-made material.

The composition of cutting tools depends on two basic criteria:

1. For the most common tools—saw blades, drill and router bits—composition depends on whether the tool is intended for hobby, craft, or industrial use. What is the selling price in relation to the intended use? What are the service-life considerations?
2. For tools intended for industrial and craft use, composition depends on the specific function of the cutter. What are the strength and performance characteristics in relation to real cost indices?

Cutting Tool Body Materials

For most cutting tools, a careful and complete description of their use will specify their metallurgical makeup. For instance, the best industrial saw blade will be made of an alloyed steel capable of withstanding high torque loads, repeated changes in temperature, and high temperatures; will resist deformation; and will exhibit characteristics of rigidity without being brittle. Also, the metal alloy must not work-harden with extensive work cycles, standing up to all the rigors of the environment, work, and temperature changes throughout its work life.

In the past 40 years, changes in the alloy content of cutting tools have come about because of the creation of man-made materials and their introduction into the woodworking industry. These materials radically changed how effectively cutting tools could function. The abrasiveness of new adhesives, higher cutting temperatures, plastic coverings and finishes, and the allowable limit in man-made panel materials of such foreign materials as sand, clay, and even metal debris have had influence on the manufacture of cutting tools. The other driving force of the changes in cutting tool makeup has been the needs created by advances in machine technology—ever larger motors, ever faster tool spindles and quicker feed speeds, and the demand for ever greater precision. It is not too surprising that cutting tools available today are unlike those of even 20 years ago.

In identifying the metal alloy contents of wood cutting tool bodies, three points need to be clear;

1. We cannot know the exact percentages of alloying materials used in each tool; we can know only what is, or should normally be, present.
2. Even knowing the elements in the metal alloy, we cannot predict totally the characteristics of any tool because subsequent heat treatment processes dictate these nearly as much as do the contents.
3. The purpose in identifying the alloy makeup of a cutting tool is to determine quality. Expensive materials, in all likelihood, will result in a well designed cutter. At least it does not make sense that a manufacturer would use expensive materials in a poorly designed or shabbily made product. Conversely, it is not possible to make a cutting tool with certain strength properties and performance characteristics without using more expensive alloyed metals.

Following is a general description of cutting tools and the alloys generally used to make them:

Saw blades for the hobby market are made primarily of plain carbon steels and mild chromium alloys.

Saw blades for the craft market vary widely in alloy content, from plain carbon steels to very specialized alloys containing such elements as chromium, tungsten or molybdenum, and vanadium.

Saw blades for industrial use vary somewhat, but nearly all are made of expensive alloys. Chromium, tungsten or molybdenum, nickel, and vanadium may be present in fairly high percentages. Most reflect the effort to support all of the task needs of industry very well.

Band saw blades for all market areas do not vary in content as much as other tools. Most are made of refined plain carbon steels and alloys containing small percentages of tungsten and vanadium. The alloying elements in these tools are not as overwhelmingly important as accurate heat treatment processes.

Drills and boring bits for the hobby market are primarily of plain carbon steels with small percentages of chromium to support heat treatment processes.

Drills for craft and industrial use vary widely in alloy content. Carbon and chromium are the primary elements used to ensure successful heat treatment, with additional tungsten, vanadium, and perhaps cobalt for strength and toughness characteristics.

Router tools for hobby use are similar to their larger counterparts for use in the craft and industrial markets, at least in strength characteristics. The alloy contents are chromium, tungsten or molybdenum, and perhaps manganese.

Router tools for craft and industrial use, though varying in makeup, are usually made of remarkably strong alloys containing high percentages of tungsten or molybdenum, chromium, vanadium, and perhaps cobalt, nickel, or manganese. The overall costs and heat treatment processes used by the particular manufacturer will dictate the alloy makeup as much as any other criteria.

Shaper and moulder heads for all market areas do not differ greatly, in that the strength features called for are fairly constant for all functions. The bodies are made primarily of mild carbon steels having moderate amounts of other elements that improve their strength, resistance to stress, and machinability. Lead added to steel improves machinability, and even small amounts of vanadium and other elements greatly increase the strength.

Shaper and moulder heads with mechanically secured knives require greater body strength than those with brazed-on knives or the entire body made of a tool steel alloy. The alloys of these tool bodies will have mild-to-medium carbon content since great hardness is not required; lead content to improve machinability; silicon, tungsten or molybdenum, and possibly other elements to increase the strength and resistance to deformation. Strength and resistance to stress and shock are the primary concerns because these tools usually have a longer intended life expectancy than other tools that perform similar tasks.

Hand-held tools for all woodworking processes—saws, chisels, planes, and others used to cut and shape solid wood—vary greatly in alloy content. Hand-held cutting tools should exhibit good wear characteristics, ease of sharpening, resistance to corrosion, and reasonable cost in proportion to intended use. These cutters do not have to withstand the great work loads and elevated temperatures of powered equipment tools. Though alloy contents vary widely, the two most common elements used are carbon and chromium, the former for hardness and the latter to resist corrosion.

Of all features we seek in cutting tools, the greatest is resistance to wear. No doubt the most popular misconception of all cutting tool features is hardness and how it relates to abrasive, thermal, and chemical wear. Wear resistance has more to do with the microscopic structure of the metal alloy than nearly any hardness criteria.

Hand-held tools, depending on their particular function, will be made of small percentages of tungsten or molybdenum, vanadium, manganese, silicon, and nickel, in addition to carbon and perhaps chromium. Though many of these tools have been produced for hundreds of years, only recently have they been made of modern alloys.

Chisels today are made of a wide range of carbon steels, tool steels, and, in a few instances, even high-speed steels.

Wood cutting files and rasps need only to be very hard, as there is no particular need for other features. Files and similar tools are made of high carbon (water hardening) steels.

Hand-held saws vary widely in alloy content, also in price, features, and workmanship. Most are made of mild carbon steels with varying amounts of chromium and other elements that support strength, wear, and corrosion resistance.

Hand-held tools are used by every woodworker, including, students, artisans, hobby workers, home owners, craftsmen, and all segments of industry. Given their widespread use, the differences in quality is to be expected. Though appearance alone is not an accurate or complete indicator of any quality, with regard to metallurgical content it is not likely that any manufacturer would invest the labor and cost to provide excellent details to a tool made of poor-quality materials. Brand name recognition is a guide to selecting hand-held tools that are common to all of us, and even there careful scrutiny is advised.

The metallurgical quality of industrial tools is not often in question because if they are made of inferior materials, they will not survive and will quickly be replaced. The means tool manufacturers utilize to bring cutting tools to industry is much different from those for other markets. Industrial tool users most often have a direct line of communication to tool makers. The sales and distribution of industrial cutting tools is a much smaller and more compact network than for other market segments. Several of the largest tool manufacturers worldwide have their own distribution systems in place, with some even having complete means to service and sharpen the tools they sell. For the tool user and manufacturer alike, there are advantages to having this close relationship. The tool users are assured of the technical support of the toolmaker, and the toolmakers are assured a firm, steady market for their products.

For all tool users, student, hobby worker, artisan, craftsman, and industrial worker, it is vitally important to learn everything possible about the cutting tools that are so necessary in their work. In the marketplace, most cutter bodies are made of steel alloys having the correct features of strength, toughness, and other needed properties, but we must understand that there are no inexpensive, well-made cutting tools.

Cutting Knife Materials

The next topic to examine is the materials that make up cutting tool knives, including:

- Steel alloys,
- Patented cast nonferrous alloys,
- Cemented carbides such as tungsten carbide and other alloys,
- Ceramic elements,
- Polycrystalline diamond.

The knives of wood cutting tools are either a part of the tool body itself, attached to the tool body by fusion, i.e.,welding; attached chemically, i.e., by brazing; or attached by a wide variety of mechanical means.

Primarily, cutter knife materials need resistance to wear and to the effects of high temperatures of the cutting process. That may appear simple, but the problems of tool wear are much more complex; what seems to be simple abrasive wear may actually be the effects of a chemical reaction or thermal corrosion (as with, perhaps, cutting a wet material with certain alloys of tungsten carbide). Numerous cutter knife materials have been developed in recent years, each attempting to solve the many wear problems encountered by woodworkers.

Earlier, we described three steels commonly used as cutting knives—*plain carbon tool steels*, *mild alloy tool steels*, and *high-speed steels*. The three differ primarily in cost (ascending as listed) and effectiveness. All are widely used in different applications where appropriate, that is, any of the three would function well for most cutting process involving any solid-wood species without glue lines. An adhesive, however, adds an element to the process that would prohibit their use; the tools would dull very quickly. Other alloy steels, such as those having high percentages of manganese, silicon, or other metal, are not as widespread because they are too expensive and are more appropriate for cutting metal where the environment, especially temperature and cutting speed, is better controlled.

Patented nonferrous metal alloys are finding more widespread use as wood cutting tool knives when they offer an advantage, usually in a specific process or range of processes such as cutting green or wet wood materials. These alloys are made from tungsten, tantalum, chromium, cobalt, and carbon. Other elements are also used to produce a temperature-resistant and wear-resistant material. These alloys cannot be softened by any heat treatment process and must be cast and ground into the needed shape. Room temperature hardness of the nonferrous alloys is lower than high speed steel, but the wear and temperature resistance is retained over a much greater range of temperatures—even red hot. As a starting point, the recommended feed speeds for these alloys can be 20 to 50 percent higher than those of high-speed steels.

Trade names of these alloys include Stellite, Crobalt, and Tantung. All can be easily braze-attached to any carrier body material such as a saw blade, planer knife, or shaper cutter. All three are gaining favor for manufacturing processes involving solid woods, especially where high moisture contents elevate the cutting tool temperatures dramatically.

Cemented carbides are the product of a unique area of metallurgy called powdered metals. The construction of metal products from powders is not a recent discovery. In fact, some of the processes involved date back nearly 1,600 years. Metals can be shaped into useful products by starting with a powder of small particles of the metal needed, pressing it into a form, then fusing it together by any of several means. There are five primary groups of metal products currently produced by these processes:

- Cemented carbides,
- Refractory metals,
- Oil-impregnated bearings,
- Porous metal filters, and
- Machine parts.

Of interest as cutting tool knife material are *cemented carbides*: tungsten carbide, titanium carbide, and boron carbide. A critically important feature of these alloys is the size of the particles in the mixture. There are several hundred grades among these alloys that primarily relate to mixture percentages and grain, or particle, size. These alloys are much harder than high-speed steels and retain their hardness and temperature resistance up to 1,400° F and higher.

A detrimental feature of carbides is that they are brittle and have to be handled carefully to avoid impact shock. Correct alloy makeup and selection make carbides an ideal knife material for most applications. Overall, carbides have brought many advances to cutting tool technology in the past 25 years and are the most widespread knife material in the woodworking industry today.

Cemented carbide means simply that the elements are bonded together with a binder or metal matrix material. Most common of the carbides is tungsten carbide, a mixture of tungsten and carbon held together with a binder of either cobalt or nickel (most commonly cobalt because nickel is too costly).

The process of making these remarkably hard metal alloys begins by mixing the elements together mechanically for an extensive time to assure uniformity, pressing the mixture into the specific shape and size (with pressure of 12 to 20 tons/sq. in.), and then baking the parts in an atmospherically controlled oven (without the presence of oxygen) at 2,400 to 2,700° F. Though this is greatly simplified, it does portray how these alloys are produced. While most carbide cutting tools were developed for metal industries, a number of grades find excellent and extensive use in woodworking.

A recent carbide development being studied for woodworking applications uses two layers of different materials sandwiched together. A patented tool knife material with the trade name Borazon consists of a thin (0.020" thick) layer of polycrystalline cubic boron nitride (CBN) that is fused to a thicker layer of tungsten carbide. This unique alloy is the hardest material known, excluding only diamond. CBN can cut effectively at temperatures of 1,800° F and higher. While still too expensive for most woodworking, it continues to be researched for use on many materials, such as thermal setting plastics and fiberglass products, now cut with what are actually wood cutting tools.

The grading identification system used for carbide alloys relates to percentages of the elements present, size of particles in the mixture, weight, strength properties, and resulting hardness. For instance, the amount of tungsten carbide in a particular alloy with a cobalt binder will vary from 97 percent tungsten carbide and 3 percent cobalt down to about 76 percent tungsten carbide and 24 percent cobalt. With more cobalt present, the alloy is less subject to impact fracture but more subject to wear.

The cobalt makes it possible to braze-attach the knife material to a carrier body. With less than 3 percent cobalt, the tungsten carbide knife material has to be secured to a cutting tool mechanically, as a chemical bond is not possible. A tungsten carbide alloy with 24 percent cobalt has the same wear characteristics as high-speed steel, providing no advantage at all while being more expensive. For the woodworker and toolmaker alike, the carbide alloy has to be selected to provide the best wear characteristics and resistance to impact fracture by relating to the specific process and the material to be cut. Most carbide knives that are attached mechanically are in the 3 percent cobalt content range. Those brazed on are in the 4 to 12 percent cobalt range.

While there are many limitations and expense problems to overcome in powder metallurgy, it will continue to be researched more than any other because of the advantages it offers for so many metal products. The methods of creating the metal powders continue to improve their quality, quantity, purity, and, most important, particle size.

Metals that have been successfully produced in powder form include aluminum, antimony, boron, brass, bronze, cadmium, cobalt, columbium, copper, gold, iron, lead, manganese, molybdenum, nickel, palladium, silicon, silver, tantalum, tungsten, vanadium, zinc, several carbides, and others.

An interesting final note about powder metallurgy is that some years ago, it was discovered that one of the most precious metals, platinum, was being formed into a forgeable solid from powder by the Incas many years before the process was discovered in Europe, which was in 1829. Prior to that, it was not known how to make a metal solid of platinum ores by any method, fusible or other. It remains good advice, perhaps, not to take ourselves or all of our modern technology too seriously.

Ceramic alloys differ from cemented alloys in that they are fused together without a binder or matrix material. The most common is an alloy of aluminum oxide that is prepared in a very fine powder, pressed into the needed shape and size, then sintered (baked/fused) into a hard, dense material. What makes this attractive, especially for metal working industries, is that the material is a very poor conductor of heat, and, therefore, can attain very fast cutting speeds without damage from high temperatures. Additionally, the hardness and wear resistance are superior to many other cutting knife materials, especially regarding finish quality of the cut surface.

Cermet is similar to the aluminum oxide ceramic, but it is a composite that joins titanium carbide into the mixture. This material can handle larger work loads because it is somewhat less brittle than aluminum oxide ceramic.

In metal industries, cermet and ceramic are gaining more widespread use. In wood cutting, however, they still present many problems. Both materials are very brittle and require a vibration-free cutting process. The inherent composition of wood and wood-based materials makes this almost impossible. Further, metal cutting can use liquid cooling to protect tools and provide lubrication that is not possible when cutting wood. Research indicates that ceramic and cermet alloys can be used to cut homogeneous wood-based materials; but in most wood cutting, the cutting process is interrupted, the cutting load varies greatly because of the makeup of the material, the process is hot cutting, and the machine processes are much faster than most in metal working, creating a wide range of vibration mechanics that ceramic or cermet alloys cannot withstand.

The next cutting tool material is unique—it is *man-made diamond.* In its naturally occurring state, diamond is the hardest material known. Natural diamond is believed to be formed within the earth, as beneath ancient volcanoes, where pure carbon is fused under great pressure and heat. Diamonds cherished as precious stones are usually not completely pure, but are mixed with a very small amount of other elements that give them their distinctive tint and array of colors. We can use natural diamond as a cutter knife material, but this remarkable material is brittle and can fracture into ever smaller pieces. An ideal material would be as hard as diamond but made of many small crystals that could somehow be held together.

Here in the U.S., in 1951, the General Electric company first created a man-made diamond material. What has later become the Speciality Products Division of GE worked for several years finding a way to fuse pure graphitic carbon into a small mass of individual diamond crystals. To accomplish this, they compressed the carbon to over 1,000,000 psi and heated it to over 2,400° F. But the problem remained of how to attach it to a cutting tool. That problem was solved by fusing the diamond material to a tungsten carbide substrate, which could then be brazed to any tool body.

Man-made polycrystalline diamond tool blanks are black, are not translucent, and have a dull luster. As they come out of the press, the tool knife blanks have the size and shape of a large tablet. Only four manufacturers worldwide currently produce polycrystalline diamond tool blanks, which are available in several sizes (8.5, 12.5, 13.2, and 30 mm in diameter and either 1.5 or 3.2 mm thick). The diamond layer is only about

0.020" thick. For use as cutting knives, these tool blanks are cut into other shapes and sizes, which we will explain later.

As with the materials of cemented carbides and ceramic tool knives, the size of the individual diamond crystal particles dictates the strengths and physical properties of diamond knives. The diamond crystals created are remarkably small, varying in diameter from 0.0000397" to 0.0003176", with the ideal size being about 0.000158".

Diamond crystals are remarkably hard, strong, and resistant to thermal and chemical effects. The forces that hold the crystals together are both physical and electromagnetic. During heating and pressing of the graphitic carbon, individual crystals are formed, and the carbon that is not made completely into crystals tends to fuse the mass together by the attraction that individual atoms of carbon have for each other when in close proximity.

Man-made polycrystalline diamond is nearly as hard as natural diamond, and its most significant characteristic is that it does not have the natural planes of cleavage that result from being composed of many crystals instead of just one. Cutting tools of both man-made and natural diamond offer remarkable advantages of high feed rates and good finish quality. Metal, mining, and other industries make the most extensive use of diamonds as cutting tool material. Woodworking industries are just beginning to do so.

Man-made diamond does have drawbacks as cutting knife material. First, it is brittle and can fracture if subjected to impact shock, such as striking a nail or small rock while cutting wood material. Second, if the cutting environment creates high temperatures and contains carbon, the diamond will turn back into graphitic carbon. Vibration can have detrimental effects on man-made diamond, but not nearly as much as with ceramic or even some tungsten carbide alloys.

As mentioned earlier, polycrystalline diamond tool blanks are cut into more usable shapes and sizes to attach them to tool cutter bodies. In constructing a cutting tool, the initial consideration is how to arrange the tool blanks, which are sharpened only on the edge surface. The next problem is how to arrange the tool knives to make the best use of the very expensive material. (See Illus. 3-2 and 3-3.)

In the past several years, diamond cutting tools have found many applications that are easily justifiable, even with the considerable cost of purchase and subsequent service/sharpenings. The first were for cutting materials found in woodworking but not made entirely of wood. In Europe, the manufacture of skis involved cutting thin wood strips laminated with fiberglass. The technology available at that time—tungsten carbide—could cut only 200' to 300' of this abrasive material before being entirely dulled. In the first tests, the new diamond cutting tools cut over 30,000' before

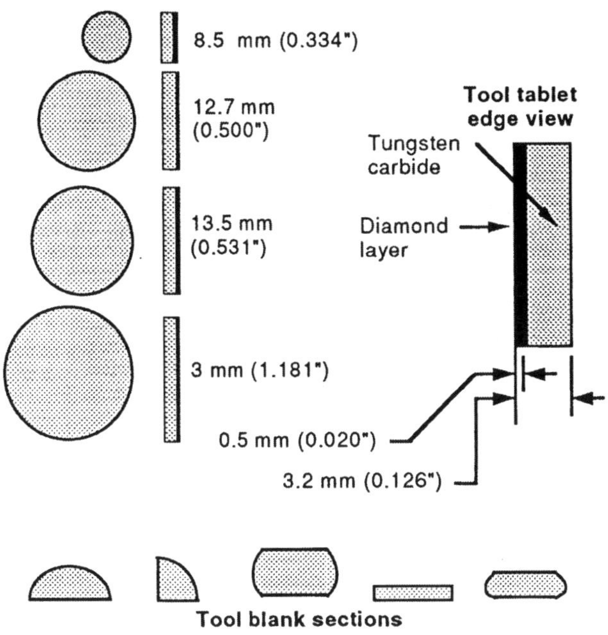

Illus. 3-2—Polycrystalline diamond cutting tool blanks

dulling. With changes in the diamond tools and material being cut, over 100,000' of production became possible.

Tungsten carbide alloys had comparable success when first used to cut man-made wood-based panel materials years earlier, but fiberglass is not anything like any wood-based material known. In the past 75 years, numerous examples of new cutting knife alloys have been developed to cut difficult materials, though

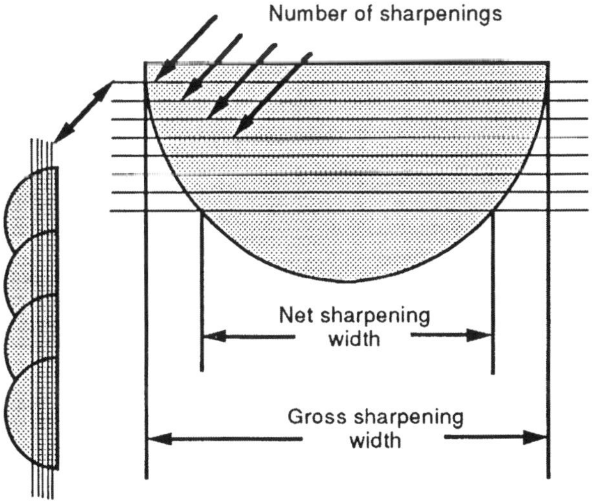

Illus. 3-3—Sharpening sequence and necessary overlap

it is not often clear which came first—the difficult material or the cutting tool that can make a useful product of it.

Polycrystalline diamond cutting tools are finding more applications in woodworking industries each year, cutting panel materials and even solid wood species that demand the advantages they offer. The first tools to make use of man-made diamond were router cutters that work in an environment that is as abusive as any possible, with heavy work loads, high temperatures, and tough, abrasive materials. As more diamond cutting tools are manufactured, geometric features are becoming further refined. Cutting, relief, and clearance angles for diamond tools are similar to those for tungsten carbide and other more brittle knife materials. (For information on use and maintenance of diamond cutting tools, see Illus. 3-4 and 3-5.)

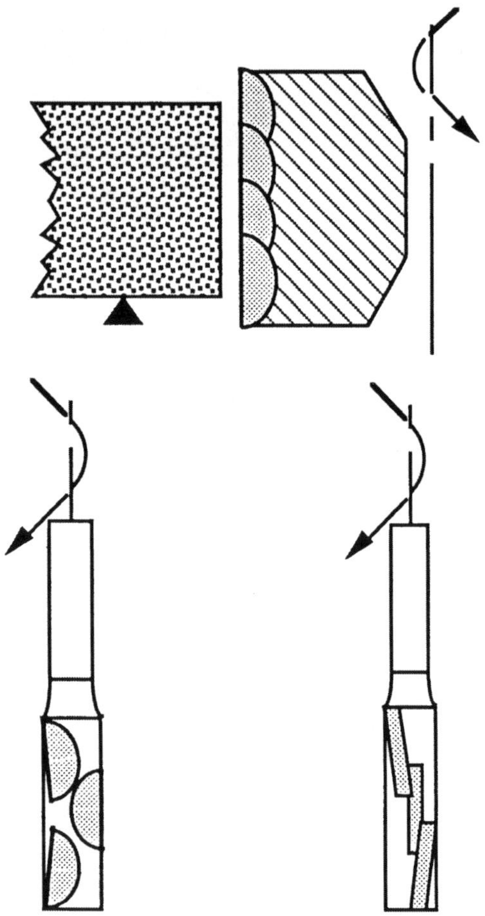

Illus. 3-4—Polyscrystalline diamond knife arranangement

The tool blanks are cut into usable shapes and sizes most often by micro-wire electrical discharge machining (EDM). This cutting system uses a very small wire filament that is drawn across the material to be cut while charged with low-voltage, high-amperage electricity. Micro-wire EDM systems cut a very narrow path through the material to be cut—as small as 0.010" wide. EDM machines are becoming more common, especially for cutting difficult, thick metals such as those used in tool and die work.

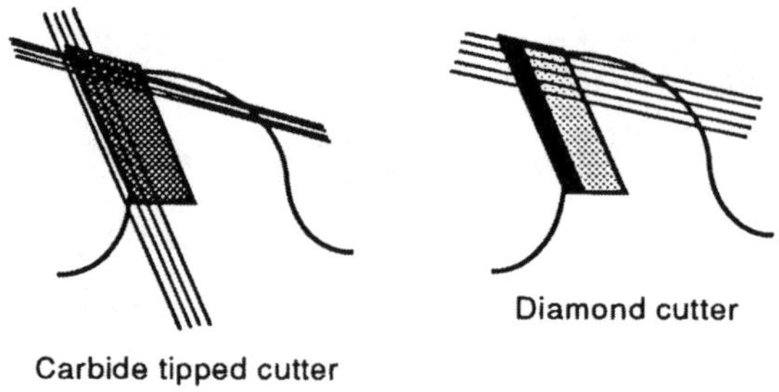

Illus. 3-5—Diamond tool sharpening

It is also possible to produce single-crystal pure diamond in a process in which carbon is melted in an electric furnace and the material is "grown" into a solidified crystal mass. The resulting material can be cut into gem stone pieces or shaped and sized for use as cutting tool knives, though the most common application would be as an abrasive, embedding the particles in a carrier body.

An interesting aside of the man-made single crystal diamonds is that they are too perfect to have great commercial value as gems, since they lack color and luster. Further, the crystals created are as difficult and expensive to cut as natural diamond. Having less commercial value as gems, most are cut for use as industrial cutting tools.

Tungsten Carbide Disposable-Knife Tools

Tungsten carbide cutting knives that are secured to the tool bodies mechanically have had profound effect on the woodworking industry in recent years. These tools offer the advantages of extended wear, concentricity, and balance. Furthermore, the knives are not sharpened; they are merely changed when dulled and therefore do not change dimensionally as do knives that are sharpened. Initially, all tools with disposable carbide knives were considerably more costly than cutters that were just tipped with similar material. As more of these tools have been produced, however, for more cutting functions and by more manufacturers, costs have diminished dramatically.

The use of disposable tungsten carbide knives has created an entire collection of wood cutting tools, from small router bits to very large cutters for tenoners and other industrial machines. The obvious advantages offered are:

1. There is no change to any dimension of the tool; the knives are changed when worn.
2. There is no change to features of balance or concentricity.
3. The strength of disposable knife tools is comparable to that of cutters with fixed knives, though somewhat less for certain processes.

4. For many cutting tool functions, there is often no need for multiple, back-up cutters because there is no service/sharpening required.
5. Many of these tools have been designed to perform a number of functions, e.g., adjustable width grooving cutters, variable bevel/chamfer cutters, and tools that can perform several different profile shapes.
6. There is an extended life expectancy for disposable knife cutters because the only parts subjected to wear (i.e., the knives) are constantly replaced when worn.
7. As more are developed by more manufacturers, both the initial cost and the cost of replacement continues to decline. Today, given the advantages overall, most mass-produced tools can easily be cost justified for any wood cutting process.
8. Unique, one of a kind cutting tools can be manufactured, especially for larger woodworking companies.

The first development of disposable-knife cutting tools was directed at machine processes that involved cutting very difficult, abrasive material, which resulted in loss of production time for tool maintenance. Cutting tools with permanently attached knives are costly and difficult to sharpen and keep in balance, and so manufacturers sought a solution to these problems. Conventional fixed-knife cutters also get smaller with each sharpening and require machine adjustments after each sharpening. If several identical cutters are needed to sustain manufacturing production, each may have a different diameter due to sharpening, which aggravates the problems.

These tools were developed to cut man-made panels and other abrasive materials being brought into the woodworking industries. Prior to the introduction of these materials, cutting any natural wood specie could be done easily with any steel alloy knife. But even the best high speed steel will cut only a few feet of the most abrasive man-made materials that are common today in most segments of woodworking.

Much development of cutting tools for woodworking began in metal industries. There are unique and different problems in each of these industries that often are solved with similar cutting tool systems and metal alloys. The basic differences are that metal cutting tools have much higher work loads, slower feed speeds, and a means to cool the process by flooding with water or other liquid. Wood cutting takes place with lighter work loads, at much higher feed speeds, and in hotter conditions. Other differences for wood cutting tool manufacturers to deal with are that wood and wood-based materials are not homogeneous; they vary greatly even within a single cutting process. In natural woods, grain orientation and defects such as knots and burls affect strength and hardness. Man-made materials, though more abrasive than nearly any natural wood, are more uniform.

Tungsten carbide knife inserts for wood cutting tools generally have thinner cross sections than those for metal work because work loads are lighter. The alloy content of wood cutting tool inserts are not as varied, with most containing only 3 or 4 percent cobalt, which makes them very hard.

Because carbide inserts are held mechanically, and most often with the designed aid of centrifugal force, and because the tool bodies are very well balanced, the cutters can be safely operated at higher rpm, giving higher rim speed. The design features in the manufacture of disposable/turn blade-insert cutting tools include:

1. Use of thinner insert knives, accomplished by essentially trading heavy work loads on individual knives for higher rim speeds made possible by the improved balance and concentricity of the tool;
2. Tool bodies manufactured to more precise tolerances and of better alloy steels that will sustain the needed work loads for longer periods of time;
3. Accurate details for chip and dust evacuation that will support the specified process and material to be cut (i.e., gullet shape and size will sustain the process and help keep the tool from overheating or plugging);
4. The best possible geometric features of all cutting and relief angles for specific materials and machine processes.

Disposable insert knives have made numerous cutting tool innovations possible, and many have made their way even to hobbyists, artisans, students, and craftspeople who could not earlier justify their cost. Among the first cutting tools to use disposable insert knives were router bits. Before disposable knives, routers exemplified best the problems of small cutting circles, high work loads, relatively short sharpening lives, and the loss of diameter with each sharpening. That is, like all other fixed knife cutters, router tools are initially expensive, run only a short period before becoming dulled, and lose diameter with each sharpening. A 1/2" router tool will cut a 1/2" groove only during it's first run time, when new. After sharpening, and with each subsequent sharpening, the tool will be narrower. If a 1/2" dimension is critical to the manufacture of a given product, that process can be very expensive, especially if the material is at all abrasive, further reducing run time. The advantages of insert knives for router tools were immediately apparent, though replacement knives were at first expensive. (See Illus. 3-6 on next page.)

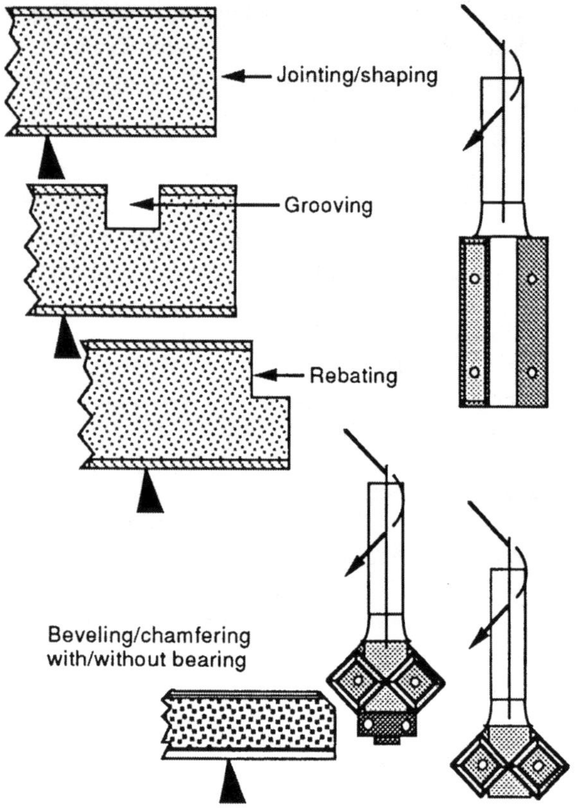

Illus. 3-6—Common router cutting fuctions

Router tools that are used to profile, shape, joint, groove, and rebate find excellent use for disposable insert knives. For router tools and all others, the least expensive are, of course, those that are sufficiently in demand to be mass produced by tool manufacturers. In all manufacturing areas, regardless of the product, the true cost is in direct opposite proportion to the number of the given product produced. More simply put, the more of a given thing that you produce, the less it costs. Development, raw materials, research, tooling, dedicated machinery—all of those costs are spread over a larger amount of capital flow as you are able to make and sell more and more of a given product.

The most expensive disposable-knife cutting tool is one in which both the cutter body and replacement knives are totally unique and no standard parts can be used at all. In today's market, the cost of even such a unique tool is less than a quarter of what it may have been just a few years ago, because more manufacturers are involved and the technology has developed dramatically in only the past 10 years. True cost has to be weighed against relative benefit. If a tool can solve a problem, even one seemingly distant from the manufacturing process, this has to be calculated into the equation. For instance, if a particular cutting tool improves the machined surface to such an extent that subsequent sanding can be reduced or eliminated, should that cost savings be calculated into the relative cost of the cutting tool that gave the savings? In another instance, a cutting tool affects how well the staining and finish operations are accomplished. Where does that cost improvement fit into what we decide the cutting tool is worth?

The point is that there are numerous costs to the implementation and use of any cutting tool, and they all have to be examined and evaluated before an accurate determination can be made. If a particular cutter costs twice the one it replaces, but is faster, improves finish quality, requires less maintenance, will perform more work before becoming dulled, and lasts longer, how will we actually calculate the true cost of the new tool? (The characteristics just described are a part of what industry seeks to evaluate in cutting tool costs.)

The carbide insert cutting tools available today are more costly than tipped or solid cutters, but the cost difference is negligible when examined against the performance abilities of each. Examples of tools with insert knives and the processes that they are used for appear in Illustrations 3-7 through 3-13.

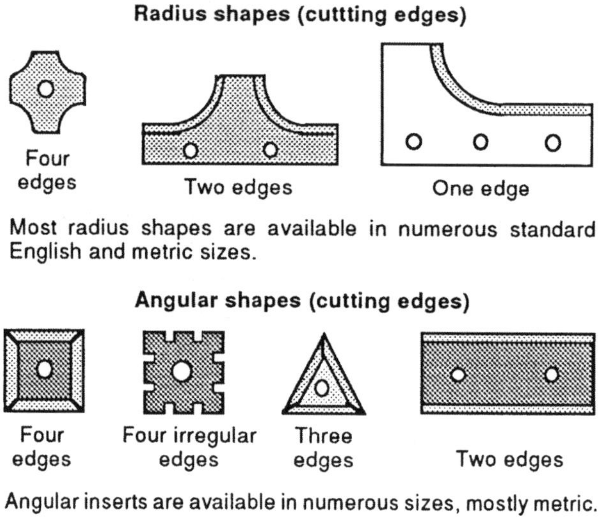

Illus. 3-7—Insert knife shapes

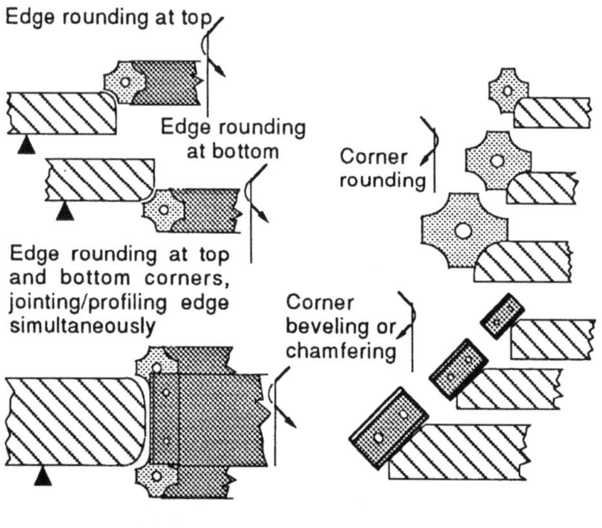

Illus. 3-8—Insert knife processes

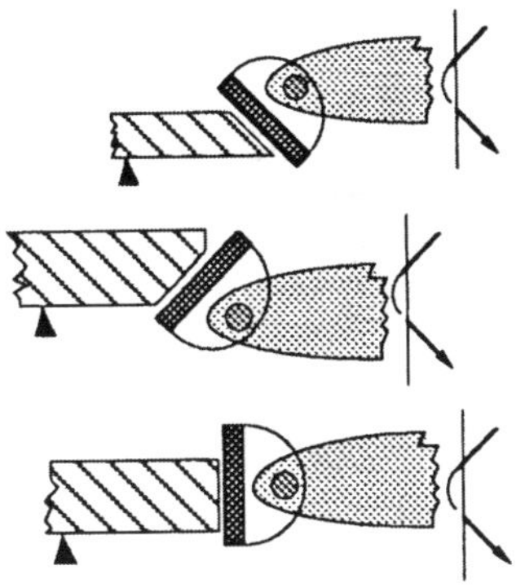

Exchangeable knives with a fixed cutting circle lend themselves to use with several unique cutting tools, among them the variable-angle cutter. This tool can cut any angle within 180° of adjustment, i.e., 90° up or down from vertical. Without disposable knives, balance, concentricity, and accuracy of such a cutter would be very difficult to build and maintain.

Illus. 3-9—Variable-angle cutter

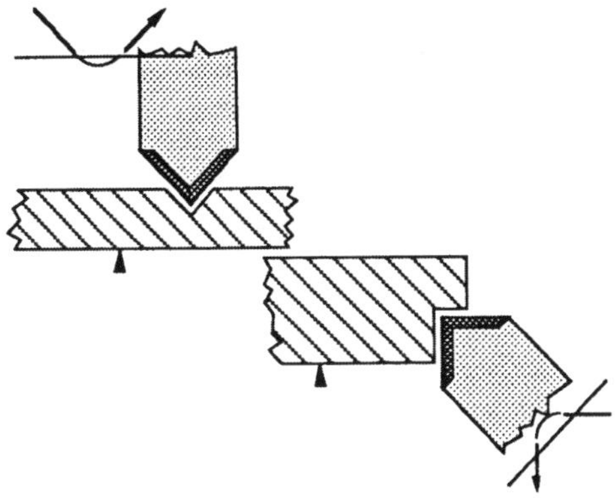

Illus. 3-10—V-groove cutter

Physically, the tungsten carbide for use as insert knives is manufactured much as similar alloys are: Powdered elements are carefully mixed, pressed into the needed size and shape, then sintered to fuse the contents together. The high cost of unique shapes and sizes (that is, those that are not mass produced) result from the need to produce either a unique pressing form or the expense of grinding the unique shape from a larger blank.

For most woodworking applications, the thickness of the insert knives will vary from 1.1 to 2 mm (0.45" to 0.80"). While that may seem small and frail, the knife/insert is usually supported and nested to the tool body or it rests against a wedge or backer plate used to secure it to the carrier body. What is most important to all insert cutting tools is that the details of the tool bodies be manufactured to very close tolerances; any irregular surfaces would fracture the brittle, frail carbide knives.

The width of the insert knives for planing, jointing, and shaping vary from 7.5 to 18 mm (0.30" to 0.70"). Even for the largest-diameter cutting tools, there is no need for a wider knife because only the leading edge protrudes from the tool body, and the knife material is very expensive. The thickness, width, and height of carbide insert tool blanks for shaping large, complex profiles vary widely, though there is always an effort to conserve the costly knife material. Since knives are supported either by a feature of the tool body itself or a backer plate, the thickness does not often exceed 2 mm. Very large, complex profile shapes are often divided into smaller segments or cut in multiple processes as a cost saving measure.

Given the cost of knife material, design criteria, and manufacturing processes, producing two smaller cutting tools rather than a single large one may cost only one-fourth as much. The manufacturing cost elements for cutting tools are similar to those for any other product: machinery, engineering , labor, etc. Tungsten carbide can be cut and shaped, though not quickly, easily, or inexpensively.

The use of disposable tungsten carbide knife tools continues to spread to more areas of woodworking, from raw lumber production to finished case goods and nearly everything in between. The benefits of extended wear characteristics, fixed cutting circle, concentricity, and balance are difficult to ignore, especially when most can provide a machining cost savings.

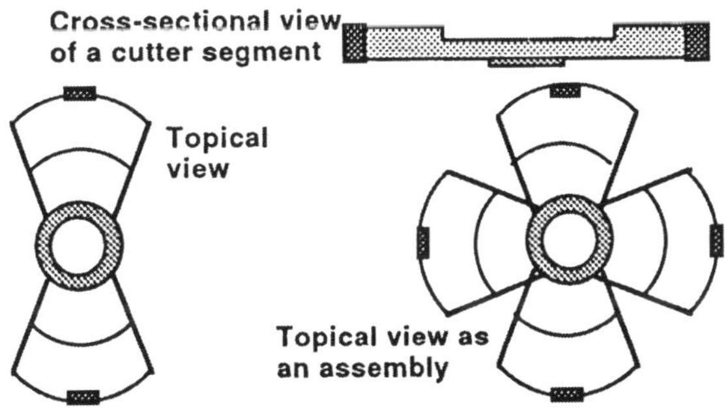

To expand the capabilities of disposable-knife cutting tools, several manufacturers now offer adjustable groove cutters that nest together as an assembly. The tools will cut grooves and rebates to twice the base width of the assembled body. Width is controlled by very precise spacers added between the sections of the cutter segments to extend the width. The body segments may or may not be attached, being stacked on a machine spindle. The tools remain well balanced and keep the same cutting circle.

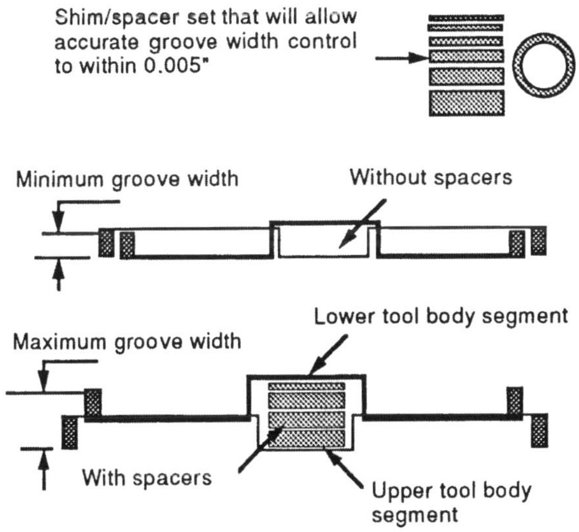

Illus. 3-11—Adjustable groove cutters

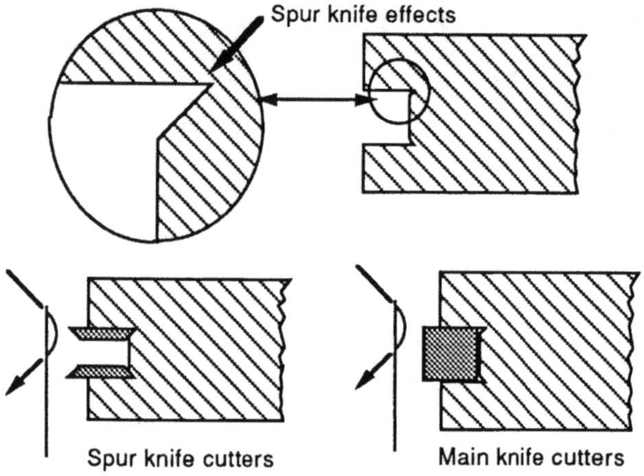

With disposable knives, tools can cut grooves and rebates with sharp, clean inside corners, even in dense and abrasive materials, by using a combination of spur and main knives.

Illus. 3-12—Using spur knives

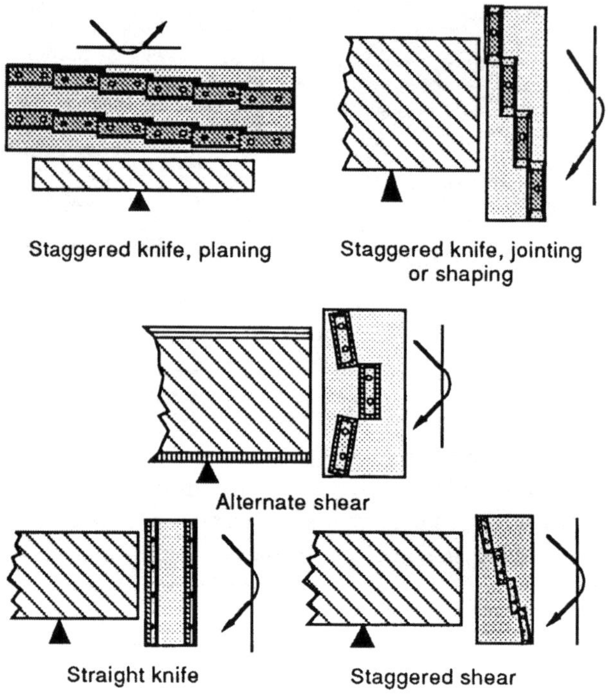

Illus. 3-13—Knife arrangements for shaping, jointing, profiling, and planing

Summary

The steel alloys that make up cutting tool bodies are carefully chosen to fulfill all the needs of toolmakers and tool users (though their needs may not be exactly alike nor in the same order of importance). While a competent toolmaker will give every effort to provide the tool user with all the needed strength and performance characteristics, cost considerations of materials, manufacturing processes, and pressures of the marketplace make the choice of metal alloy a critical first concern. For the tool user, the needs are driven by the performance characteristics of the cutting tools and their overall operating costs, though in industry the initial cost (of an individual cutter) is usually not the first concern.

Most woodworkers purchasing cutting tools from a competent industrial supplier would seldom need to question the integrity of the metals in the tools. A very old marketing axiom makes the observation, however, that "the farther the marketing means is removed from the manufacturer, the greater is the likelihood that someone will try to take greater profits from lesser-quality products—in direct proportion, likelihood to distance." While this unfortunate observation is surely not universal, given the range of cutting tools available to the hobby, craft, and industrial market segments, it does prompt the idea that caution is called for. This is especially true considering all of the possible things a poorly made cutting tool can (or cannot) do to the machine operator, the machine, or the material being cut.

In the manufacture of all metals and their respective heat treatment processes, the marketable features of each are closely protected. A part of what the consumer of any product sees is the effect of marketing strategy and advertising verbiage. Given the commercial value of a particular metal alloy over another in the marketplace, it is not too surprising that some of the claims made seem exorbitant at best.

The variety of knife materials for cutting tools will continue to expand in the future with the continuing development of more rare earth alloys, ceramics, and others. The most outstanding problems of wood cutting are abrasiveness, high heat, and impact shock. None of the materials available today completely resolves all of the problems simultaneously, but research will, no doubt, continue.

Chapter 4
Circular Saw Blades

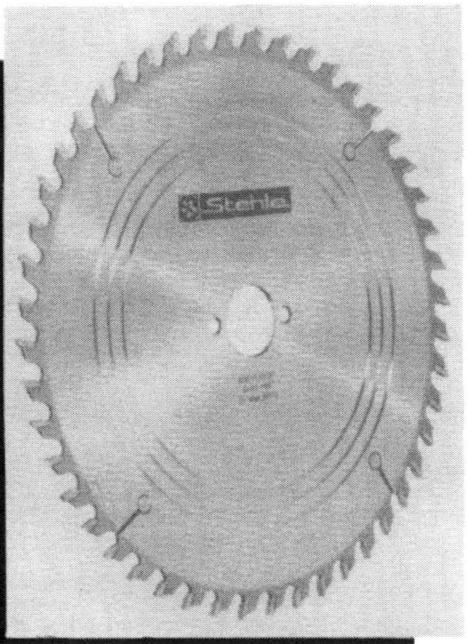

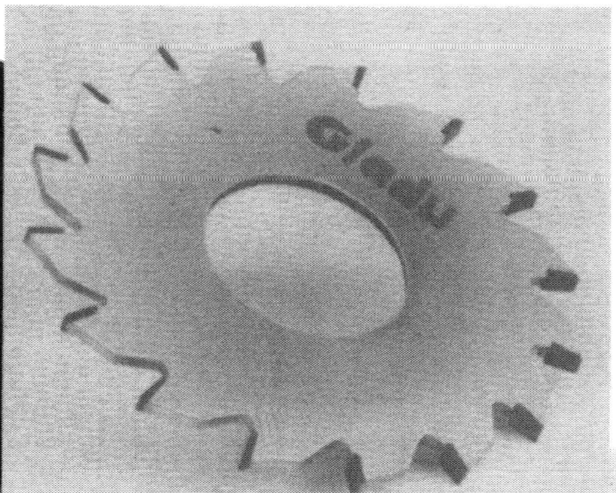

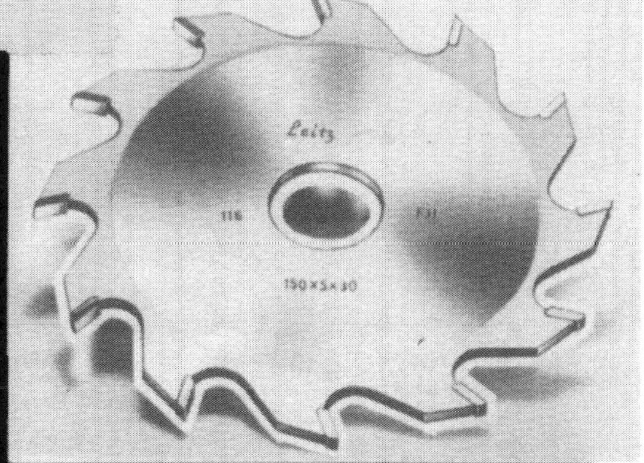

Photos courtesy of Stehle GmbH & Co., Leitz Tooling Systems, Inc., and Gladu Tools, Inc.

Circular Saw Blades

Saw blades are the best example of "chisels on a wheel." The knife is turned upside down, but, otherwise, the teeth behave precisely as individual chisels. Each tooth segment cuts the material, gathers the waste-chips into the gullet space, then empties itself to repeat the process in the next revolution. Except for several variables that need to be explained, qualified, and quantified, the discussion could perhaps stop there.

Saw blades are the most complex cutting tools that we commonly use in woodworking. They have more unusual tooth/knife configurations than any others. There are numerous sizes and shapes of cutting and relief angles and many tooth shapes for different materials. Machine adaptations are similarly diverse, with nearly 30 types of saw machines. Though these are not greatly different, they are sufficiently so to require consideration of the variations in tool geometry. A saw blade is not just a saw blade.

Saw blades perform tremendous amounts of work, driven, in some instances, by motors of more than 100 horsepower. And yet saw blades have very little lateral strength. Their greatest mechanical strength is the rotating force that places them in motion.

Circular saw blades were invented at least a couple of hundred years ago and have been under constant study and refinement ever since. With ever-growing numbers of man-made materials and a vast array of natural woods to be cut, tool manufacturers are kept busy. The combined effects of new materials and larger, faster machines have made numerous changes in all cutting tools necessary. Many of the tools in use in industry today did not exist even 35 years ago. The basic mechanical functions of cutting and shaping are the same, but tools are not. Saw blades appear, in a very general sense, as they always have, but all their features have been refined. Even the materials that make up saw bodies are relatively new. Tooth/knife materials—e.g., tungsten carbide, diamond, Borazon, Stellite, and other patented alloys—vary also, from one saw blade to the next, depending upon intended use.

Today, most saw blades, especially those intended for industrial or craft use, have metallurgically well designed bodies with teeth made of another alloyed metal attached. The earliest saw blades were made of various steel alloys, and geometric features were ground or shaped into the metal in various ways. (See Illus. 4-1.) The first circular saws were essentially round versions of hand tools

Close examination of a hand saw will reveal the same detail features that are designed into powered tools. The teeth are bent sideways to provide lateral clearance, and the cutting angle is ground into the face area. (See Illus. 4-1, A-D) The geometry of the earliest tools was not nearly as complex as modern tools for three very basic reasons:

- The materials being cut were mostly wood solids.
- The machine processes were neither fast nor complex compared to what is commonplace today.
- Other technology was not available yet.

There are, of course, basic principals of physics and mechanics that have not changed, but modern cutting tools are quite unlike their predecessors. In

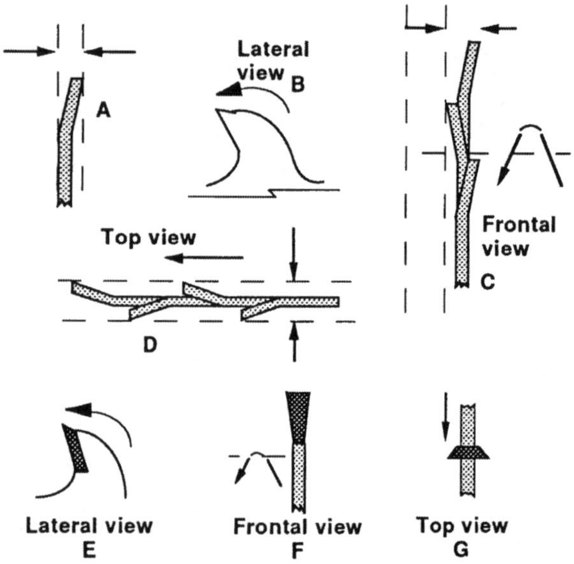

Illus. 2-1—Saw blade detail

Illustration 2-1, Figs. E, F, and G depict modern attached-tooth tools.

Physical Features

Illustration 4-2 shows the larger features of a modern saw blade, some of which were not features of earlier tools:

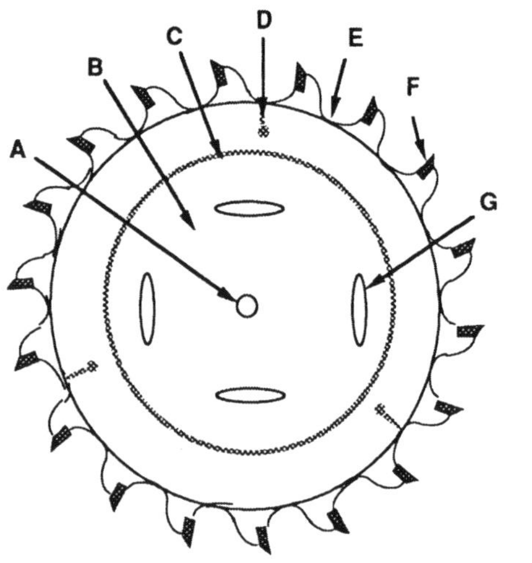

A. Arbor hole (bore)
B. Saw body (plate)
C. Tension ring
D. Expansion slot
E. Gullet space
F. Tooth/knife face
G. Plate/body cooling slot

Illus. 4-2—Physical features of saw blade

A. The *arbor hole*, or *bore*, is at the precise center of the tool body. Its size depends on the tool and the shaft it will be placed on. The fit between the tool and the shaft should be as close as possible and still allow the tool to be placed on and removed from the machine. A note about fit and tolerance that we will address later in more detail is that regardless of how close in size the bore is to the machine arbor, the tool will turn eccentrically by the dimension of the difference in their sizes. Even if the fit tolerance is only 0.001", the cutting circle is not perfect.

We will also later examine methods of holding and driving cutting tools to minimize fit tolerance. The likelihood that the bore-to-arbor size ratio will present significant problems is remote, but we should not disregard it. Mass-produced tools are manufactured for the most popular machines, though bore sizes can be adjusted. Use of bushings to reduce the size, however, should be avoided.

B. The thickness of the *saw body*, or *plate*, just as the number and shape of the teeth, is a design parameter defined by the tool's intended range of uses. A saw blade designed for cutting solid woods will have different plate thickness from one of the same diameter designed for cutting material such as particleboard. Natural solid wood has much more strength than any man-made material, and so the tools to cut it need to be stronger. We do not absolutely need to have a different saw blade for every application or material, but there are significant advantages to having a particleboard saw if we are cutting thousands of sheets every day with a tool that will not cut anything else. Primarily, plate thickness is a relationship between the anticipated work load on the tool and its inherent strength. Today, intensive research seeks ways to use much thinner saw plates with narrower kerfs and, consequently, less waste.

C. The *tension ring* is an area in the saw body that has been altered mechanically to help the tool turn without wobbling and to make it more rigid. There are two commonly used methods to accomplish this: by anvil hammering and by pinch-rolling. In mass production, few blades are still tensioned by anvil hammering.

Illustration 4-3 shows how saw blades are tensioned by pinch-rolling. As the saw plate is held and rotated, two hydraulically powered wheels are pressed against the surface. The wheels have a rounded peripheral edge that presses against the saw body hard enough to often leave a small depression. The metal in the area of this process is harder than anywhere else in the saw blade. The hydraulic pinch-roller places from 2,500 to over 8,000 psi pressure against the saw blade surface in a

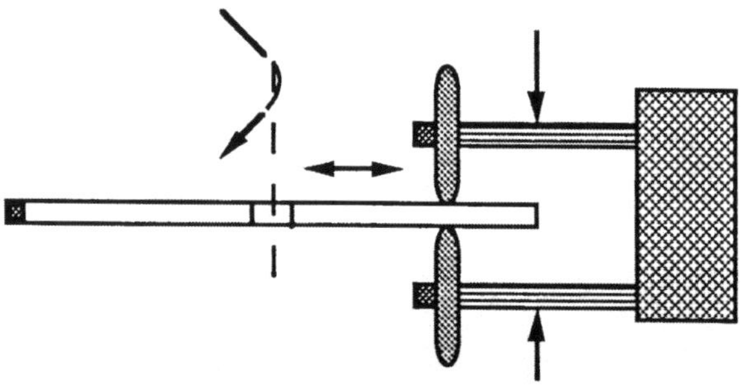

Illus. 4-3—Hydraulic pinch-roller tensioning

mechanical hardening process. The hardened area can be located variously between the bore and periphery, depending on the work load to be placed on the tool. Most often, the tension ring is in the innermost area where the work load is placed. On heavy rip saws, the tension ring may be closer to the bore than for a lighter cross-cut saw.

Anvil hammer tensioning is still a viable tensioning method, though mostly used today to straighten warped or bent saws. Anvil hammersmiths, like blacksmiths and others, have all but vanished as technologies have advanced. Most quality saws 12" and larger are tensioned, though there may not be clear evidence of it.

D. As explained in Chapter 1, cutting tools undergo dramatic and constant temperature changes. The *expansion slots* at the periphery of a saw blade allow it to expand without wobbling. Without a means to change dimensionally at the outer edge, the blade would distort and become unstable. There are zones of temperature in the working blade that may differ by hundreds of degrees. At the bore, the blade is probably the same temperature as the motor driving it, usually from 140° to 180° F. While in contact with the material being cut, the blade's leading edges may reach temperatures of 800° F and more.

The saw plate has to endure this great range of temperature, which, furthermore, changes constantly during each cutting process. The extremes of temperature are not sustained, and after a period of time, the rate of feed, the rotation, and the nature of the material all stabilize the saw blade into a given range of temperature for any particular process. The area of the saw blade between the bore and periphery often has temperature in the area of 200° to 350° F, cooler toward the bore and hotter toward the outer edge. Well made saw blades 10" and larger will have expansion slots as will smaller special purpose blades, such as those for edge banding machines and others.

E. The *gullet space* of a saw blade may appear to be simply the space between the teeth, but it is actually another feature of tool design. A collection of new man-made panel materials have come into the woodworking industry. They display many of the same cutting characteristics of the wood that they were derived from, plus a few of their own. Research in industry can lead or be driven by technological advances; gullet design has been driven by safety laws that prescribe allowable noise levels and by the search for more effective ways to cut man-made panel materials.

Gullet space has essentially two functions in the cutting process: The first is to gather waste chips, carry them out of the contact zone, and discharge them fully to return and repeat on the next revolution. The second and less obvious function is to control noise level. Without consideration of several factors, a saw can generate noise beyond any legal, or bearable, limit.

In the marketplace today there are distinct gullet designs to produce as little sound as possible. Legal sound levels are described in terms of amplitude or loudness, not frequency, so tool designers and engineers attempt to mix frequencies and not exceed a particular loudness.

F. The *tooth face* was detailed in Chapter 1.

G. *Plate/body cooling slots*, while not as elaborate or complex, are just as important as other saw blade features. As mentioned earlier, heat build-up is the most significant problem encountered in any wood cutting procedure. Simply stated, cooling slots are unusually shaped holes in the saw body that keep heat from building up.

The most prevalent shapes are an elongated oval or a figure eight. These shapes are effective in controlling both movement of air and sound. Cooling slots are necessary for large-diameter saws (16" and larger) that are subject to very heavy work loads. In industry, for instance, it is not uncommon to cut stacks of panel materials of 6 inches and more. There are specialty, small-diameter tools for unique machine processes or unusual materials that may require tool bodies that are not mass produced because of singularly shaped cooling slots, and, hence, not common to the marketplace.

We covered in Chapter 1 most of the details depicted in Illustration 4-4 in the context of how the basic geometric features are common to all cutting tools. The details of cutting, hook, back relief, lateral relief, and lateral relief to the periphery angles differ from one saw blade to another, depending on intended use. That is, a saw blade designed to cut only solid wood along the grain—i.e., to rip cut—will have those five angles arranged for the best efficiency. Another saw designed to cut hardboard will no doubt have a different arrangement of angles.

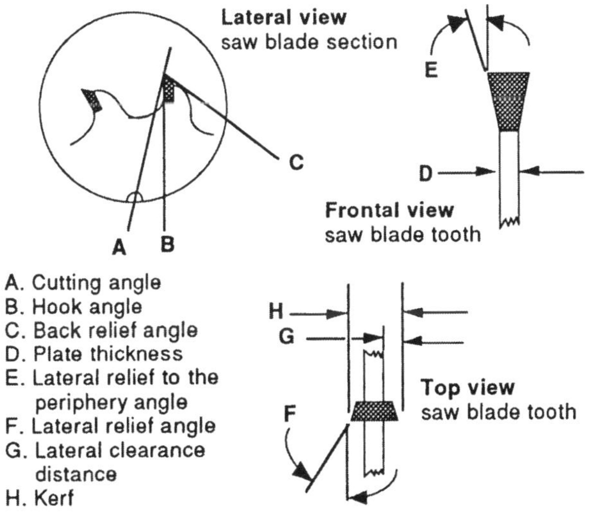

Illus. 4-4—Saw blade peripheral features

Lateral clearance, not covered earlier, is another feature of tool design that is determined by the material being cut or by a particular machine process. Generally, strength and moisture content of the material to be cut will dictate a particular clearance distance. In the last chapter, we will provide lists of recommend features for various materials.

The last feature in Illustration 4-4 identifies the *kerf*, or cutting width. Mass-production manufacturing produces tools to satisfy general needs. That is, toolmakers attempt to produce tools that will address most machine size limitations and the widest range of materials to be cut. Blades for 10" or 12" table saws, for instance, will have a 5/8" bore and a particular range of tooth number, shapes, kerfs, and plate thicknesses. Kerf size relates to the relative strength of the tool and to the horsepower needed to accomplish the work. Finally, we need to understand that all design features of a cutting tool are interrelated.

Designing Tools to Overcome Problems

Woodworking, like all areas of study, has what seem to be exceptions to the rules. An example is construction of tools for cutting grooves. At question here are two of the basic geometric features—the angles of lateral relief and lateral relief to the periphery. In Illustration 4-5, A and B will cut a groove of exactly the same width. After the first sharpening, however, the tool labeled B will be narrower and of no use as a groove cutter. The design and construction of A increase the lateral relief angle and eliminate the lateral relief to the periphery angle—that is, the angle measures 0°.

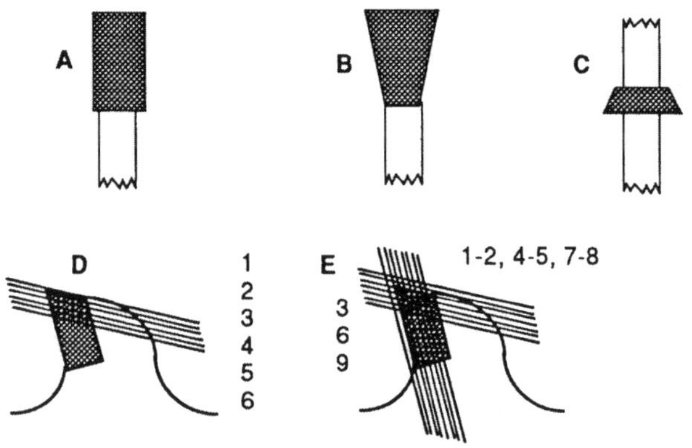

A. Frontal view of groove cutter tooth/knife
B. Frontal view of saw blade tooth/knife
C. Top view of either saw blade or groove cutter tooth/knife
D. Sharpening sequence of groove cutter—lateral view
E. Sharpening sequence of saw blade—lateral view

Illus. 4-5—Saw blade to groove cutter comparison

This tool design allows the tool to cut throughout its entire useful life without any deviation of groove cutting width. There are several other such applications in the other families of tools that we will examine in later chapters.

Another problem that woodworkers face regularly with circular saws is the finish quality of the surface edges. For instance, if we are cutting a panel, sheet, or even rail part with a table saw and want both upper and lower edges to be as smooth as possible, we have a problem. Circular saw blades are at the same time, fast, strong, and efficient. They are not, without some consideration, generally able to cut through a material and leave both top and bottom edges smooth and free of tearing. What is present is another geometry problem.

As a circular saw blade turns, the peripheral angles (cutting angle being of most concern) change with even 1° of turn in relation to the material being cut. (See Illus. 4-6.) The cutting angle on the saw may be, say, a positive 15°, but the actual angle of the cutting teeth as they enter the workpiece depends on several criteria. And the same angle will be different again as it leaves the material at the bottom.

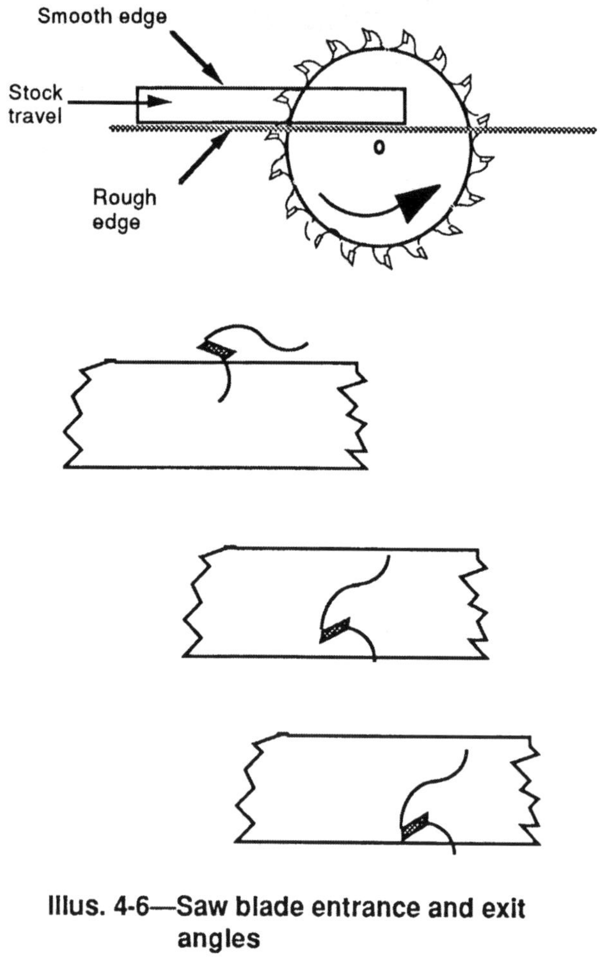

Illus. 4-6—Saw blade entrance and exit angles

The cutting angle is only a single part of a larger, more complex set of features that generally define what the cutter can do. In the case of a saw blade, we can orient the tool and the material to be cut at any location in the circular path of the cutter, with respect given to being able to hold the stock. And other variables must be considered and controlled with respect to any particular process:

1. Stock thickness,
2. Blade elevation,
3. Tool cutting angle,
4. Feed rate and tool rpm,
5. Strength of the material being cut,
6. Possible surface covering such as veneer or laminate,
7. Number of teeth on the blade.

Illustration 4-6 shows the path of a saw tooth through the work material. Note that the inclined angle of the cutting tooth changes constantly, becoming more acute as it progresses. By the time the tooth leaves the material, it may be pushing the waste material ahead of itself—a worst possible case, but not uncommon. This too often causes the lower edge surface to tear, and there are several ways to reduce that effect.

The first thing to examine is the machine process itself. Illustration 4-7 A shows the same material being cut by the same tool but at three different blade elevations. If the entrance and exit angles result from three variables—cutting angle, material thickness, and location of the material in the cutting circle—we can easily affect one of the variables by raising or lowering the blade height. This changes the entrance and exit angles dramatically, by as much as 30° in a best case.

Surface-edge quality can also be affected by changing the cutting angle of the tool. For most applications, saw blades have a cutting angle between 0°, for hard, dense materials or those that have very little surface strength, to 20° or more, for rip-cutting solid woods. Cutting man-made panels with very frail or brittle coverings, may necessitate reducing the cutting angle to a negative 10°. (See Illus. 4-7 B.) Altering the blade height or the tool cutting angle can reduce many bottom chipping problems significantly or eliminate them completely. Later in this chapter, we will explain more causes for irregular surfaces created by saw blades.

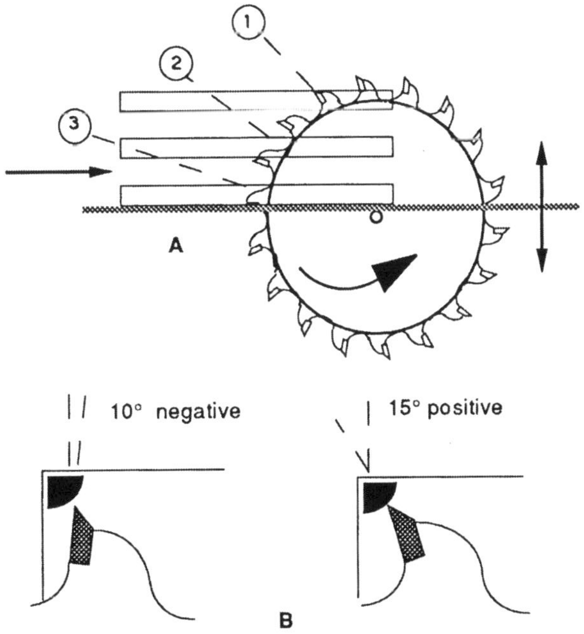

Illus. 4-7—Blade elevation

Scoring Saws and Strobe Saws

In industry, the solution to the bottom chipping problem is more costly than just manipulating tool geometry or the machine process, but it is also more effective. The process involves two saw blades—a scoring blade and a main blade—held in precise alignment to each other and turning in opposite directions. (See Illus. 4-8.) In variations of this process, either the saw blades travel to the material to be cut, or the material moves to the saws. What is common to all such processes is that the lower surface is scored—i.e., cut by a smaller blade ahead of the larger main blade that completes the parting cut. Thus, the outer surface has already been cut and cannot chip. The opposite rotation of the scoring blade protects the lateral walls and trailing edge of the material from tearout.

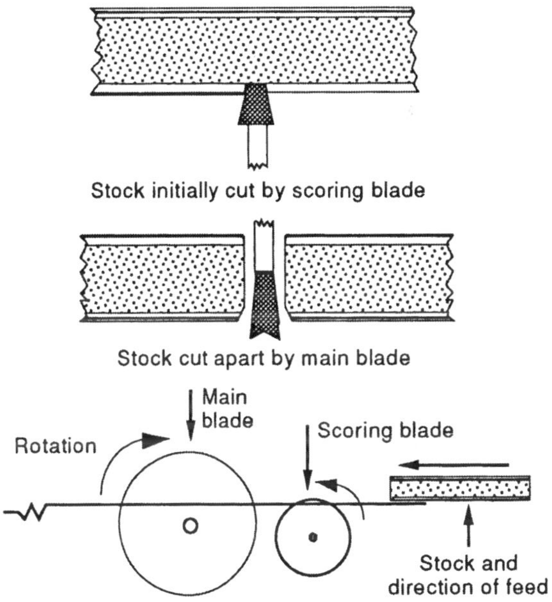

Illus. 4-8—Basic principles of main and scoring blades

This process, while completely effective, is costly and requires constant attention to details. These details involve alignment on three axes, main-saw kerf, scoring-saw kerf, and scoring blade type. (See Illus. 4-9.)

There are two types of scoring blades. In Illustration 4-9 A we have illustrated the inverted-V scoring saw. Here we have to refer back to Chapter 1, to explain the function of the kerf relationship between the scoring and main blades. As described earlier, a saw blade tooth

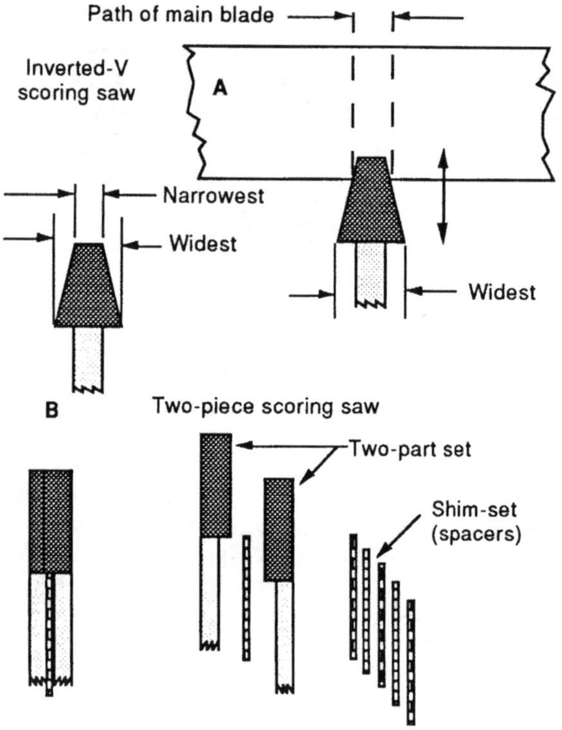

Illus. 4-9—Scoring blade types

is widest at the outer perimeter and leading edge and narrowest at the bottom rear of the tooth segment.

This is important because sharpening will change the dimensions of the main blade; each tooth segment becomes shorter and narrower with each service cycle. Yet, the scoring blade must cut a kerf of precisely the same width as the main blade kerf. Consequently, the scoring-saw kerf has to be altered. The inverted-V scoring saw, when first put in use, cuts a deeper groove to match the wide main blade. As the main saw is sharpened and becomes narrower, the scoring saw must cut shallower to match perfectly with the main saw.The mechanism that alters the height of the scoring blade has micrometer precision. Most often, the main and scoring saws are put into service as a set and kept together for their entire useful lives.

The second type of scoring saw is actually a multi-part set, consisting of two thin saw blades and several very thin spacers. (See Illus. 4-9 B.) The need of matching the kerf of the scoring saw to that of the main saw remains the same, though the mechanics are different. When the scoring saw is put into service, spacers are added to make it as wide as the new main saw. As the main saw becomes narrower with repeated sharpenings, spacers are removed from the scoring saw to keep it at the width of the main saw.

This system has some advantages over the V type. The depth of cut for the two part saw is not critical, as width is maintained by precise spacers. Further, the saw set does not have to be paired with a particular main saw, and, usually, the tool can be sharpened more times. Another advantage occurs when cutting sheet or panel material with an unusually thick lamination or veneer covering that must be score-cut through. The V type may not be able to accomplish this and still match with the main blade. The multi-part scoring saws are, however, more expensive, in both purchase price and maintenance. Both types have been successful in industry for over 35 years.

As mentioned earlier, there are a number of cutting tools that seem to break the rules. But when dealing with a substance as obstinate as wood, the rules need to be flexible, allowing for tools such as the *strobe saw*, which has extra cutting teeth/knives in what would normally be expansion slots.(See Illus. 4-10.)

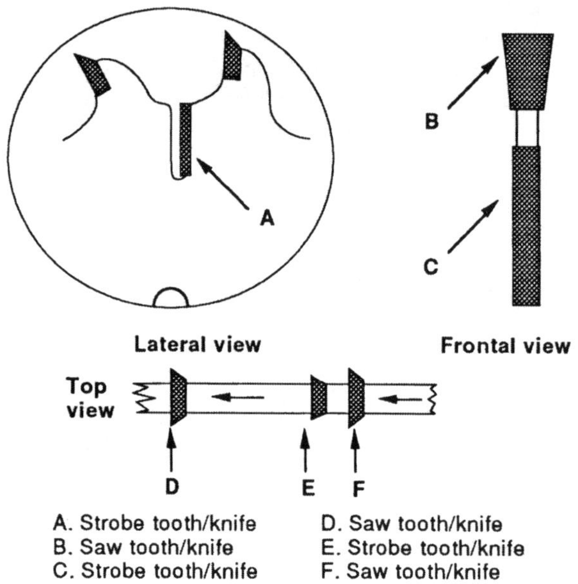

Illus. 4-10—Strobe saw features

Freshly harvested wood is heavy with water and remarkably strong. While in this green-wet state, wood being cut can pinch the saw blade, in the worse case, with sufficient strength to stall the motor driving it. At the very least, the binding and pinching can generate enough heat to destroy the saw blade. This is not common, but was such an ongoing problem for the timber industry that tool designers sought a solution. The answer was to actually recut the wood with a long, narrow tooth segment below the peripheral teeth. After careful study, it became clear that the strobe knives need not be as wide as the teeth, and they are most often 70 percent as wide.

The strobe saw takes advantage of an often overlooked aspect of cutting with a circular saw, i.e., that the teeth pass through the cut material twice. (See Illus. 4-11.) The periods of time a saw-blade is in contact with the material are the *descending period of contact* (or cutting period of contact), during which the saw teeth enter the material, and the *ascending period of contact.*

During the ascending period, contact with the material should be minimal. Only the natural elasticity of the material should affect what would otherwise be an idle pass through the already-cut tool path. When cutting green-wet wood, however, the saw blade can bind and be pinched during this period, especially if a lot of twisted grain is present. The strobe saw can help alleviate this situation.

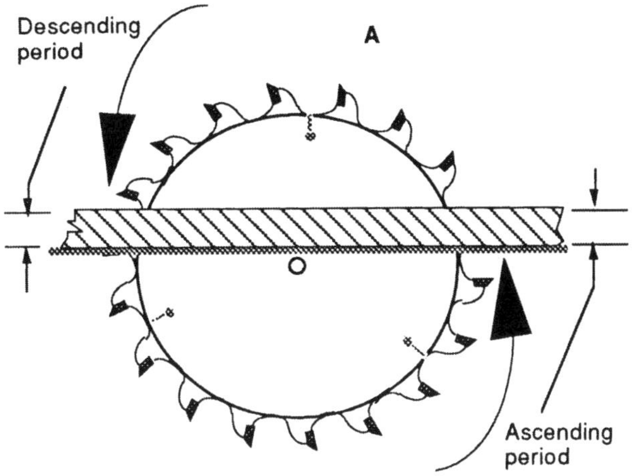

If there is visible marking of the cut surface, it should be a faint cross-hatched, radiused tooth path created by both periods of contact of equal depth.

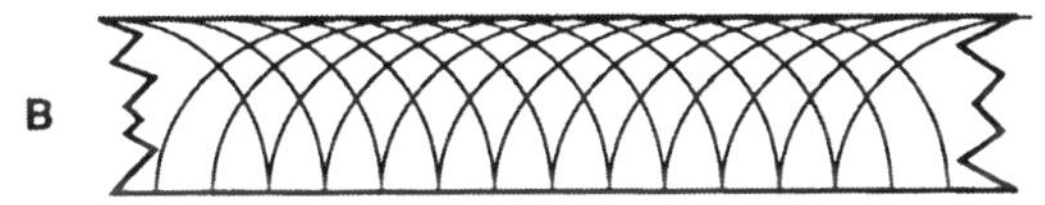

Ilus. 4-11—Circular saw double cut

Saw Blade Mechanics

To understand the mechanics of the cutting process of wood and wood-based materials better, it is also necessary to study wood morphology. Wood is remarkably complex in that all wood materials, regardless of specie, have internal stresses present, and as a piece of wood is cut, it reacts to its own stresses simultaneously in all directions. That best describes (though does not solve) most difficulties woodworkers encounter.

Because individual cutting teeth and knives work at such remarkable speed, precisely what they do is difficult to visualize and comprehend. This seems particularly true of saw blades because they are so common to us, and we have accepted the idea that they really are just "chisels on a wheel." Furthermore, with so many teeth moving so quickly, the action of a single tooth does not seem that important. To understand and make better use of all cutting tools, however, we have to examine each individual tooth in the cutting process.

A topic of relevance here is *contact surface pressure*, which, for our purposes, is dissipation of forces that result from the leading edge of a tool cutting some material. That would be simple enough for most materials, but wood reacts differently in all directions to forces that are placed on it or that originate from within. This is what has made the design and manufacturing of wood cutting tools the subject of much ongoing study. The number of variables, which change almost constantly, reduces most of the "rules" to general guidelines. Cutting metal, though it requires more horsepower, is more straightforward, constant, and generally easier, once the make-up of the metal is known.

Contact surface pressure is important to comprehend not just for saw blades but for all cutting tools. Tool teeth or knives have various shapes to overcome the effects of contact surface pressure. Recently developed and marketed tooth shapes are generally intended to solve problems with newer man-made materials.

To explain further, Illustration 4-12 shows that as an individual V-shaped tooth enters the piece being cut, the material compresses slightly ahead of the tool. As the process proceeds to C, the tooth, by virtue of its shape, forces the material to compress away from the leading pointed edge. By the time the tooth has proceeded to D, the chip offal is being forced onto the tooth face and into the gullet. In E, which indicates the pressure zones toward the end of the chip-formation area and the relative size of the tool path, the chip mass is fully formed, and the compression is nearly complete as the tooth is about to leave the descending contact period area. In A, we trace the tool path and indicate the size and shape of the chip-formation area.

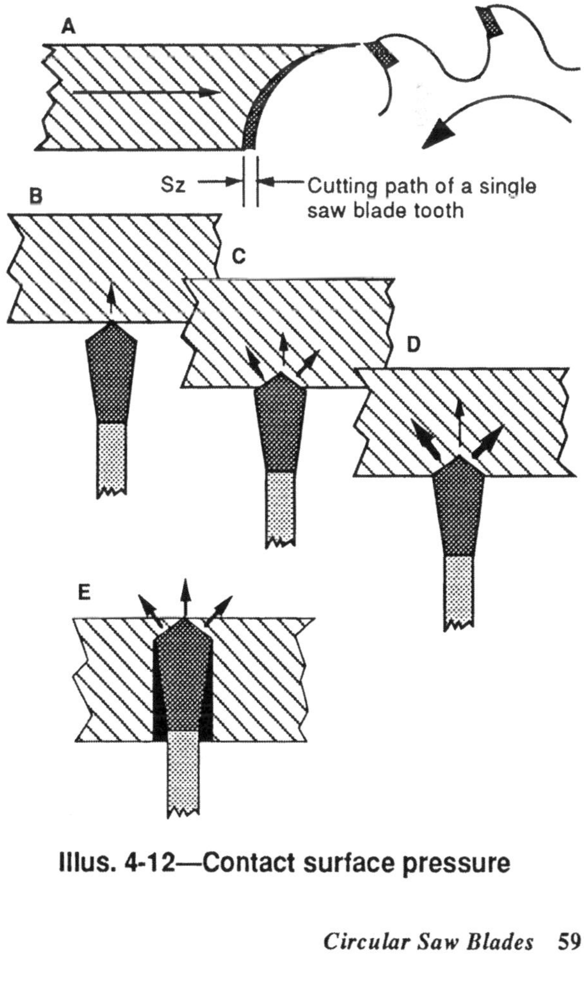

Illus. 4-12—Contact surface pressure

Saw Tooth Shapes

Illustration 4-13 more specifically shows contact surface pressure. The five tooth shapes have distinct effects on the material being cut and the forces in the process. Contact surface pressure is not a force that we need to measure in specific terms such as pounds per square inch, but we must understand that it is present.

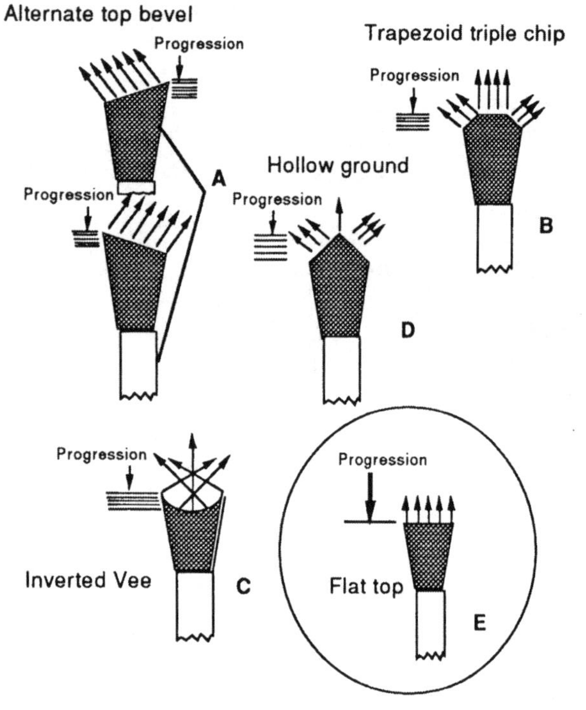

Illus. 4-13—Managing contact surface pressure

Contact pressure between the cutting tool and material being cut will vary greatly in one wood specie and even in a single process. This will be clearly evident as a single sample of wood is cut, for instance, in three different directions, as might be done to rip cut to width, cross cut to length, and later shape into a profile. The contact surface pressure of each cut will be different, and the primary determining factor in this instance is grain orientation.

Illustration 4-13 A shows two teeth in what, except for the flat top, is the most common form. This two-tooth combination—called *alternate top bevel*—is the most versatile and successful of tooth shapes because altering the angle of contact in every other tooth divides the contact surface pressure. Further, in the period of time it takes the tooth to enter the material being cut, the forces build gradually, at least relative to what happens with the flat-top tooth in E.

To say that an alternate top-bevel saw tooth builds its contact surface pressure gradually (because of its shape) as it enters the material has to be understood in proper perspective. Let's return for a moment to the effective speed at which saw blades operate—our earlier example of a 12" saw with 72 teeth, turning at 4,000 rpm. The teeth enter the workpiece at a rate of 288,000 per minute, or 4,800 per second. The time it takes a single tooth to complete a cutting process is then 1/4800 of a second.

But various shapes of saw blade teeth do each have different contact surface pressures and will have varying effects on the materials being cut. The shapes, excluding the flat-top tooth (E), build pressure forces differently and successfully divide those forces along the plane of the leading edge of the tooth. That is, no exact point on the leading edge enters the workpiece at precisely the same time as another. With the alternate top tooth, the leading edge enters the workpiece in an angular progression, rather than abruptly as does the flat-top tooth. The effect, when cutting a material that is at the same time very strong, elastic, and fibrous, is a reduction of the horsepower required and smoother cut surfaces. The wood fiber cells are less crushed and cut more cleanly, and the finish appearance is smoother.

In Illustration 4-13 (A, B, C, and D), each saw blade tooth divides the contact surface pressure along its leading edges a bit differently. Often, two tooth forms are used on the same tool to satisfy the requirements of cutting a particular material. As mentioned earlier, the alternate top bevel (A) tooth is the most widely used, especially cutting solid wood materials—cross cutting, bevel cutting, etc. In B and C, the *inverted V* and *triple chip* were developed to deal with the unusual properties of man-made panel materials, especially those laminated with brittle or frail coverings.

The *hollow-ground* tooth (D) requires special explanation.The complex and more complete distribution of contact surface pressure that this tooth form achieves provides the best surface quality of any saw tooth shape. There are reservations regarding its use in industry, however, that do not relate to the effectiveness. The primary concerns relate to cost and will be addressed later.

The hollow-ground tooth shape handles contact surface pressure uniquely. As the radiused surface enters the material, the forces created along the leading edge appear to "shear" when there is no shear angle present. During the entire period of time that the saw tooth is in contact with the material, the effects of the hollow tooth continues as though there were many shear angles present on a single cutter segment. The lateral view of a hollow-ground tooth appears the same as others. (See Illus. 4-14 D.) Additionally the sharpening sequence is the same as for other tooth forms, though grinding equipment and procedures are different.

The *flat-top* saw blade tooth (Illus. 4-13 E) creates the harshest and most abrupt contact surface pressure of all tooth forms. The entire leading edge enters the material at the same moment, building the chip mass across the face width and forcing the severed offal to

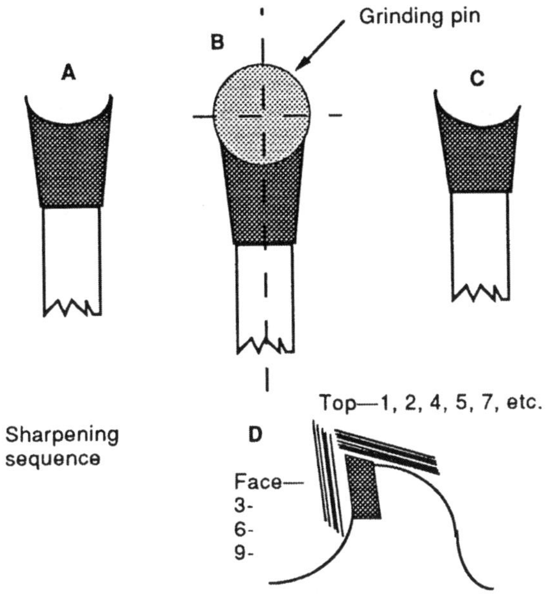

Illus. 4-14—Hollow ground tooth

roll into the gullet space. The forces created in this manner of progression crush and tear the wood fibers ahead of the tool and laterally. This is more pronounced as the cutting edge becomes dulled or when the cutting angle is incorrect for the material.

Flat-top saw blade teeth, however, excel in several cutting processes and have two cost advantages over other shapes: They are the least expensive of all to manufacture and, further, are sharpened very easily, requiring only simple face and top grinding. Saw blades with flat-top teeth are perhaps best at rip cutting solid wood. In only this instance does the abrupt contact surface pressure have a dramatic benefit. As the flat leading edge of the tool progresses into the wood, the surface pressure created tends to pre-split the material ahead of the tool with the net effect of requiring less energy. The contact pressure works much like driving a wedge into the wood ahead of the cutter. This does present disadvantages with very dry, brittle wood materials.

Saw Tooth Applications

Contact surface pressure is not really anything that we can measure; in a single process, such as cutting a large piece of lumber into several smaller ones, all the forces present will vary somewhat in each cut and from one to the next. But it is important to understand that we can alter and manipulate tool geometry and features, processes, and machine characteristics to seek the best advantage. We have to know what is available to us.

Illustration 4-15 shows 12 of the most common saw blade tooth shapes—common to industry, that is, and not necessarily what is available in all market segments. The marketing of cutting tools is such that local sources, such as building supply outlets and hardware stores, perhaps have never seen many of the tools that are readily available to industry. In addition to the 12 illustrated, there are perhaps three times more examples when the basic shapes are grouped together in various configurations. There are also saw blades with a wide range of other features directed at cutting unique materials that are only remotely related to the woodworking industry, e.g., plastics, soft metals, and fiberglass.

Refinement of cutting tool characteristics advances to satisfy product manufacturers' needs. For example, in years past, kitchen cabinets were made entirely of solid woods and plywoods; today, they are made of fiberboard, particleboard, plywood, and solid woods and covered with wood veneers, plastic laminations, impregnated papers, etc. The cutting requirements of all of these materials are different from one to another, and the driving forces of tool design and refinement for these processes are effectiveness and costs.

Illustration 4-15 provides a brief view of saw blade tooth shapes. A review of their particular uses follows:

A—*Flat-top* saw teeth are most commonly used to rip cut solid woods, taking advantage of the "pre-splitting" effect of the contact surface pressure. Also, groove saws and scoring saws may have flat-top teeth. One advantage of this tooth shape is the relatively low cost of manufacture and maintenance. The disadvantage of the flat top for other cutting processes is rough surface appearance and the horsepower required to cross cut, for instance, solid wood materials.

B—*Alternate top bevel* is the most widely used tooth shape, especially with solid woods.

Saw Tooth Applications

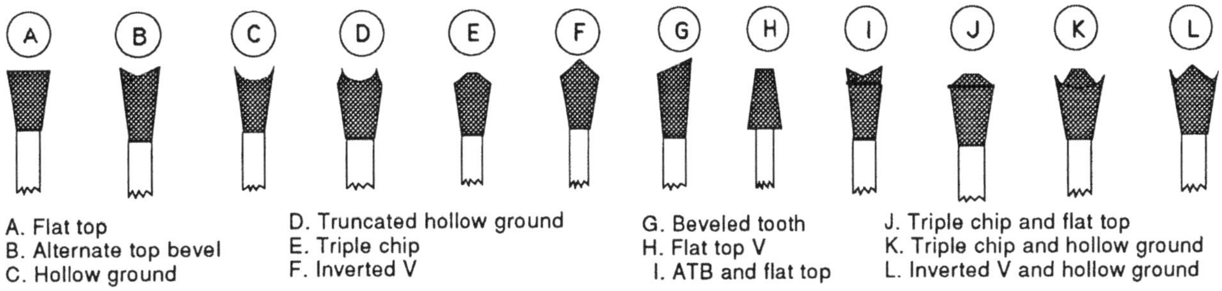

Illus. 4-15—Saw blade tooth profiles

This tooth shape leaves a surface that is cleanly cut and uses a relatively small amount of energy, depending upon the wood-grain orientation in the process.

C—*Hollow-ground* saw blade teeth provide the best surface quality of any tooth form. The contact surface pressure is dissipated along the radiused leading edge, leaving the cut surface smooth and relatively free from tears and other imperfections. For industrial use, however, there are three disadvantages that cannot be tolerated, and all relate to cost.

First, the cost of manufacturing hollow-ground teeth is high. Second, the leading outer edges of each tooth are relatively frail and prone to impact fracture, the tips actually breaking off, especially in man-made panel materials or knots in solid woods. Last, the maintenance cost is high, requiring unusually precise machinery and grinding wheels. (See Illus. 4-14 B and C.) To grind and maintain the small radius feature requires a small "pin" tool having the same radius as in the tooth top and perhaps face. This process, even with automated equipment, is slow and expensive. Additionally, the sharpening process has to be done with exact precision or the tool will not perform well at all. Illustration 4-14 C shows the necessity of precision; if the radius is not exactly centered, the saw blade will not cut in a straight line and will wander. Where the hollow ground tooth has its greatest advantage is cutting thin plywood panel materials, such as wall coverings, covered with decorative wood veneers or frail laminations.

D—*Truncated hollow-ground* teeth have the outer edges of the radius beveled, moving the more frail points away from the outer rim. This is not a common tooth form but illustrates an attempt to provide the benefits of a hollow ground tooth without one of the disadvantages, i.e., the frail, outside leading tips. In an industrial saw blade, the truncated hollow ground tooth will most often be used together with another shape such as a V or flat top, again in the attempt to dissipate the effects of contact surface pressure.

E—The *triple chip* or *trapezoid* tooth shape is finding wider use in all areas of woodworking. The effective distribution of contact surface pressure along the three plane surfaces is successful over a wide range of materials and processes. From rip cutting solid woods to parting cuts in man-made panels, the triple-chip tooth shape is proving to be cost effective and successful. This is especially true as automated sharpening machinery becomes more available and efficient. Prior to recent developments in tool grinding machinery, the sharpening of a triple-chip tooth saw blade would have required completely different set-up and operational procedures for the three peripheral edge angles and another for the face of each tooth. With current technology, the entire sharpening process is accomplished in a single operation. Additionally, with electronic computer controls, newer saw grinding machines can duplicate manufactured details with precision not possible even a few years ago.

F—The *inverted V* is another attempt to dissipate the contact surface pressure. This tooth form was developed and manufactured to cut particleboard in Europe, where the material was developed. The V tooth is used with a flat top, in a somewhat unusual configuration, that is, the flat-top tooth does not cut in exactly the same cutting circle as the V tooth, but rather below it by as much as 0.015". The V is allowed to protrude further to stabilize the process and take better advantage of the contact surface pressure benefits. The flat top lessens the total load on the V tooth and, in effect, cleans up behind it and aligns the lateral walls of the tool path. The tooth shape and arrangement was for a number of years patent protected.

G—A *beveled tooth* may be used with another, opposing bevel tooth shape in an alternate top bevel arrangement (B), or used alone in very light trimming/beveling processes. Regarding the use of teeth beveled in one direction only, great care must be given to such a process, or the saw will not cut in a straight line. Saw blades with ordinary body/plate thicknesses do not have sufficient lateral strength to withstand the deflection of cutting with a beveled tooth. Except for very light cutting processes, where the tool can completely overwhelm the material being cut, it should be avoided. This is done successfully in special machining processes, such as edge banding, where it is necessary to trim and bevel the extra width of an applied, relatively frail edge-banding material. If a heavy load were placed on a saw blade beveled in only direction, there would be no sure method to control the cutting alignment without providing a body/plate of extraordinary thickness.

H—A *flat-top inverted V* is a scoring blade such as described earlier in this chapter.

I—*Alternate top-bevel (ATB)* and *flat top* is a grouping of teeth designed (without hard supporting research) for use in cutting solid woods, attempting to take advantage of the effects of contact surface pressure of both flat-top and alternate bevel teeth. The results may be beneficial in unique conditions, but the advantages are not outweighed by the additional costs of sharpening and manufacturing the tool.

There are several other tooth shape groupings produced and marketed with claims of improved performance for particular materials or unique processes—every fourth tooth having a different shape, three or four different shapes, and so on—somewhat like the vast array of fishing lures, whose first objective is to catch the fisherman.

The last three tooth groupings have had wide success and acceptance in cutting man-made panel materials of different make-ups:

J—A *flat top* is combined with a *triple-chip* tooth, as described in E. The triple-chip is gaining more widespread use than all others, especially when it is paired with a flat top for a very wide range of materials and processes.

K—The pairing of *triple-chip* and *hollow-ground* teeth—and L—the pairing of *inverted V* and *hollow-ground* teeth—are finding wider use cutting man-made panels with frail or brittle laminations. The unique tooth configurations have solved surface chipping problems that have been prevalent with these materials. In this instance, the panel coverings were perhaps developed before an adequate means to cut them was widely available, even though there is constant communication among all of the segments of industry involved in this kind of progression.

Woodworkers at all levels of participation and involvement are being affected and influenced by events and conditions far removed from their environment. Many developments in the design and use of cutting tools relate to solving problems of working with materials that are only wood based or are used in the construction of a given product.

It is not difficult to conceive that most of the man-made materials were developed in Europe where there is not the abundant supply of solid woods that we have (for the moment) here in the U.S. Europe needed a panel material that could be easily produced at low cost from low quality, readily available materials.

Hence, in 1945 when particleboard was developed in Germany, there was a need to cut and shape it effectively. The needs of cutting the new materials have directed the research and manufacturing of most cutting tools that are common today, to the benefit of all.

Saw Blade Performance

Studying the results of the cutting process to see if the tool is preforming correctly is an important process. Too often, cutting tools are blamed for any number of problems that are not of their doing at all. There is a wide range of variables to observe and control in every cutting process, many only in part directly related to the tool itself. The tool has to have the correct geometric features, sizes, and shapes, but there are two more complete sets of variables in the process:

- Variables inherent in the material to be cut;
- Variables in the machine that will hold and drive the cutting tool.

In each of these, the number of possibilities, though large and diverse, can be reduced to a manageable set of guidelines. Further, as we review and execute set-up procedures prior to attempting a particular cutting process, we can eliminate many problems— certainly the larger, most obvious ones. That is, as a normal procedure, prior to putting tools and machines into motion, we need to develop a check list that will force us to examine what we are attempting.

This list need not be long nor terribly complex. There are only two major groups of criteria to address, and they become easier to manage each time the procedure is used. The criteria groups that need to be reviewed prior to a cutting process are the following:

- Is the tool correct for what is known about the material?
- Is the machine process correct for the procedure? And are there any mechanical problems with the machine that will affect the performance?

All of this, of course, demands that we understand the materials we are working with and that our machines are in working order. Failure to do this might explain some of the problems that are blamed on the performance of the cutting tools. We often make an incorrect tool selection or fail to detect mechanical problems with the machine we are using.

Illustrations 4-16, 17, and 18 depict the most common saw blade problems that relate to tools or machines. We will address others later in this chapter. Illustration 4-16 A shows saw blade *axial misalignment*, viewing from above the machine. This problem will be further illustrated because it is perhaps the most common and pervasive. Axial alignment problems can result from misalignment of the arbor shaft that supports and drives the saw blade, worn bearings that support the arbor shaft, and so on.

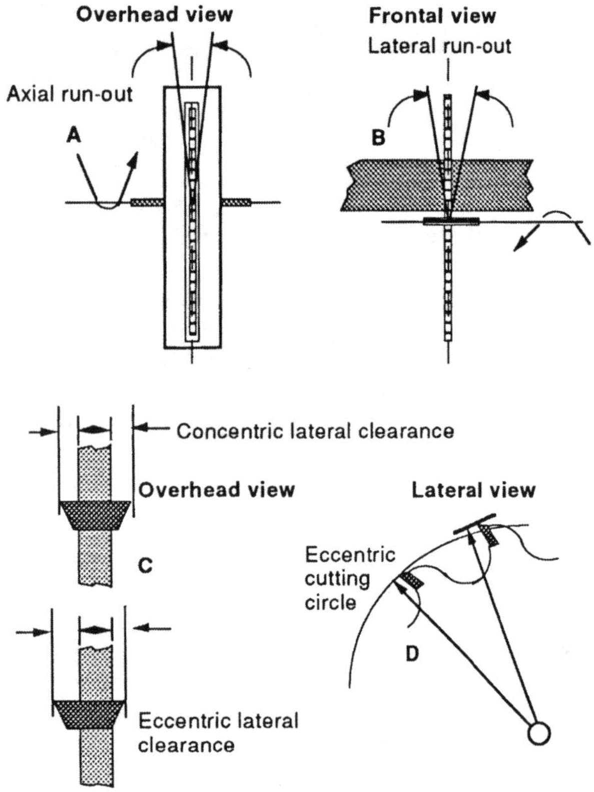

Illus. 4-16—Common circular saw problems

Axial misalignment can also be in the tool itself, though this is not likely because the tool would quickly overheat. An instance when the saw blade could cause this and still run, albeit not well, would be if only half the teeth in an alternate top-bevel tooth configuration were sharpened. I have seen this, but only once. Another source of axial run-out can be warped saw-blade stiffening collars. If saw collars are warped only a little, near the center, the distortion on the blade at the periphery can be several times as bad. As cutting tools become dull, all mechanical forces and stresses are multiplied so that even minor inconsistencies become more apparent and threatening to the success of the procedure. This is an excellent reason to understand the expected wear characteristics of cutting tools that will be used regularly.

Illustration 4-16 B is a frontal view of *lateral run-out*, that is, as though we were cutting a very small-angled bevel. If the machine is capable of cutting a bevel, an adjustment will take care of the problem. But a machine with a fixed arbor presents another problem. A saw blade cutting any degree of angularity, as in a bevel, requires considerably more energy, and this may cause what appears to be more lateral run-out. Part of this is because a saw blade in motion creates a gyroscopic force that works to return the tool to an upright position. Additionally, worn or damaged machine parts such as bearings and shafts can contribute to this problem.

In Illustration 4-16 C, the two saw blade teeth viewed from overhead show the uncommon problem of *eccentric lateral clearance*. A quick inspection with an accurate measuring device will accurately diagnose this. The device needed is a lateral clearance micrometer and should be a standard tool for any serious woodworker. Teeth that protrude laterally in an uneven manner can create several problems. If the deviation is beyond an allowed tolerance (the normal maximum lateral clearance tolerance is 0.003" on each side of the blade), the mechanical effects can be dramatic on alignment, finish quality, and horsepower consumption. The tool may be manufactured within tolerance, and subsequent errors during service may create the condition during sharpening or in a retipping repair procedure. Again, this is not common.

Illustration 4-16 D depicts another uncommon tool problem that woodworkers should understand. An *eccentric cutting circle* would in most instances be immediately apparent. If the bore is not perfectly centered, the vibration will be tremendous. It is possible, however, to create an eccentricity through improper service/grinding procedures that remove irregular amounts of tooth material. Saw blades are less prone to the problem of eccentric balance than other types of cutting tools. Tools that have adjustable or replaceable knife segments are much more likely to have out-of-round and out-of-balance problems.

Illustration 4-17 (B, C, and D) shows drawings of wood samples cut with a circular saw blade. Each represents half a sample that has just been cut. In B, the tool-path marking indicates that the saw blade was tracking squarely, as there are equal tooth marks left during both the descending and ascending periods of contact. If any tooth marks are deeper than others, it would indicate either teeth of unequal lateral placement or some movement of the saw blade during each rotation. A part of the appearance from even a perfectly aligned saw blade and machine will vary somewhat in response to the strength of the material being cut. Grain deviation, moisture content, and the composition of the man-made panel materials all influence the cutting process. Inspection of these results is an important practice in all woodworking procedures, providing important knowledge of tool performance characteristics that will help with subsequent procedures.

The samples in C and D could very well be opposite sides of the same cutting procedure. In C, there is clear evidence that the saw blade was misaligned and making primary contact during the descending period of contact and no contact at all during the ascending period.

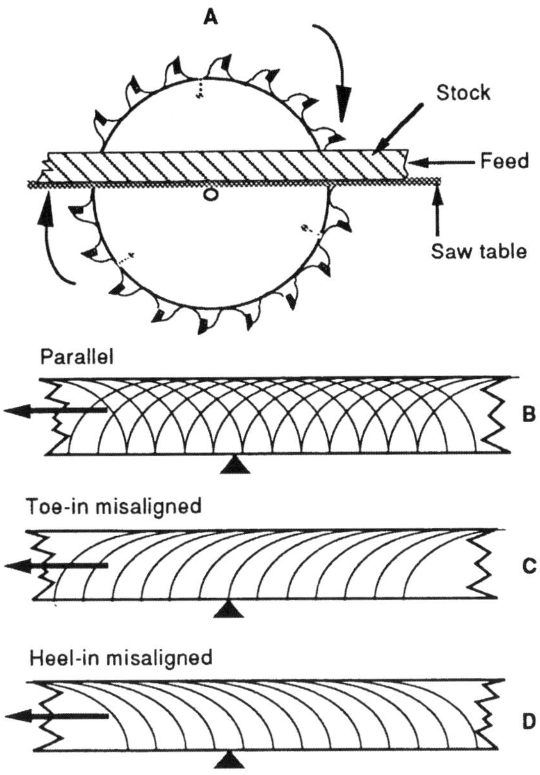

Ilus. 4-17—Tooth-mark saw alignment

Another explanation for these markings would be that all teeth were protruding far out to the left side of the tool, a condition that is possible but unlikely. Misalignment that would produce such marks would present several dangers to the tool, the procedure, and the machine operator. The initial danger is to the saw blade itself because the stress of cutting in this position affects all the tool's mechanical strengths and will surely shorten its useful life. The temperatures at the cutting interface will be elevated, dramatically increasing the danger of permanent thermal damage to the saw body. If the toe-in (or heel-in) is too great, there is also the possibility of distorting or warping the tool permanently.

Danger to the operator exists in the possibility of the material being thrown backwards. The binding effect requires more feeding effort. That is, on a table saw, if the blade is not properly aligned the effort necessary to push the material into the saw blade will be greater. Moreover, if the saw blade wobbles or otherwise becomes unstable, the machine can use all of its available horsepower to throw the material backward. Unfortunately, this is not an uncommon occurrence.

A similar though opposite misalignment of the saw blade in relation to the feed-travel of the material being cut is depicted in Illustration 4-17 D. In this instance, the tooth marks are left by the teeth in the ascending period of contact when there is essentially no cutting taking place. The dangers to the tool and operator are similar to those of the opposite condition, with only a minor difference when a table saw is involved. With misalignment to the right, or heel-in, the danger is more that the saw blade will lift the material being cut and throw it in the direction of tool rotation. Throwing the material either toward the operator or upward and away from the operator could bring his hands into contact with the moving tool.

In all misalignment conditions, there will be poor surface quality and dangers to the tool, the machine, and the operator. Although wood cutting procedures seem deceptively simple and harmless, they are neither. We have designed and produced powerful and effective machines and cutting tools. Yet cutting procedures still require attention to detail and work habits that do not ignore correct machine maintenance, such as keeping tool arbors in correct alignment, free of vibration and movement. Wood is a remarkably strong material, and the tools and machines we use to cut and shape it are stronger yet. They must be handled with care for our own safety. Illustration 4-18 further illustrates saw blade alignment.

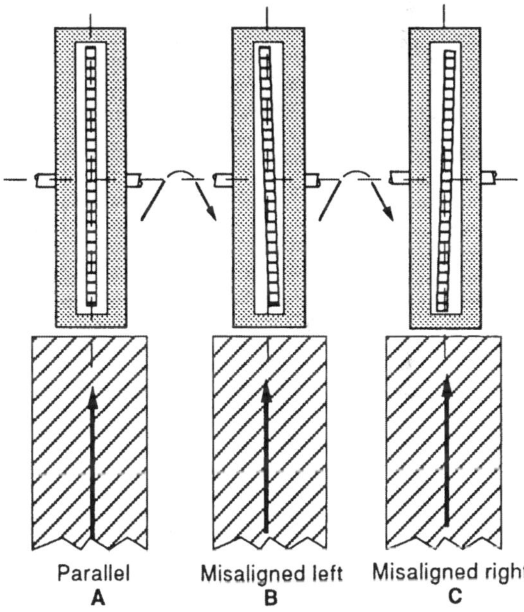

Illus. 4-18—Saw blade alignment

Saw Blade Stiffening Collars

The saw blade stiffening collar is essentially a machine part, though produced by both toolmakers and machine manufacturers. It looks like a huge, thick washer, as we might use with a bolt and nut, except that it is not quite flat. (See Illus. 4-19.) Saw collars add to the rigidity of the saw blade. They are an important accessory to circular saw blades that is too often overlooked. Stiffening collars are as much precision tools as the saw blades they support. Like saw blades, stiffening collars have a predictable life expectancy, though it is much longer than the cutting tool's. Saw collars should be periodically inspected for warp and distortion. With repeated working cycles and tensioning against the blade, collars can be fatigued and bent. If the saw collar, which is most often thicker and stronger than the saw blade, becomes warped, it will transmit deformation to the saw blade and cause it to wobble. Periodic inspection of saw collars is not complex or expensive and requires only a light source and a straight-edge ruler.

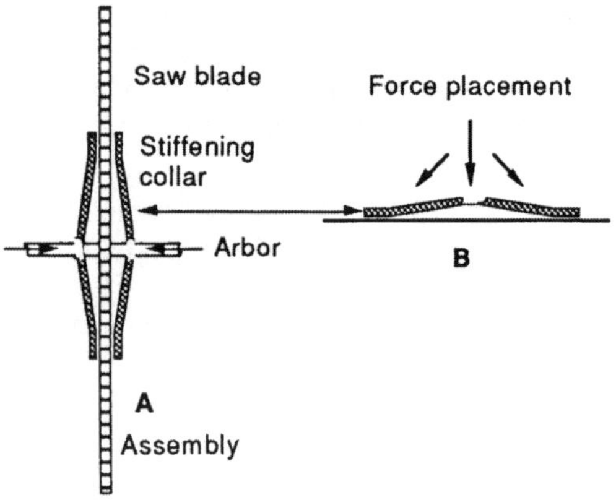

Illus. 4-18—Saw blade stiffening collar

A good saw stiffening collar must have sufficient strength to add support to the saw blade during the cutting process. Illustration 4-19 B depicts how the saw collar is tensioned against the saw blade. The shape of the collar allows an air space beneath most of its diameter, dissipating heat. The area about the bore (hub) and the periphery are nearly parallel. When tensioned against the saw blade and tool arbor, the saw collar first makes contact with the saw blade at its periphery. As it is tightened, the hub area at the center makes contact. The saw collar is designed to flex slightly as it is secured against the saw blade to assure support of the tool. It is this flex that eventually fatigues, and the tool accessory should then be replaced.

Saw collars should made of tool steel, machined accurately, heat-treated, and precision ground. Unfortunately, there are no hard, accurate guidelines regarding life expectancy of saw collars, as the factors that determine this are the amount and type of work the collars are subjected to. A basic rule may be that good ones usually cost more, do a better job, and last longer. Tool collars are a necessary accessory for many processes and need to be maintained.

Saw Blade Service and Sharpening

Our examination of sharpening, handling, and storage of saw blades will emphasize blades with attached teeth of tungsten carbide, Stellite, and similar materials, rather than those of high-speed steels, though many of the conditions are the same. For more recently developed metals and metal alloys, see Chapter 3.

The maintenance and sharpening of saw blades and other cutting tools is too often not clearly understood nor performed well enough to fully take advantage of the tool life they were designed for. The basic premises of tool maintenance include the following:

1. Well-made cutting tools can be sharpened repeatedly, and after each cycle perform nearly as well as when new.
2. Sharpening procedures should duplicate in every detail the geometric features of the tool when new.
3. The finish quality of the tool teeth/knives during manufacture can be duplicated during every sharpening process.
4. The sharpening procedures have to be correct to get the most of the tool life designed into the tools.

Exactly what constitutes sharpness or dullness to determine when a tool needs to be sharpened will vary according to specific processes and the cost of maintenance service. That is, the sharpening cycle and life of a tool in a given process will differ from one to another, based on different criteria, those being, in no particular order:

- The required finish quality of the cutting process,
- Total cost of the sharpening service,
- Cost of lost production time for sharpening,
- Increasing energy cost during the sharpening life cycle,
- Cost of the material being cut,
- Overall effectiveness of the cutting process,
- The relationship of the given cutting process to the chain of events en route to a finished product,
- Initial and replacement costs of the cutting tool,
- Production and scheduling requirements of the process, and
- Comprehension of tool sharpening practices.

While all ten criteria are probably not examined for each cutting tool and process, they are what has to be pursued to make the best use of a very costly resource.

Perhaps the best indicator for determining that a particular cutting tool needs sharpening is the finish quality of the cut surface. In a hand-fed operation, such as on a table saw, diminished finish quality will be followed by a second indicator—the additional effort required to push the material into the cutting tool. In this particular progression, it is not necessary to look for a third indicator.

All cutting tools, from the moment they come into contact with the material being cut, become duller, and the rate at which this takes place results from many factors.

The briefest explanation of *sharpening cycle* is the period of time or amount of material a tool cuts before the decision is made to sharpen it. Further, *sharpening life*, or *tool life*, is generally the total number of times a tool can be sharpened, as dictated by the amount of tooth/knife material present when the tool is new and how much is removed during each sharpening process. The importance of this is easily understood in industry, where tool expenditures are a very important part of manufacturing cost and are a consideration in the initial purchase of all cutting tools.

The rate at which cutting tools become dulled varies widely from one process and material to another. For instance, a rip-cutting process in clear solid woods may require thousands of linear feet of production over several days, while a similar parting cut in particleboard may completely dull a similar tool with a tenth of that production in four hours. Admittedly, this may not be a fair comparison, but the point is that there are wide ranges of sharpening life. Factors that affect the rate at which tools become dull would include:

1. Makeup of the material being cut—chemical content, moisture content, and strength;
2. Effectiveness of chip and dust removal;
3. Mechanical conditions of the process;
4. Feed speed and rim speed of the tool process; and
5. Correctness and appropriateness of tool geometry.

There are several more factors that relate to either normal or abnormal progression of tool wear. We need to pursue the factors we can alter or control that will most improve how and at what rate the cutting tools become dulled. Often, causes of unusual tool wear are simply that we have overlooked something very basic. The two most common causes that are overlooked are moisture content and chip/dust evacuation.

The surfaces of cutting tools that receive the most wear are those in most immediate contact during the process. (See Illus. 4-20.) In A, B, and C, we have three views of the same cutter tooth, which could be for any type of tool—saw, shaper cutter, router, etc. The surface edge that makes contact in the cutting process will show 90 percent or more of all wear to the tool. This 90 percent figure is important because it dictates sharpening procedures and underscores the fact the other surfaces away from the contact point also wear, but dramatically less. Wear on the edges and surfaces away from the contact point result from heat, abrasion, and other factors inherent in removing the chip/dust offal.

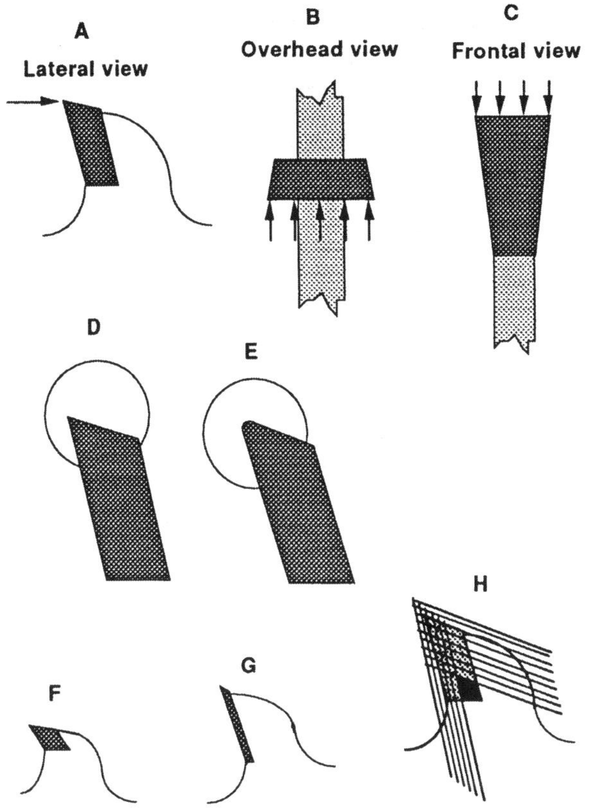

Ilus. 4-20—Saw blade sharpening

Illustration 4-20 (D and E) shows the relative size and shape of a tool tooth/knife as it wears down—D being sharp, and E being dulled. The progression of this wear will vary with the material being cut and the metallic makeup of the tool/knife material. The size and shape of the tooth/knife as it wears will be relatively the same, regardless of other factors, and will require repetition of the same sharpening procedures to return the tool to serviceability.

In Illustration 4-20 F and G appear two of the most common tool sharpening errors. In F, the tool tooth/knife has been sharpened only on the top, effectively wasting tool life. G illustrates the opposite problem. The tool has been sharpened only on the face, again wasting tool life and certainly increasing the overall cost of the cutting procedures.

In H, we illustrate the accepted sharpening progression that will yield the best tool life for saw blades and most other cutting tools. As important as the selection of overall tool configuration is—diameter, number of teeth, tooth shape, and so on—we can waste half of the tool's effectiveness by not carefully maintaining it. This may seem unimportant to most woodworkers, who replace tools infrequently. But in industry, where yearly tool costs may run into hundreds of thousands of dollars, effective maintenance is crucial.

In Illustration 4-21, we have a more complete depiction of the most commonly accepted progression and method of sharpening saw blades. The three drawings at the top of the illustration show the size and shape relationships of a single saw blade tooth.

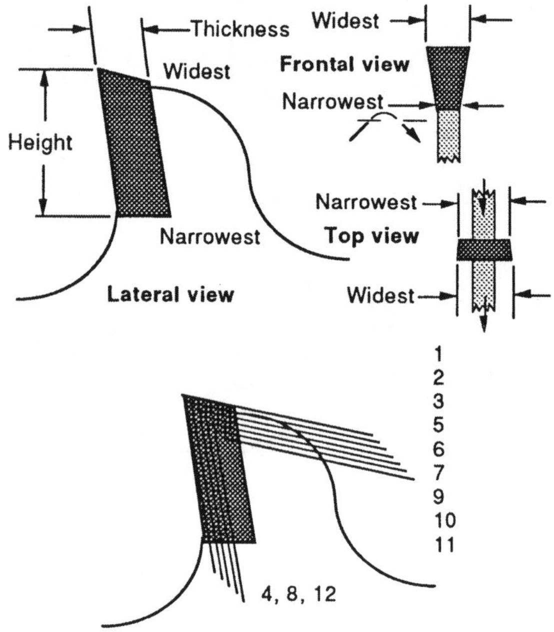

Illus. 4-21—Saw blade sharpening sequence

Given that the tooth material is widest at the outer leading edge and narrowest at the rear lower edge, and that the knife/tooth is in some ratio considerably taller than thick, the sharpening sequence should be a progression of cycles to take advantage of the shape and size. Too often, the useful life of saw blades is reduced considerably by incorrect service practices. Most commonly, during each sharpening, material is removed from both the face and top surfaces. Unless the tool has been run excessively, this is unnecessary and a waste of tooth material.

Another wasteful practice may remove an amount of material from the tool well beyond what is required to return the cutting edge to sharpness. Depending on the size of the tooth when new, most sharpening procedures will remove only a few thousandths of an inch of material per cycle (0.004" to 0.012"). Another waste of tool life occurs when the amount of tooth material removed varies from one tooth to another, ensuring that the tool will have an irregular cutting circle and that its useful life will be shortened. This is not common, though it is a mistake to be watchful for.

In Illustration 4-22, we show perhaps the most common mistake in the use of cutting tools: the idea that we can somehow get more tool life by extending the normal sharpening cycle time or amount of production.

A indicates the tool tooth/knife when sharp. B shows the tool at the end of a reasonable sharpening cycle. C shows the relative condition of a tool tooth that has been run too long and will be costly to repair.

D and E show the amounts of tooth material that will have to be removed to return each tooth to sharpness. The false economy is concluding that we can simply run the tools longer, and the sharpening procedure will correct it. It will, but at the loss of perhaps several sharpenings worth of tooth material. In E, it will be necessary to remove considerable material to correct the damage, no matter the sequence of grinding procedures. This will be true of all cutting tools, and for all cutting tooth/knife materials. There are simply no bargains to be sought in regard to tool wear characteristics.

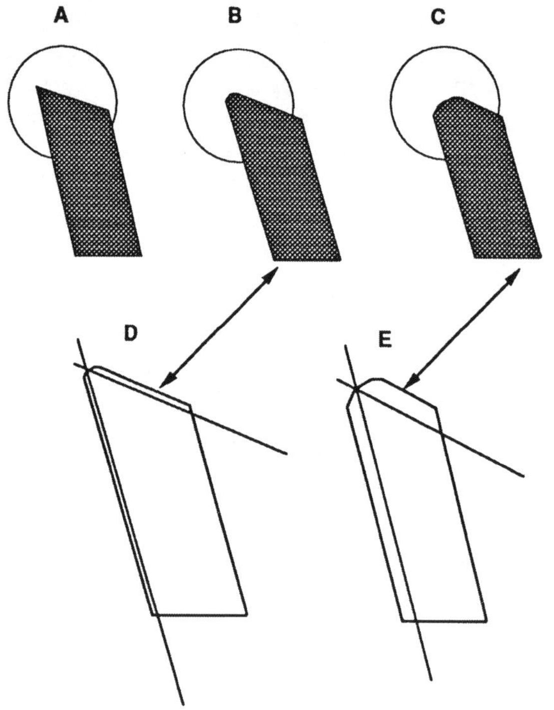

Illus. 4-22—Overextending the sharpening cycle of any cutting tool

As a precaution in tool service, it must be clear that all types of dull tools are dangerous; use more energy to do the same work; generate more noise, heat, and vibration; and leave cut surfaces of diminishing quality.

When a saw blade or other cutting tool is new, it should be carefully inspected down to every tooth. Tooth shape/configuration, how teeth are attached to the body, and the grinding finish quality are all features that must be duplicated in every detail during each subsequent sharpening cycle. There is no reason for tool performance to diminish significantly over its life. If it does, there are problems to solve, not specifically with the tool itself but rather with what has been done to it.

A last point regarding the service and sharpening of

saw blades is the practice of replacing all of the teeth after a complete sharpening life. Given the enormous number of work cycles that the blade has performed, the body plate has no doubt been stressed and fatigued to such a condition that it has become unpredictable and dangerous. The best advice would be to buy a new one.

Cleaning, Handling, and Storing Saw Blades

Often, a cutting tool's serviceable life can be extended with consideration given to its overall care, including cleaning, handling, and storage. The need to clean cutting tools does not occur to most woodworkers, but removing any build-up on the lateral portions of a saw blade can improve its operating efficiency by as much as 15 percent or more in some instances. That saving can extend into both the horsepower needed to perform the work and, by reducing tip temperature, the sharpening life.

Saw blades can be cleaned by any of several means. A couple of oven cleaning chemicals work very well, as does washing the tool with nearly any good detergent, such as for an automatic dish washer, followed by a thorough rinsing and application of a very light oil film to eliminate rust. Before cleaning by a mechanical method, such as wire wheel or abrasive sanding, a woodworker should contact the manufacturer or distributor of that tool regarding the advisability of such methods.

A number of materials that woodworkers may have the occasion to cut will very quickly leave deposits on any cutting tool. Solid woods that are high in rosin or pitch content will create a heavy gum-like deposit on cutters, especially if the wood contains an abundance of moisture. Many man-made panels with large amounts of adhesive can cause a build-up problem that has to be addressed.

Cutting tools, in addition to being expensive, are in many aspects very frail. The materials that make up cutting teeth/knives are at the same time very hard and very brittle, especially the most expensive alloys such as tungsten carbide, polycrystalline boron nitride (Borazon), Stellite, and high speed steel. The most common error in handling is to allow the tools to come into contact with other metals, such as by setting the tool on a metal-covered work table or stacking saw blades on top of each other. It may be possible to stack saws on top of each other and not inflict any damage, but it is not likely—they are that frail. The best rule to observe is not to allow any metal-to-metal contact.

As regards storage, saw blades or any other cutting tools have to be protected from any eventuality. Coverings, such as individual containers, should be used to protect the cutting edges. Saw blades should not be stacked unless means have been very carefully provided to prevent any distortion of the tool. As mentioned earlier, saw blades do not have great lateral strength. What is overlooked is that the frail teeth are wider than the body is thick. Consequently, whatever weight is put on the tool is placed on the sides of the teeth. The preferred method to stack saw blades, if necessary, would be to place a plywood or similar softwood panel between each pair of tools, ensuring that the stack is even and parallel.

General guidelines to tool maintenance is more a matter of paying attention to details than any complexity:

1. Inspect the cutter after each use.
2. Sharpen as needed.
3. Keep all surfaces of the tool free from build-up materials.
4. Always handle with great care, not allowing any metal-to-metal contact.
5. Make certain that the sharpening procedures are correct in every detail—finish, procedure, and sequence.
6. Also inspect the "mating" surfaces of the machine(s) that the tool(s) are used on.
7. Store the tools in a manner that will not in any way harm them.

Guidelines for Selection and Use of Saw Blades

The number of variables that we have to consider when selecting the best tool to perform a particular cutting process is getting more complex as woodworkers are introduced to more man-made materials in addition to the more than a hundred native natural wood species. Add to that the numerous machine processes and possible grain configurations in the natural woods, the possibilities of all of the panel materials, and their different veneer and laminate coverings, and we can begin to understand why it is not reasonable to assume that a single cutter of any type can do everything: *There is really no general purpose saw blade.*

There are tooth shapes, sizes, and numbers that preform certain tasks better than others, but none do everything well. There are variables in the processes that we can manipulate toward the desired outcome, i.e., rpm, feed speed, saw blade elevation, or tool approach. We have to begin, however, with selecting the best possible tool for the process, for the material, and for the machine that will hold and drive it. Earlier, we described the specifics of how each feature performed in the cutting process; what follows are general guidelines for selection.

By way of general use guidelines for saw blades, there a few very basic premises to observe. The first consideration is to select a saw blade with the correct tooth spacing for the material to be cut. For instance, it would be very difficult and perhaps dangerous to cut very thin materials with a saw blade with very few teeth. The saw blade in Illustration 4-23 A indicates one tooth in the material as a basic guideline. The formula to identify the spacing is to simply divide the tool's circumference by the number of teeth present, e.g.:

- A 10" saw with 40 teeth has a tooth spacing of 0.785".
- A 12" saw with 60 teeth has a tooth spacing of 0.628".
- A 7.25" saw with 24 teeth hlas a tooth spacing of 0.948".

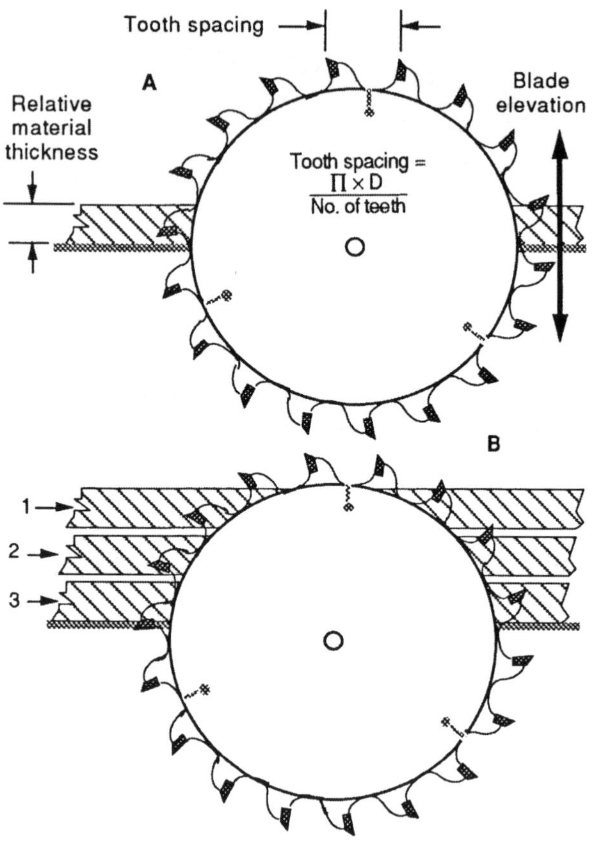

Illus. 4-23—Application guide for (tungsten carbide tipped) saw blades

To cut materials that are thinner than the tooth spacing without dramatically affecting the finish quality will require altering the means that bring the cutter and material together. We have two ways to ensure that a cutter tooth is always in the material being cut: by using saw blades with more or fewer teeth for any particular process or by changing the elevation of the blade in relation to the material. With respect to altering the blade elevation, it is best not to cover the saw blade gullets. To do so would interfere with the necessary air movement about the saw.

It is possible with changing blade elevation to change the number of teeth involved in the cutting process from one to four or five, depending on the material and particular saw to be used. (See Illus. 4-23.) The other advantage of changing elevation is the effect of entrance and exit angles, seeking the ideal finish quality.

After selecting tooth spacing, the other geometric features will have to be reviewed with respect to the process and material to be cut. Rip cutting solid wood, parting cuts in plywood or another man-made panel material, cross cutting a very dense hardwood—each has a completely different set of criteria to examine with the desired outcome in mind. For instance, there is a clear difference between rough-dimension cutting of frame lumber in construction and the final-size cutting of expensive hardwood for a furniture piece.

Knowing the performance characteristics of the tool's geometric features, the machine's capabilities, and the makeup of the material to be cut can make tool selection more sure and accurate.

General use guidelines:

1. Always at least one tooth in the material;
2. For hard or strong materials, more teeth;
3. For soft or less-strong materials, fewer teeth;
4. For thick materials, fewer teeth;
5. For thin materials, more teeth.

Considerations prior to purchase:

1. Nature of the material to be cut—strength, thickness, makeup;
2. Machine features—spindle/arbor capability, size, etc.;
3. Cutting direction, i.e., grain orientation;
4. Requirements of the process—feed rate, rpm, etc.;
5. Required quality of finish.

Purchase considerations:

"Buyer, beware" is still good advice. While most tool manufacturers are technically competent and their products are well made, it is still best to know as much as possible about what we are purchasing, especially considering the complexity of cutting tools. That is, not many of us are automotive engineers, but most of us purchase and drive cars, and we are much more intimately involved with our tools than with our cars. A last word—the larger tool manufacturers are probably more able to completely support the products that they have developed and sell.

Examine all purchased cutting tools for precision, attention to detail, and workmanship. If it is available, seek point-of-purchase information that will identify the intended-use criteria for the particular tool. Most industrial sales organizations have technical personnel who can help consumers with unusual needs and inquiries. While we woodworkers can be thoroughly competent as woodworkers, we still need the support of our toolmakers.

Summary

The saw blades produced today are in many ways unlike the tools manufactured even ten years ago. The degree of precision in every feature and detail has become possible only in the past few years with the development of refined automated machinery. Progress in the metal working industries is in many ways well ahead of that found in woodworking though the gap is closing quickly.

Perhaps some clarification is needed here: In woodworking, we shape wood and wood-based materials into finished products; in metal working we shape metals into finished products. Both groups use cutting tools made of metals. Toolmakers are then obviously metal workers providing the tool needs of woodworkers. The metal working industry has developed the technology of computer-controlled machinery over the past 40 years that is responsible for many of the advances seen in woodworking tools.

Mass production of quality cutting tools at a reasonable cost has been possible only in this manner. Modern manufacturing processes have provided the means to automate nearly every procedure, from cutting the plate features to putting saw blades in a box for shipment and nearly everything in between. Electronic and computer controls on nearly every conceivable machining process have made the accuracy and duplication of critical details possible as they have never been before.

For the woodworker, the appreciable effects of leaps of technology and competition in the marketplace have provided for our needs at a reasonable cost. That point may be arguable, but not if we completely understand the tools we use. As we examine other families of cutting tools in later chapters, just how far we have come can be better appreciated.

A last point is that many of the cutting process details that apply to saw blades, apply also to other tools. How chips are formed, compressed, and later discharged, for instance, is similar for all cutters. Where there are differences, they will be explained and qualified in detail in relation to the unique geometric features of the tool or process.

Chapter 5

Shaft-Mounted Cutting Tools I: Drills

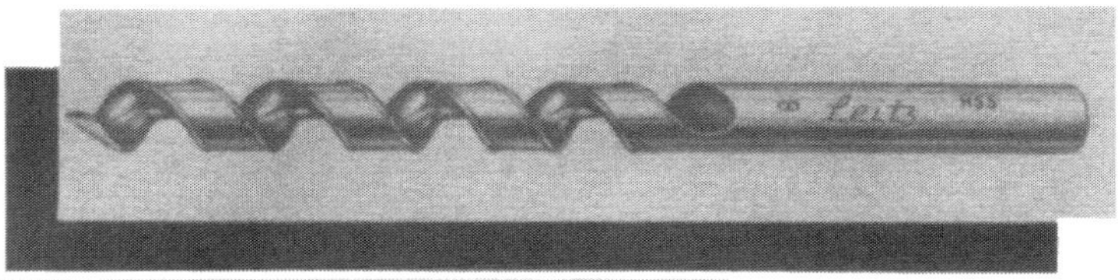

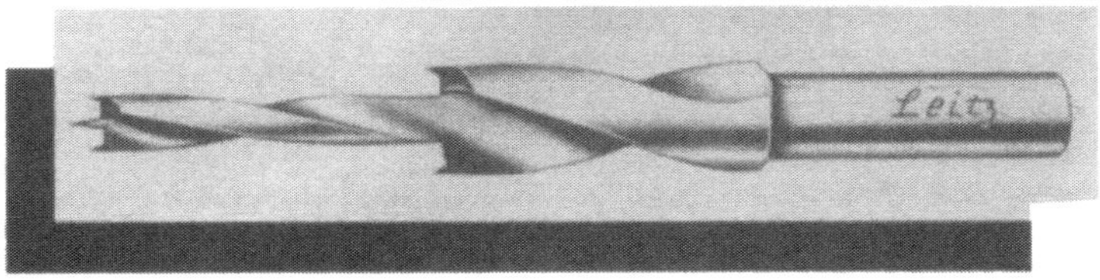

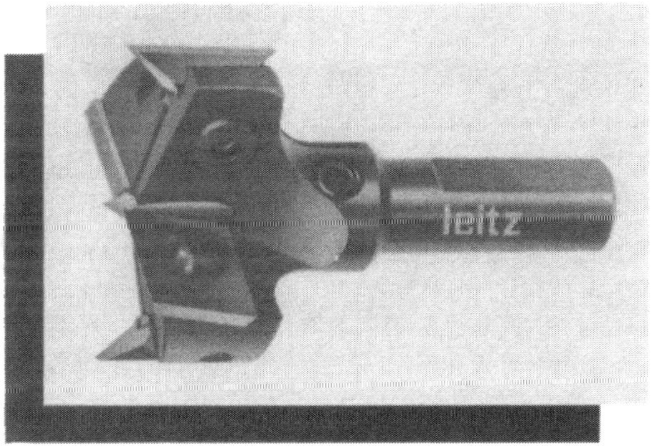

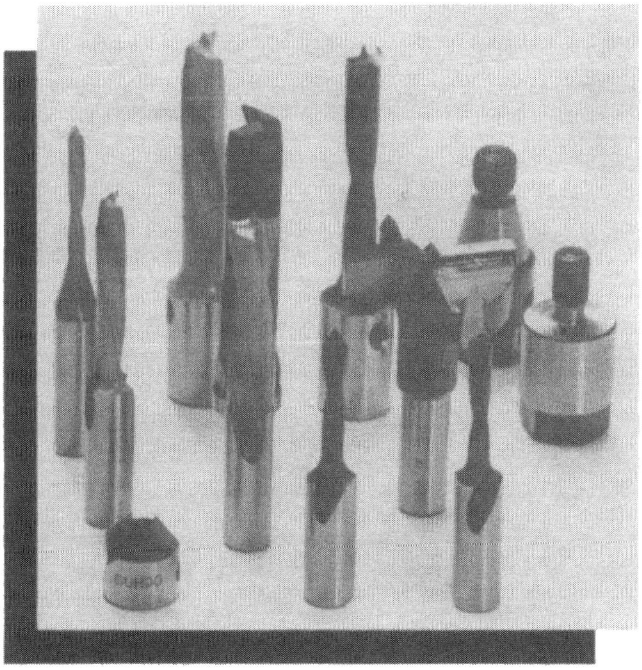

Photos courtesy of Leitz Tooling Systems, Inc., Guhdo USA, Inc., and Leuco Tool Corp.

Shaft-Mounted Cutting Tools I: Drills

It is not really possible to discuss a single technology without acknowledging that there are interrelationships among several technologies and historical progressions to examine before any of it makes much sense at all. That is, modern cutting tools did not simply arrive in our local hardware stores as we see them today.

The basic principles for using shaft-mounted cutting tools date back many thousands of years. The first tools developed as weapons and later were adapted for chopping, boring holes, and sawing—as best as can be determined, in that order. Chopping tools evolved into chipping, scraping, and sawing tools and later into others. Boring bits/drills began as knives and spears and later were adapted to pierce animal skins and other materials. All tools evolved essentially as an extension of the human hand.

The family of tools broadly identified as shaft-mounted cutters is large and includes tools for a wide range of work tasks—from simple spade drills to complex profile-cutting router tools to spindle carving and mortising cutters. The first and perhaps most basic similarity among them is that their body shape is essentially cylindrical and the horsepower necessary to perform the work is transmitted by the shaft portion of the tool body itself. But even the means to hold and drive shaft-mounted cutters varies greatly; at least six methods are commonly employed. The machines and machine adaptations are similarly diverse and complex, from manual, hand-held drill machines to huge computer-controlled router machines as large as a small house to multiple-head carving machines that are larger than an ordinary house. Shaft-mounted cutters include:

- Drills and boring bits of many sizes, styles, and shapes;
- Router tools of similar variety;
- Mortising cutters, both reciprocating and rotating;
- Spindle-carving tools of many types and sizes;
- Abrasive, shaft-mounted cutters of large variety.

The first shaft cutters were primitive drills and piercing tools. For the thousands of years before other forms of energy were developed, all cutting tools were held and driven by hand.

It is important to understand these two concepts, i.e., "held" and "driven." To hold and turn a drill bit in your hand, for instance, would be very difficult. Consequently, numerous clever devices have been developed to hold shaft tools, even if the effort to turn it was supplied by a human hand. Among the first refinements of drilling tools was a handle on the tool itself. Later, this idea evolved into an interchangeable handle that could be used on several tools. Much later, an offset crank was developed to allow even greater efficiency. It is interesting that many of the most primitive means to hold and drive shaft tools are still employed in some form, probably in your own shop. (See Illus. 5-1 on following page.)

With the advent of other means to hold tools and provide power to drive machines, such as water, steam, and electric motors, technological advance accelerated dramatically, as woodworkers sought ever-greater efficiencies of speed for a wider range of tasks and materials. The interrelationships mentioned earlier are those among the advances in the other technologies necessary for advancement in woodworking tools—metallurgy, mechanics, and power generation.

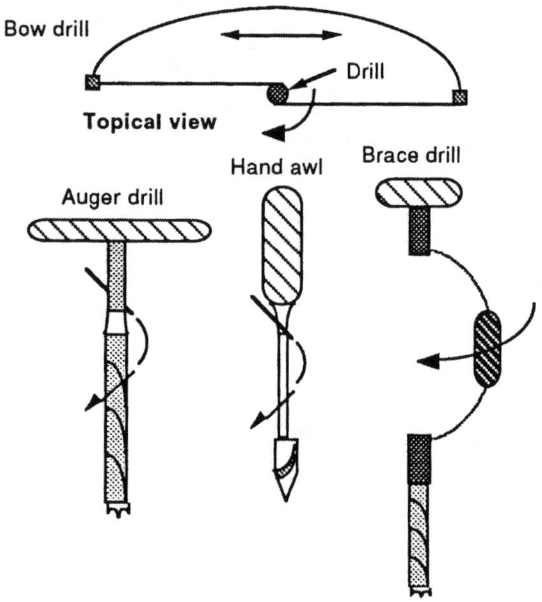

Illus. 5-1—Primitive drilling methods

Performance Characteristics

It is difficult to conceive and understand what drills and boring bits are doing in the course of their work because the processes are upside down to the perspective we are accustomed to; the work is hidden from view; and the tool knives/teeth are not on the periphery of the cutter but rather on the end of a shaft. (See Illus. 5-2.) In A, a saw blade, and C, a shaper cutter, the tools turn in a circular motion and contact the material being cut with only the outer periphery of the tool bodies.

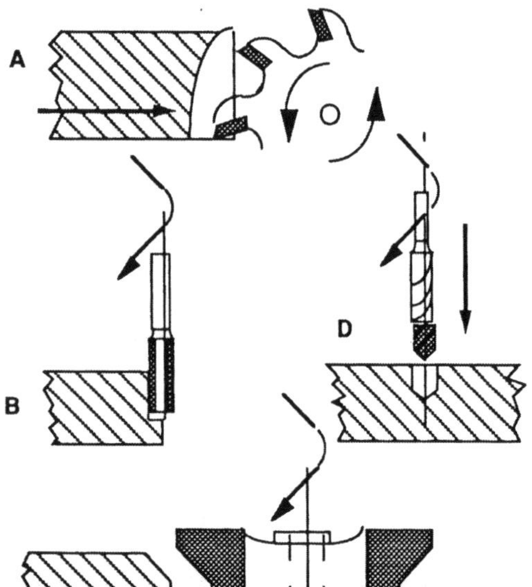

Illus. 5-2—Tool cutting motion

In B, the tool is a router bit, and it turns in a rotary motion, contacting the material with the periphery of the body; that is, it travels laterally (though router tools can have features to plunge cut, also). In D, the drill motion is rotary, and the contact is on the end of the tool only. The geometric features and cutting motion of these tools are unlike those of all other cutting tools in several ways: We must use unusual elements of circle geometry to describe what the tools are doing in the course of their work.

Perhaps the most difficult concept to understand is rim speed, also called surface speed, and sometimes peripheral speed. All of these terms identify the rate of travel of a tool as it makes contact with the workpiece in a cutting process. With all cutting tools that have teeth/knives on the periphery, rim speed, usually expressed in feet/second, can be modified to improve performance by altering either the rpm or the number of teeth/knives. With drills and boring bits of all types, however, we use approaches to processes and tasks that are different from those we use for other tools because having the teeth/knives on the end of a shaft creates somewhat paradoxical performance.

Illustration 5-3 shows the unusual surface speed characteristic of drills and boring bits: At the center of the tool, there is no surface speed, and as the cutting circle becomes larger the surface speed increases. The entire tool body is turning at 1,500 rpm, but the tooth/knife portions at the outermost part of the tool travel farther with each revolution, do more work, and remove more material. The fact that different parts of the cutting knife move at different velocities does not greatly affect tool performance, but it does limit the rate at which the tool can work, that is, how much material can be cut in each revolution.

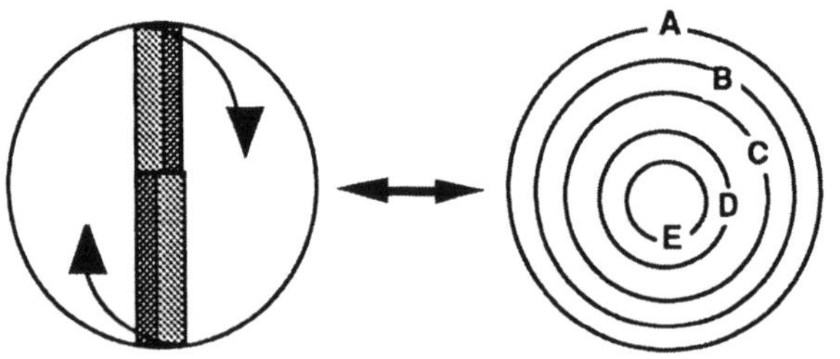

Above is an overhead view of a drill and five of the cutting circles in it. The seeming paradox of drills and boring tools is that across the diameter of the cutter, the cutting knives are traveling at different speeds. This tool is revolving at 1,500 rpm; the diameter is 2.25 inches. Referring to the rim speed calculation in Chapter 1, surface speeds are:

A—14.47 ft./sec. (dia. = 2.25")
B—13.08 ft./sec. (dia. = 2.00")
C— 9.79 ft./sec. (dia. = 1.50")
D— 6.54 ft./sec. (dia. = 1.00")
E— 3.27 ft./sec. (dia. = 0.50")

Illus. 5-3—Drill/boring bit, rim-surface speed

While it is true that drills and boring bits usually have small diameters and require little horsepower, the overall performance must still be as efficient as possible. Most critical to the success of drilling tools is how well they can endure the thermal conditions of the drilling/boring processes. Unlike most other tools, drills are constantly in contact with the workpiece until the process is finished; other tools, cutting on the periphery, are not. We will explain how woodworkers can address this problem later in this chapter.

The oddity of rim speed in drills and boring tools creates a necessity for other geometric features, such as unusual cutting angles, flute angles, spur knives, and center points. Further, all drills and boring bits must have great precision and concentricity to perform very well. Accurate balance and precise details in all features are necessary.

Basic Tool Geometry

The cutting and relief angles necessary on drill bits and boring tools are similar to those on other tools, except that on shaft-mounted tools the physical dimensions are smaller, which means the details must be even more precise, balanced, and concentric than for other cutters.

Illustration 5-4 (A-G) shows the basic angles of a boring tool by comparing them to features of a circular saw blade, which is much easier to visualize and more

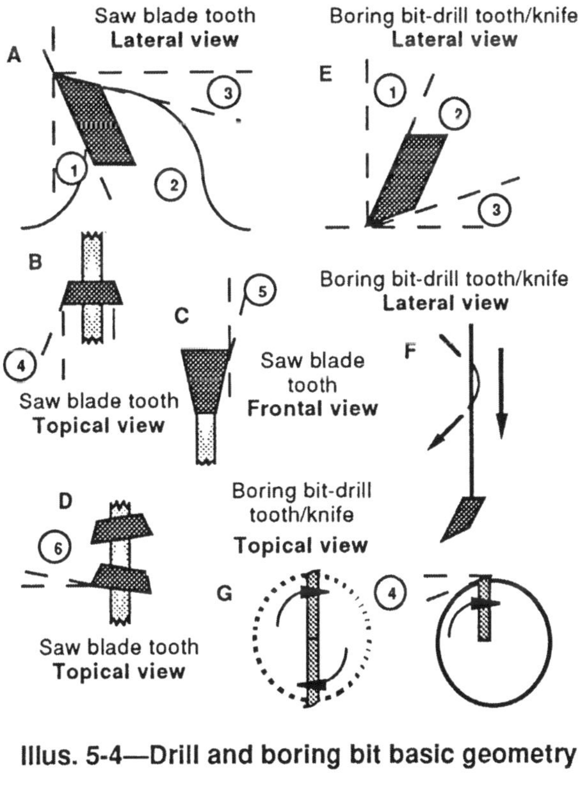

Illus. 5-4—Drill and boring bit basic geometry

common to us. Angles are illustrated and numbered 1 through 6.

The angles provided for cutting and relief on drills and boring bits are similar to those on other tools:

1. *Cutting angle*;
2. *Hook angle*;
3. *Back relief angle*;
4. *Lateral relief angle*—B for a saw-blade, and G for a drill/boring bit; the drill and saw-blade teeth, of course, operate on different planes—the drill is rotary, the saw is circular.
5. *Angle of lateral relief to the periphery*, as drawn in C for a saw blade, does not apply to drills or boring bits because those tools are sharpened on the end only. Otherwise, the tool would become narrower as it is sharpened. In most instances, shaft-mounted tools do not have lateral relief either, except in the sense that the outer peripheral end of the tooth/knife is rounded to the dimension of the tool's diameter as shown in G.
6. *Shear angle*, or *decline to axle*, as illustrated in D, does not apply to most drilling tools, though in unusual applications there would be an advantage to providing shear when cutting especially dense and strong materials. Manufacturing a shaft cutter with shear would be more difficult and expensive, and the tool would require special procedures to sharpen, adding to its operational costs. Shear angle may be present in shaft tools that have a large number of cutting teeth, such as multi-spur bits produced for unusual applications such as drilling very large holes. Shear angle is advantageous in any cutting process, helping to distribute contact surface pressure along the plane of the cutting knife edge.

Design Characteristics

As mentioned, it is difficult to conceive what drills and boring tools are doing in the course of their work, but a cross-sectional view of the process shows that drills perform very much like other tools in cutting the material, severing the waste into chips and discharging them. The revolving motion with the teeth/knives turning about a center point, however, creates a problem with the last two elements of this process. Most other cutting tools have teeth/knives arranged about the periphery and cut on the periphery so that contact with the workpiece is interrupted during each revolution. Most waste discharge is thus accomplished by centrifugal force.

For all the tools illustrated in Illustration 5-5 except the drill, the cutting process is interrupted for a part of each revolution allowing the waste to be discharged. But for drills and boring bits of all kinds, a means must be provided for the chips to gather in the gullet/flute

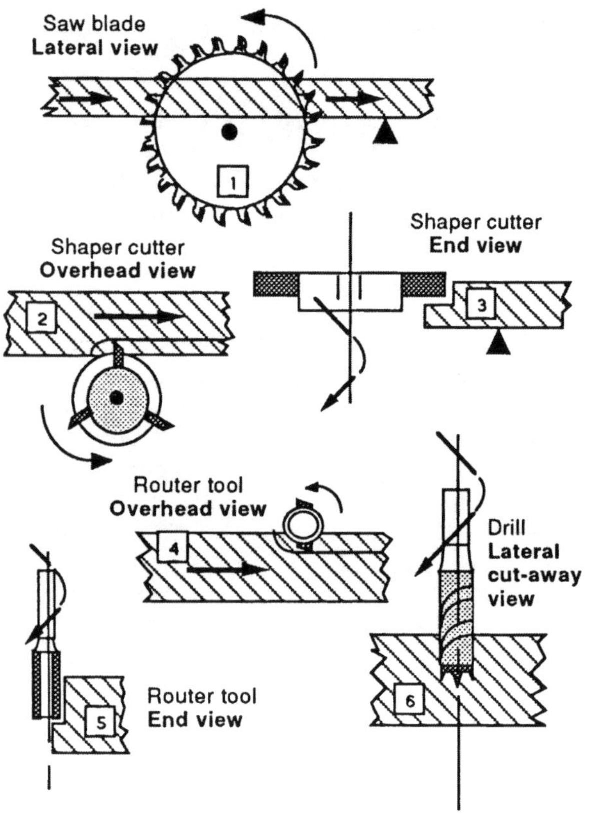

Illus. 5-5—Waste discharge

space and then be discharged in the direction opposite from the tool feed. The gullet/flute has to be large enough to accumulate the waste as it is created and shaped so as not to become plugged. Drills are manufactured with different flute inclination angles and shapes to suit the material to be cut.

For drilling tools, like other cutters, mass production and competition in the marketplace have reduced end-user cost and, to a great extent, standardized drill shapes and sizes and the materials that the tools are made of. The tools we purchase are designed to perform well over a wide range of applications and materials, but we may find that a standard tool will not perform a given task as it should (assuming we have correctly diagnosed the inherent problems in the process).

In such a case, a special tool can be produced. In industry, hundreds of special tools are produced each day, and though such tools may seem very expensive, they are not—if the tool solves a problem or increases production. The most common special tools perform more than one task in a single operation, such as drill and counter bore or drill and counter sink. (See Illus. 5-6.) It is generally not possible to mass produce such tools, given the large number of sizes that are used. Today, that is further complicated by the mix of English and metric dimensions of cutting tools found in the various areas of the woodworking industry.

All drilling processes have the same desired outcome—fast, clean cutting of the material and equally fast, clog-free evacuation of the waste-offal chips. Gullet/flute angles, shapes, and sizes are design characteristics directed at a given range of cutting processes and materials. The general criterion is that denser, stronger materials require more energy to remove chips and waste. Metal-cutting drills and boring tools have similar flute sizes to those manufactured for wood, but

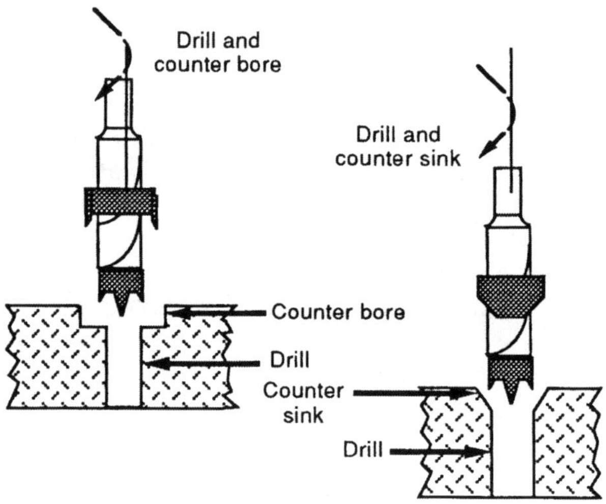

Illus. 5-6—Drill multiple function

the angle of inclination is more acute. It requires more effort to direct metal chips into the gullet space and away from the cutting process, hence the more abrupt angle of the flute. A metal-cutting drill may have more than two more complete flutes/turns over its length than a similar drill for wood would have. Wood and wood-based materials are, of course, lighter and less dense than nearly any metal, and the drill feed speeds are much faster for wood.

One of the biggest problems encountered in drilling/boring processes (and others, too) is that we overlook and take for granted the most simple aspects of our work. A last point regarding drill/boring bit gullets and flutes that is often overlooked is that the waste chips have to be directed first into the gullet/flute, then upward in the cutting process, in the opposite direction to the cutting motion. That is, as the tool progresses downward, the chips move upward. (See Illus. 5-7.)

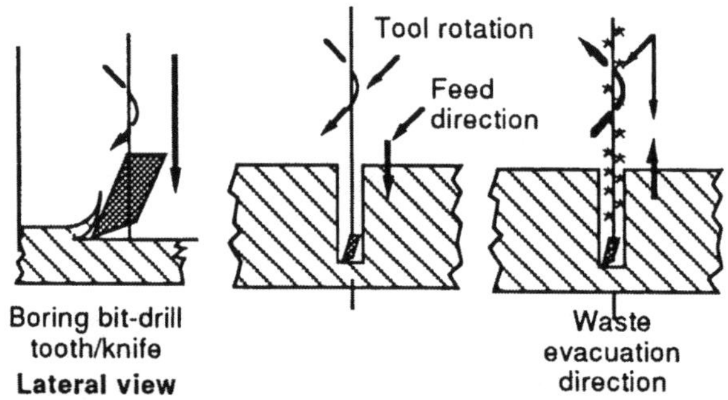

Illus. 5-7—Drill/boring bit waste evacuation

As simple and obvious as this may appear, it is not. Of the possible failures of drilling tools, failing to evacuate waste chips will probably be the most prevalent. The relationships among the material being cut, the rpm of the cutting tool, the size and shape of the gullet/flute, the cutting and relief angles, and the feed rate all must be correct for that process to be successful. If evacuation of the waste chips is restricted or the gullet/flute becomes plugged, the process will quickly overheat the tool, and it will cease to function. The basic guidelines relate primarily to rpm and feed rate, though dramatic changes in the material, such as large increases in moisture, can affect tool performance even if the process is otherwise correct.

Cutting Process Variables

As with all other cutting tools, drills rely on the chips and offal to carry away the heat generated in the cutting process. Unlike other cutters, drills are trapped in their own tool path. For this reason, they are subject to a wider range of operating temperatures, and the time in contact with the workpiece has to be more closely controlled.

Controlling the rate at which drilling/boring tools are fed into the workpiece is critical because of the revolving motion of the tool knives/teeth about a center point, the fact that the cutter is trapped in its own tool path, and the constant contact with the work process. (See Illus. 5-8.)

Knowing the machine rpm and the material to be drilled, we can, with a simple arithmetic exercise, tell how long the tool will (or should be) in contact with the workpiece during the process.

For instance:

1. To drill a 1/2" (0.500") diameter hole, 3/4" (0.750") deep, at 1,500 rpm, progressing at 0.006"/revolution will require 125 revolutions of the tool and take 5 seconds; that is:

   A. $\dfrac{0.750\text{" (hole depth)}}{0.006\text{" per rev.}} = 125 \text{ rev.}$

   B. $\dfrac{1500 \text{ rpm}}{60 \text{ sec./min}} = 25 \text{ rev./sec.}$

   C. $\dfrac{125 \text{ rev.}}{25 \text{ rev./sec.}} = 5 \text{ sec.}$

2. By the same formula, to drill a 1/4" (0.250") diameter hole, 5/8" (0.625") deep, at 2,200 rpm, progressing at 0.0025"/rev. will require 250 revolutions of the tool and take 6.83 seconds.

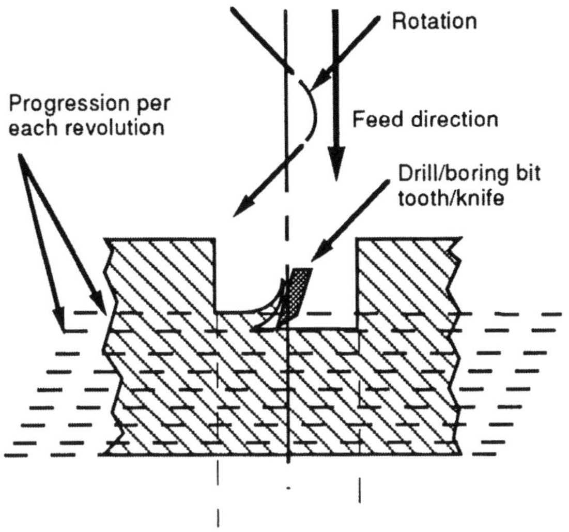

For most wood and wood-based materials the normal range of progression, which for drills is also feed rate, will vary from 0.0015 to 0.010" per revolution. The criteria are the diameter of the hole drilled, the strength and density of the material, and the rpm used to perform the work.

General Rules:
- Hard, strong materials—lower feed rates,
- Soft, weak materials—higher feed rates,
- Large drills—lower rpm,
- Small drills—higher rpm.

Range of rpm for most applications:
- 500 to 1500 for large diameter drills,
- 1500 to 2500 for medium diameter drills,
- 2500 to 4000 for small diameter drills.

Illus. 5-8—Drill/boring bit progression-feed rate

While the time that drills/boring bits are in contact with the material undergoing a cutting process is very brief, the temperature can rise dramatically. It is not uncommon for a drill to go from room temperature to 400° F and more in a single cutting process, especially if the material is wet, dense, and strong (a worst-possible case). If such a cutting process were repeated rapidly, the cutting tool would soon be ruined. Elevated tool temperatures can be avoided by following basic guidelines that relate to feed speeds:

1. Know the makeup of the material to be cut—density, strength, and moisture content.
2. Carefully select the progression rate in relation to tool diameter, machine rpm, and material makeup.
3. Make sure that the machine is as vibration free as possible.
4. When drilling especially deep holes, make sure that the waste is being freely removed. If the hole is too deep for efficient waste removal, "peck drill" the process, that is, drill only a portion at a time, back out and repeat.
5. To drill completely accurate holes, clamp the workpiece firmly. This will also prevent vibration.

6. Select the best possible tool to perform the task, one that has the best features and is made of the best materials.

In selecting the tool and machine process for a particular task, we learn to make generalizations because there are far too many variables to keep and commit to memory. However, there are a couple of such generalizations that need clarification and are more concepts of expression than general rules to memorize. These apply more particularly to drills and boring bits than to any other tools.

The description of a circular object or the size of such an object seems somehow to lose clarity. For instance, we can no doubt easily select the correct tool process to drill a 1/4" hole, but if the next task involves drilling a 1/2" hole, how many of us would recognize that we may have to change the machine process in either rpm or feed rate? A 1/2" hole is four times larger than a 1/4" hole, not two times larger! (See Illus. 5-9 and 5-10.)

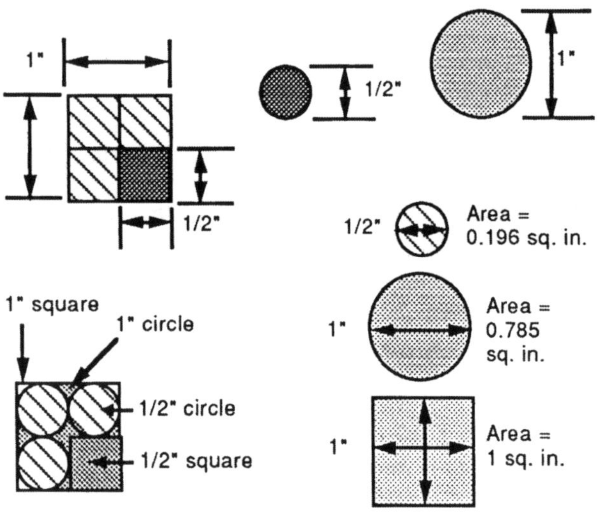

Illus. 5-9—Circle vs. square areas

Area is expressed in terms of a squared figure—inch, foot, mile, etc. A circle however, is described by its diameter, and that figure does not reflect the area contained within it. This may not seem of any great consequence, but in the selection of cutting tools, it is very important. Doubling the diameter results in removing four times the amount of material, and four times the horsepower will be necessary, even in identical processes.

The difference in area from one drill size to the next, if taken in small increments, will not be a problem to deal with; but doubling the size, as in the earlier example, will require consideration of rpm, feed rate, and possibly tool selection for any given process. It is important to remember that circular area increases geometrically in relation to circumference. The amount

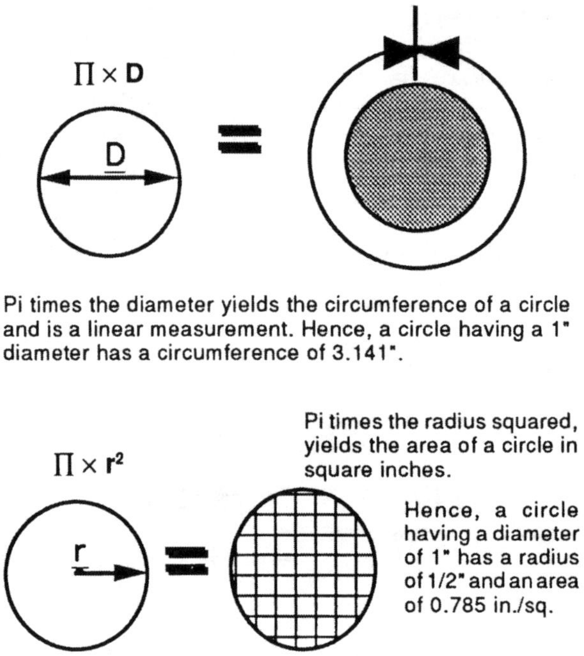

Illus. 5-10—Circular area and circumference

of energy necessary to perform the work, the amount of material removed, and the feed rate must reflect that. Most often, the difficulties occur when drilling holes of 1" in diameter and larger—rpm have to be reduced substantially, and the horsepower increases greatly.

Effects of Grain Orientation on Drilling

Another seeming paradox of the drilling process relates primarily to wood solids rather than man-made panel and composite materials. Because drill teeth/knives rotate about a fixed center, they encounter the grain orientation of the wood stock differently during each rotation. (See Illus. 5-11.) That is, during a portion of each rotation the teeth/knives cut the wood first along the grain and then across end grain. This does not cause dramatic changes in performance for the machine process or the cutting tool, but as grain orientation changes constantly in relation to the teeth/knives as they go about the cutting circle, the work load also changes slightly.

While in most cutting processes, this will not be of great concern with regard to effects on the cutting tool or the machine, the effect will be evident if large diameter holes are needed. The change of grain orientation in every drilling process will be clear, though, in the surface quality of cut edges. The along-the-grain edges will be smoother and freer of tearing; the end-grain surfaces will be rougher and more torn, as evidence of the added effort to cut these areas. It is easy to identify the growth direction in a particular wood sample by examining the edge surfaces of a hole drilled across the grain because the cutting tool passes more easily through the material when it turns in the same direction

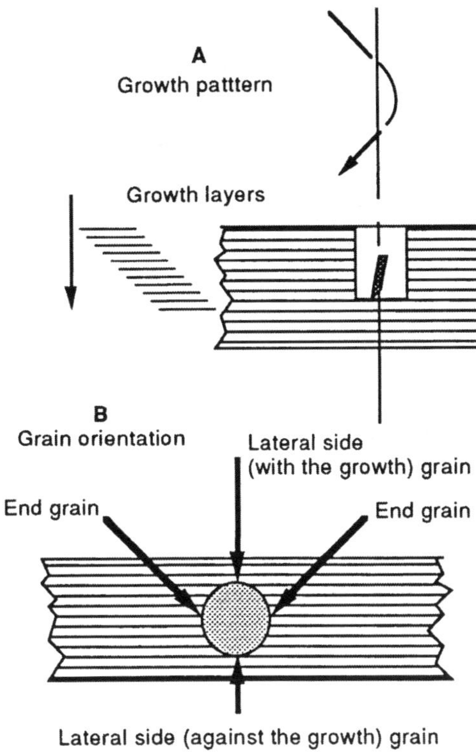

Illus. 5-11—Grain growth pattern/orientation

as the tree grew. The opposite side of the same hole will have a rougher appearance.

Another wood-grain effect is that as a drill passes through annual growth rings, the energy necessary to sever the fibers differs in each layer, but only slightly. There are very few tree species in which the density within each annual ring changes enough to affect cutting tools. We understand that in each growing season, the early wood is less dense and probably covers a wider area than the denser late-wood portion of an annual growth ring. Only in very unusual conditions is this difference in density ever obvious when using power tools, but it is more noticeable in using hand held tools, like those for carving, for instance.

In plywood and other laminated panels, there can be a noticeable difference in the energy necessary to cut the fibers in adjacent layers and especially glue lines. The alternating grain orientation will affect the amount of horsepower required, but not to a point of being a serious problem. Finish quality can be a concern, though, requiring close attention to keep cutting tools sharp. There may be a significant difference in the amount of heat-generated friction, caused by the additional abrasiveness and the increased density created by the adhesive. As with all other cutting tools, the presence of adhesive will shorten the sharpening life of drills and boring bits, regardless of the knife material used in the tool's construction. Overall, most cutting tools and machine processes can deal with a wide range of materials without radical changes, but we still have to be thoughtful and observant.

Early Drills

The first toolmakers were what we would call blacksmiths today. The tools they made were large and crude by our standards, made first of simple alloys of copper, zinc, and tin (i.e., brass and bronze), then later of iron that could be hammered and shaped easily. For instance, a toolmaker would make an auger bit heating and twisting a flat bar into spiral flutes, then hammering one end into a shaft and sharpening the other end to cut into the wood as it turned. Making such a tool would require the efforts of a skilled craftsman for perhaps an entire day. Today, using refined machinery, a skilled metal worker can produce hundreds of tools each day. (See Illus. 5-12.)

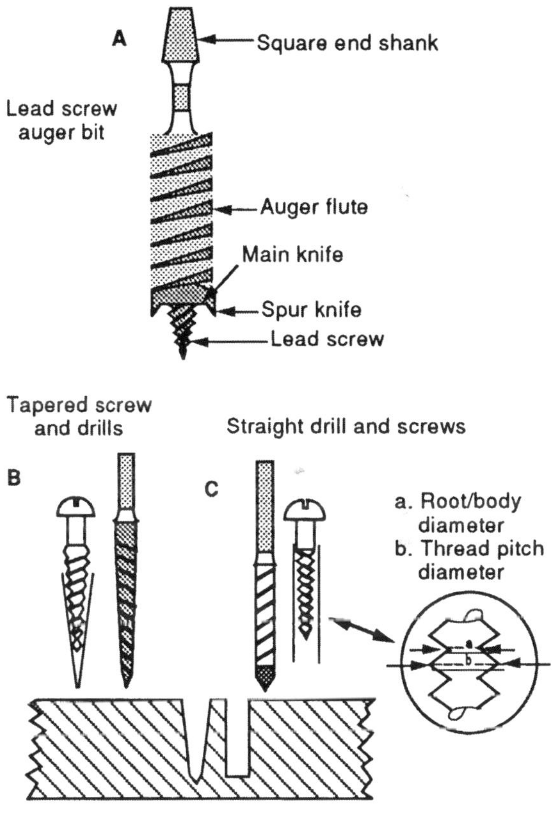

Illus. 5-12—Tool examples

The auger drill in A is not of the earliest type, in that it has a lead screw, which serves to pull the tool into the cutting process as the cutter is turned slowly by hand. The first such tools were used to drill large holes for constructing buildings with peg-tenoned beams. Many examples of pegged buildings exist all over the world—from huge cathedrals to log cabins. The drilling tools used in these were developed over hundreds of years in numerous countries.

Later, as technology developed for more refined details, the cutters became smaller and more precise. For constructing furniture, the first mechanical fastener that was developed, after different nails, was a tapered screw. The drill produced to prepare the pilot hole had the same taper as the root/body diameter of the screw itself. Hence, there was an entire collection of these drills, one for each size of screw. (See Illus. 5-12 B.) The logic was to take advantage of the wedge effect of using a tapered screw to make the attachment even stronger.

The observation was only marginally true, and the tapered screw was bothersome because the depth of the pilot hole had to be too closely controlled. If it was too shallow, the screw could split the wood; if it was too deep, the screw would not hold at all. In industry, tapered screws did not survive long, though they are still manufactured and used today.

The solution to the problems of the tapered screw came from the sheet-metal-working industry. The screw still most in use today is not tapered; the pilot hole drilled is a bit smaller than the root/body diameter of the screw itself. (See Illus. 5-12 C.) The particular sizing of the drill tool allows the screw threads to compress the wood slightly, assuring a strong mechanical attachment. Using a straight-thread screw and parallel drill tool allows for variations in the depth of the pilot hole with less chance of failure. Additionally, drills can be manufactured to do both counter-sink and counter-bore operations with a single tool in a single setup and process.

The Demands of Industry

As industry has developed tools to solve the problems of mass production, most of what is in use today has been created to meet the needs of finish quality and increased speed. Admittedly, there have been parallel developments in all industries, and there has been liberal borrowing from one area to another. The metal working industries have, in large part, always developed tools for all the others; and metal workers are in the business of making tools for woodworkers and all others, too, so the resemblance of woodworking tools and versions of metal working tools is not hard to understand. The first metal workers, blacksmiths, no doubt made more wood cutting tools than metal cutting tools for the obvious reason that wood as a building material (for everything) has been with us forever.

There are almost countless variations of basic tools produced in various parts of the world today. Most refinements are occurring in those parts of industry that are becoming more automated, i.e., more machine intense and less labor intense. Tools are designed to do a more specific range of work or for special dedicated-task machines. For instance, many tools are produced only with certain overall lengths, specific shank dimensions, or other features for a particular type or brand-name of machine. For the end user, this can work as either a detriment or a benefit.

For many years, drills produced for specific machines, such as multiple-spindle drilling machines, have been readily available, but recently, the leading technology has been imported to the U.S. from Europe and Asia. Not many places in the world outside the United States still use the English system of measurement—inch, foot, mile, and gallon, etc.—which causes another problem. Today tool manufacturers in the U.S., Europe, and Asia have to produce tools that have some of the following specifications:

- English cutting dimensions and metric shanks,
- English cutting dimensions and English shanks,
- Metric cutting dimensions and English shanks,
- Metric cutting dimensions and metric shanks,
- Smooth shanks,
- External threaded shanks,
- Internal threaded shanks.

These features, taken in whole, indicate a fairly close working relationship between the toolmakers and the machine manufacturers; otherwise there would be no tools available.

In recent years, an especially large number of European-made woodworking machines have been brought to the U.S., which created a number of problems of short-term tool availability. The English measuring system is as alien to Europeans as their metric system is for American woodworkers. The problem was that the machines imported could hold and drive drills and boring bits with metric-dimensioned shanks, and the Americans wanted to drill holes measured in fractions of inches. The problem was short lived, as the toolmakers quickly expanded inventories to accommodate the new shank and cutting dimensions.

The opposite condition also exists, i.e., the need for inch-dimensioned shanks and metric-dimensioned cutting diameters. For instance, a woodworker may have an older American or European machine that will accept only inch-dimensioned shank tools, and the product he is producing has already been converted to all metric dimensions. Again, toolmakers have risen to the occasion and provided what is needed, though by now they are no doubt hoping for eventual standardization. That will not be in the near future, as many of these machines will last for years.

In industry, tools are, as stated earlier, becoming more specialized. As we learn more about the geometric features of drills and boring bits and how each functions in the cutting process, we build features into a particular tool to perform just a narrow task or range of tasks. The speed, longevity, accuracy, and finish quality of cutting tools improves constantly, especially as newer alloys of metals, ceramics, and polychrystalline diamond are being used as knife materials. Some of the unique drills and boring bits used in industry currently include:

- Through-hole drills,
- Dowel drills,
- Hinge-hole boring bits,
- Hardware-insertion boring bits,
- Screw pilot-hole drills,
- Counter-bore drills,
- Counter-sink drills,
- Screw clearance-hole drills,
- Assembly access-hole drills,
- Service access-hole drills.

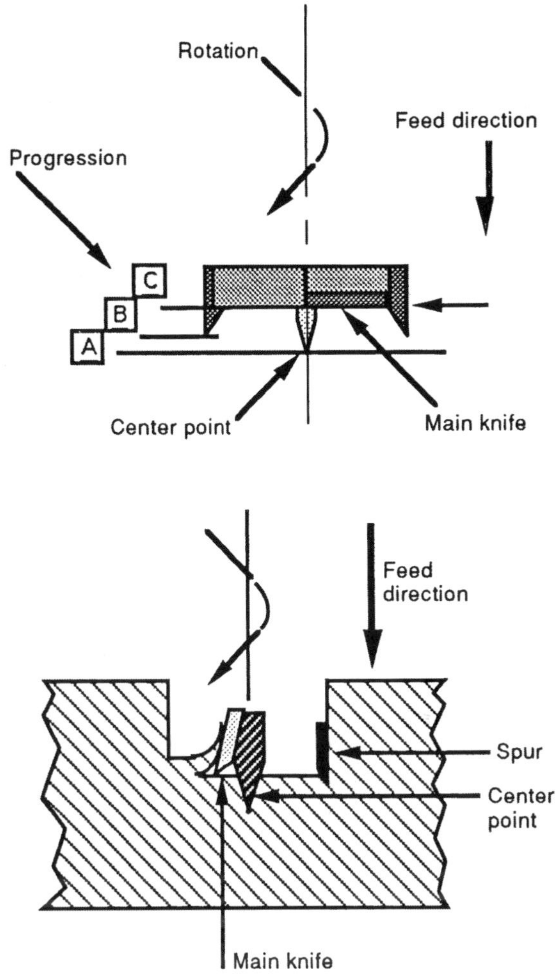

Illus. 5-13—Spur bit

This list does not suggest that each of the drills mentioned will look totally unlike another or that it will function in some dramatically different way, but rather that each has been designed to perform a process particularly well. The differences will be minor and are as much directed at the machine process as at the material to be cut .

Mechanical Features and Improvements

As in all industries, managers in woodworking seek cost reductions wherever they can be found. If a machine process can be faster, produce better quality, reduce scrap (e.g., damaged parts), increase tool operating life, reduce tool maintenance costs, run longer with less lost production time, or bring about any of perhaps dozens of other improvements, it will be pursued. This is not to say that cost reductions are necessarily directed at affecting the quality or integrity of the finished product. That may be true in some instances but is not a part of this discourse.

When the industry seeks cost reductions, cutting tools are always open to scrutiny because they are a large and obvious factor in all areas of woodworking. Improvements in even the smallest detail of any feature are pursued, tested, and implemented.

Drills and boring bits appear not to have changed in many years, but close examination shows that they are very refined. Most industrial tools are tipped with tungsten carbide and may have from two to five or more individual cutting knives. Alternatively, they may have specific features ground into a single carbide tip, with the goal of providing the features while getting the best use of the expensive carbide material. This is more common in smaller-diameter tools.

The most common drill/boring bits in use today will have some variation of the features shown in Illustration 5-13. Bits will all provide some consideration of the *center point* and, in most instances, *spur knives* and *main knife*. Details such as the length of the center point and spurs are specified for the tool's intended use, the machine process to be employed, or the material to be cut.

The center point steadies the tool as it is fed into the material. Soft material could require a longer center point, and harder material a shorter one. The need for a center point is not subtle—it is often necessary for controlling even very small amounts of machine vibration, wobble in the tool, or an eccentric feature in the method of holding or driving the tool in its rotary motion.

Most often, drilling tools are held on the end of a long, complex rotating shaft or collection of shafts inside of other shafts, as with the telescoping shaft, or *quill*, of a common drill press machine. The likelihood that all of these moving parts will be perfectly balanced, precise, and concentric diminishes with increased length. The center point increases accuracy dramatically throughout the entire process and through the workpiece.

Center points can be prepared to cut in addition to being simply forced into the material ahead of the knives, though they are generally no more than a smooth, cone-shaped protrusion. The angularity, length, and diameter of the center point will vary for both the diameter of the cutting tool and hardness of the material to be cut. There are several variations of center-point type: Some use a small, movable, solid shaft that can be removed, sharpened, and extended. Others, have features that can be sharpened in place and become shorter through sharpening in direct proportion to the other cutting features of the tool. In some, the point is ground into a part of another knife.

As the drill progresses into the material, the spur knives are the next part of the tool to make contact. They function to cut the periphery cleanly ahead of the main knives. Spur knives assist in providing speed to the process and surface-edge quality to the product. The spur knife cuts the material at the periphery and relieves the natural tension in the wood fibers. This allows the following main knife to progress faster than would be possible otherwise and increases the efficiency of the process dramatically. This is true in every instance except the most difficult of all drilling processes, that of cutting into the end grain of natural wood. (While the arrangement of center point, spur, and main knives improves this process over others, tool sharpness and geometric features are of even greater importance.)

Spurs can be individual tool segments or ground as a feature on the peripheral end portion of a main knife segment. The arrangement is dictated by the diameter of the cutter; the features of all cutting tools have to be such that they can be easily sharpened. Very small-diameter drills will have all features ground into a single tip, making best use of the material of which it is made. Large-diameter drills will generally have individual segments of carbide or other material attached to the body for the center point, spurs, and main knives. Spurs are not a recent development for drills and boring bits; they were among the first features added to improve finish quality of edge surfaces, even on hand-held and hand-driven tools in use for hundreds of years. Spurs allow increased feed rates without loss of finish quality.

Following the center point and spurs, the main knife enters the material. This can take several forms with different cutting angles, relief angles, and other features, depending on the tool's intended use or the machine process. The main knife, of course, cuts the material into chips and evacuates the waste, but it is not quite that simple. The nature of the material, its density and strength, the rate at which the tool is turned and fed into the material, and the tool features for waste removal all demand attention. While it is not necessary to examine each of these characteristics for every hole drilled, it is important to understand how the processes are accomplished so we can alter individual features to improve performance characteristics.

The main knife of a drill/boring bit initially severs the material and lifts the waste onto its face surface and into the flute/gullet much as a chisel would. (See Illus. 5-14.) The analogy of a drill main knife to a chisel holds if we picture the chisel upside down. With the chisel, the wood chip is forced onto the beveled face of the tool as it is pushed foreword. The chisel, however, is not in constant contact with the material nor in constant motion during the procedure.

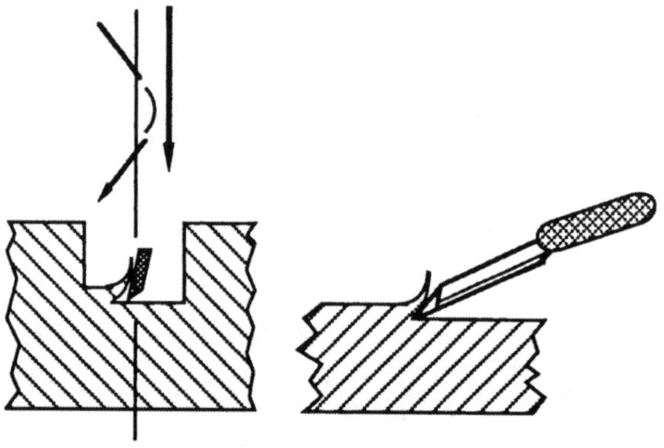

Ilus. 5-14—Drill main knife vs. chisel

Overcoming Heat in the Cutting Process

We can manipulate the chisel to any angle of incidence, and it will be fed into the work at a slow, incremental rate. Even then, the chips will form according to the nature of the material being cut. Similarly, as the drill knife rotates and progresses into the wood, the chips are forced upon the leading face surface, but the effects that the material and the process have on the tool now must be considered. If the size of the knife surface is not reduced, the tool will quickly overheat because the leading edge of the cutter is in constant contact with the material.

The surface area of the main knife is reduced by changing the relief angle and by the way it is sharpened. Only the very tip of the knife is in contact with the material, whereas in cutting tools that cut on the periphery, the relief angle allows the tool to travel in the circular motion and not rub against the material during the cutting process. (See Illus. 5-15.)

The problem to be overcome, as in all cutting processes, is heat generated by contact friction and compression. The size of the relief angle on cutting tools that cut on their periphery—saws, shaper tools, and router bits—is proportional to the feed speed they can support. Especially with router bits, which have very small cutting circles, the relief angle is critical to achieving high feed speeds while maintaining sufficient strength to preform the work. Any angle cut into the tool reduces the mass and strength of the tooth/knife.

In Illustration 5-15, drawings A and B show the function of the relief angle as the drill knife is in motion.

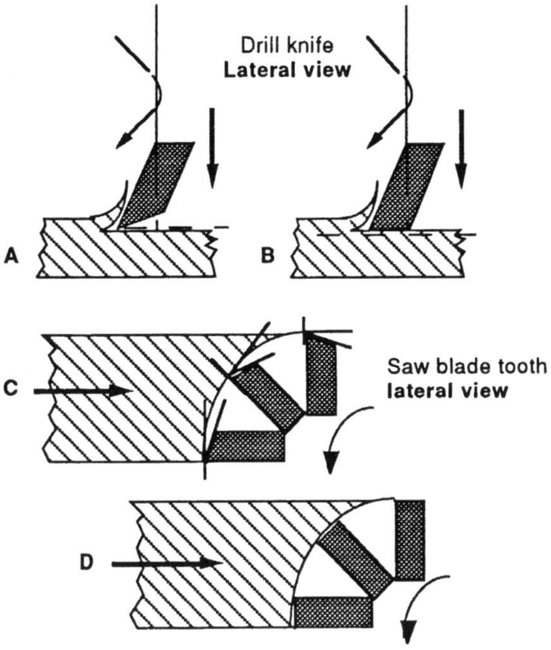

Illus. 5-15—Drill and saw blade relief angle

As the drill shaft turns about its axis, the knives are forced into the material being cut, and the waste is forced onto the face of the knife and into the gullet/flute space and away from the process. Even though the tool is in contact with the workpiece for a short period of time, if there were not adequate relief provided, the tool would quickly overheat and be ruined. The more obvious result of not having a relief angle is that with completely blunt knives, the drill could not penetrate any material. It would turn but could not even begin to cut.

To achieve needed feed speeds or to drill dense, strong material, the relief angle often must be quite severe. The effect of such a severe angle weakens the leading edge of the tool knife, making it too frail for practical use. This is overcome by grinding two angles on the back side of the knife. (See Illus. 5-16.) The secondary relief angle provides strength to the knife while meeting demands of materials or the high feed rates.

In addition to the general rules guiding feed speed and rpm, the relationship between cutting angles and relief angles for all drill/boring bits is based on the material to be cut and the machine process: (See Illus. 5-17.)

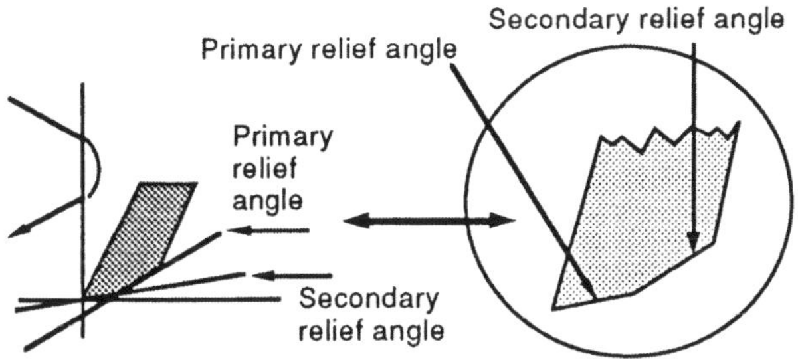

Illus. 5-16—Secondary relief angle

Cutting angles:

1. Shallow for strong, dense, and hard materials;
2. Steep for weak, less dense, and softer material.

Relief angles:

1. Shallow for slow cutting of dense, hard, heavy materials;
2. Steep for fast cutting of softer, less dense materials.

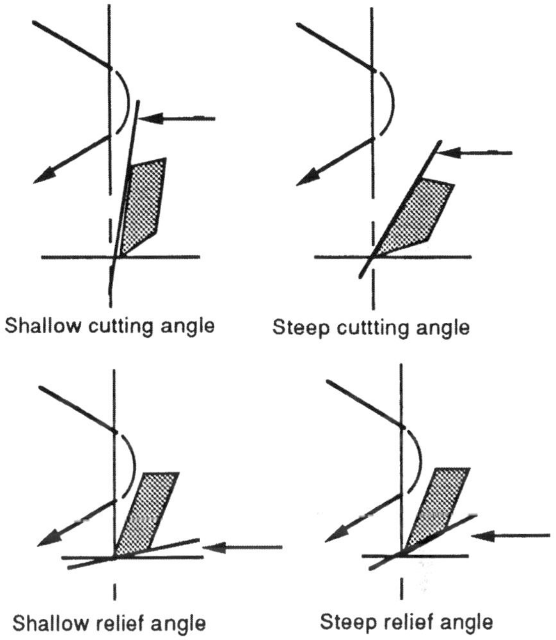

Illus. 5-17—Cutting and relief angles

Usually, woodworkers can accomplish their work with standard off-the-shelf cutting tools, but it is still important to understand what takes place in the cutting process. Knowing even general guidelines of tool performance, we can learn to solve the frequent problems of everyday work and become more proficient. With addition of new materials into an industry that in the past has dealt with only natural woods, it will be even more critical to extend our understanding of all processes.

Chucks for Holding and Driving Drill Bits

When we examine cutting tools and machine processes for effectiveness, we often overlook the fact that the tools have to be held and driven in some manner. But this is an important topic that we should address here.

The most prevalent devices to hold and provide power to drills and boring bits are *adjustable-diameter chucks.* (See Illus. 5-18 A.) With a correctly sized key wrench, the mechanism will adjust to hold several drill shaft diameters. While these devices work well over a range of sizes, they do not provide the best concentric accuracy. Without some care, they can create an out-of-round rotation as the tool is turned.

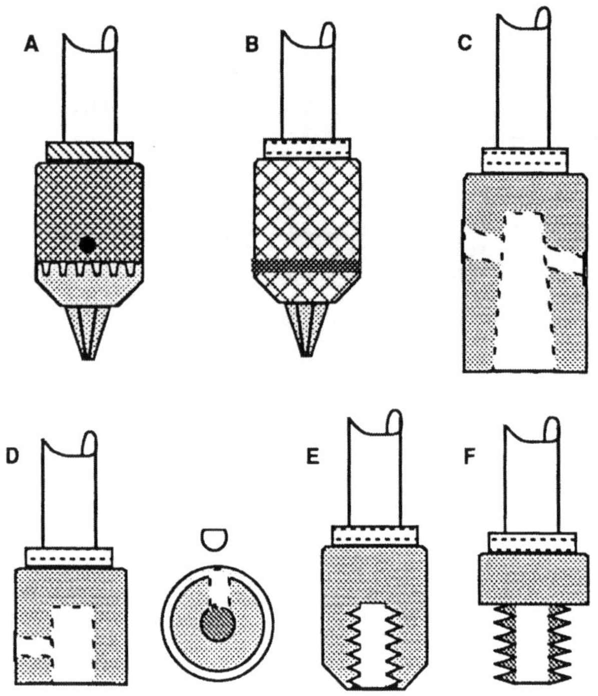

llus. 5-18—Holding and driving devices

The three jaws of the chuck move inside the housing that holds them in a spiral groove. Mass production has sacrificed some of their accuracy, and they are not intended as precision tool devices. Standard adjustable chucks will function well for most applications if we constantly monitor their use. The best practice is to rotate the drill/boring bit as the chuck is tightened, to ensure the jaws grip the tool as well as possible, and then to check for concentricity.

Precision adjustable chucks are available, though are considerably more costly than those on small hand-held electric drills, for instance. In fact, a precision chuck will cost considerably more to manufacture than the retail price of such a drill. Precision chucks ensure that the drill /boring bit will be held securely and turned concentrically with great accuracy.

Illustration 5-18 B is a *keyless chuck,* a device that, with only hand tightening, grips the shaft of a drill bit with sufficient force to drive it. Most often keyless chucks are both more expensive and more precise than any standard chuck that requires a key wrench. The jaws are held and driven in a system that allows the jaw segments to wedge against the tool shaft and the grooved-track walls that hold them. For machine processes that require quick and frequent tool changes, keyless chucks represent considerable improvement. Cost can be justified easily for a tool that gives both precision and speed, especially in a production setting.

Illustration 5-18 C depicts a *Morse taper chuck.* Not commonly used in the woodworking industry, the chuck requires drill/boring bit shafts with a dimension that mates to the chuck itself. Still common in metal working, Morse taper chucks are produced in several diameters, each tapering at a rate of 3/4" per foot. The tool shaft is held and driven simply by being tapped into the taper. The resulting wedge effect is sufficient to transmit large amounts of horsepower and hold the tool shaft with great precision. There are other taper systems in existence. The Morse is the most common and successful. The use of Morse taper chucks in metal working is more appropriate because the tools turn more slowly and require much more horsepower than in woodworking.

The most common chuck in industry use is the *smooth-shaft grub-screw chuck* in Illustration 5-18 D. All drills and boring bits for use with these chucks have the same diameter shafts. This technology was imported from Europe, arriving in the U.S. in the late 1960s. Many of the machine systems developed are for use with multiple spindles. That is, they are fitted with perhaps 50 or more drills that work at once. The need to change the tools easily is apparent. Even for more simple machines, the advantage of the smooth, common-sized shanks is also apparent.

Drilling in wood and wood-based materials usually requires relatively little energy, so transferring power by wedging the tool shaft with a grub-screw is not a problem, unless drill size is very large. Accuracy, that is concentricity, of a smooth bore chuck is equal to if not superior to most other types, and it is surely among the least expensive to produce. Drills with knives/teeth made of tungsten carbide and other hard metal alloys have extended sharpening lives, so the time it takes to change bits, though slower than with some other kinds of chucks, is not a problem.

The drill shafts have a flat area machined lengthwise onto which the grub screw is driven. The screw serves both to hold the drill bit in the chuck and to transfer the rotating motion to the tool.

A unique method has been developed to maintain consistent depth of the hole as the drill tools are sharpened. Very simply, as the drill tool is sharpened and becomes shorter, the screw in the shaft is adjusted by the same amount, until the tool is finally discarded at the end of its sharpening life. This allows a drill or a set of drill tools to continue working with great accuracy.

Illustration 5-18 E and F show *chucks for use with external- or internal-threaded tool shanks.* For many years in the U.S., it was common for multiple-spindle drill machines to use tools with external-threaded shanks, like the chuck in Illustration 5-18 E. The obvious problem is that in addition to being slow to change from one size to another, there was no way to control the length except to sharpen them all exactly alike. The cost of producing all of the necessary sizes of diameters with threaded shanks was considerable.

The chuck in Illustration 5-18 F requires a drill bit with a threaded socket base opposite of that for the chuck in E. Both the male and female threaded drill shaft chucks can be found today, but they are rapidly falling out of favor. These holding devices have been given extended life by using the threaded base to accept a quick change device of which there are several types now marketed.

To summarize the importance of carefully using the devices that hold and drive drills and boring bits, it must be remembered that the cutting tools themselves have no moving parts. Given all the other moving parts of even simple machines, there is a greater likelihood of inaccuracy in a part of the machine than in the cutting tool being held and driven. The cutting tools are at the end of a long chain of events and movements and consequently are the first thing to be examined and blamed. This is not to say that all cutting tools are without fault, but there are many other elements in the process to consider when seeking the cause of a problem.

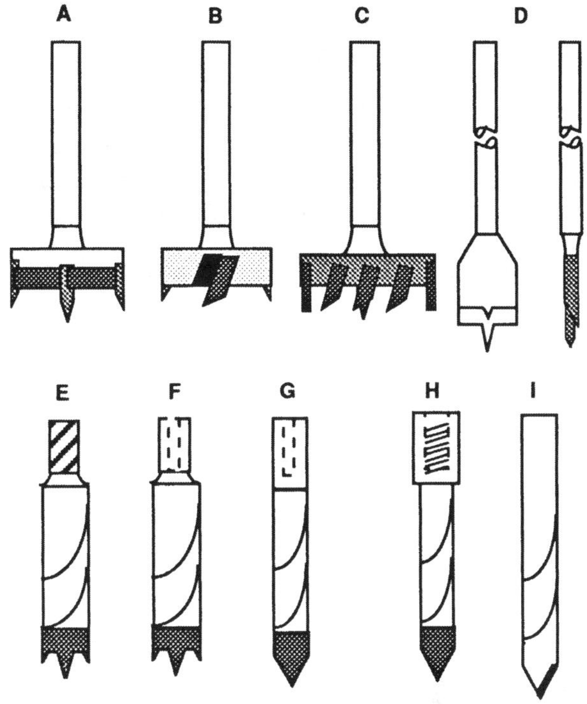

Illus. 5-19—Common drill/boring bit forms

Types of Drills and Boring Bits

On the market today is a wide range of cutting tools that we describe alternately as drills or boring bits. A drill has spiral flutes over its entire length to carry away waste chips; a boring bit does not. Even so, the terms drill and boring bit are, and will continue to be, used interchangeably at no particular loss to either.

Drill and boring bit tools have evolved into current shapes, sizes, and efficiencies for specific areas of woodworking. For instance, tools used in construction trades probably will not look or perform like their counterparts in the furniture industry. Even tools common in furniture manufacture, such as for bedroom or living room case goods, may not have an exact counterpart in the production of kitchen and bath case goods because the manufacture of kitchen and bathroom case goods has become more standardized and automated. The machine technology and tooling systems for much of this automation were imported from Europe and Asia; hence the cutting tools are somewhat different.

Illustration 5-19 depicts several of the most common drill and boring bits:

- A—the *spur bit* (see also Illus. 5-13) has been developed in several versions to cut clean holes in a wide range of materials, even when drilling completely through the workpiece. The configuration of spurs, center point, and main knives effectively cuts wood and wood-based materials and removes the waste. The cutter is particularly effective in drilling larger holes in the furniture industries and others. The spur bit finds widespread use for many processes.
- B—the *Forstner bit*, is a predecessor to the spur bit. The tool body and cutting edges are made of a tool steel alloy, and the bit was intended for use only on solid, natural woods. In fact, it was developed long before any of the man-made materials. The tool is still manufactured and popular today, but it is very expensive to produce, having all of its complex features cut from a large solid bar of steel. This cutter is most effective in drilling large diameter holes in furniture construction, producing clean, smooth edges. The outer periphery acts as a spur to cut the edge cleanly ahead of the single main knife. Like the spur bit, the Forstner has a leading center point to hold the tool as the other cutting edges enter the material. Current manufacturing methods are such that a modern spur bit with tungsten carbide cutting edges is less costly to produce than many of the older steel alloy tools. The latter may have historical value but are quickly falling out of favor.
- C—the *multi-spur bit* is similar to the spur bit in A. The name is a misnomer because the cutter knives at the periphery are not spurs but cutting teeth. The idea behind the design of this arrangement is that more teeth will be more effective. In very hard, dense materials that must be drilled very slowly, this may be helpful. In most applications, however, the extra teeth are not needed, tend to slow the drilling process, and frequently plug with waste offal.

• D—the *spade bit* comes in several variations for use mostly in woodworking in the construction trades. The tool functions more to scrape than to cut as other drills and boring bits. Holes produced by spade bits are rough with torn peripheral edges. They are used most commonly for clearance holes in construction members for electrical wires and plumbing. Spade drills are made with features to drill and counter sink and drill and counter bore where finish quality can be hidden or is of no consequence. Screw clearance and screw starting holes are among the processes suited to these tools. The single greatest virtue of spade drills is their low cost. They are "disposable," in that their cost is so low that if they are damaged by rough use, the loss is negligible. Though they can be sharpened several times, most are discarded. Of all drilling tools, spade drill types probably make up the largest number because they find use in so many areas, from the hobby market to the construction trades to limited use in industry.

• E and F—*dowel drills* have many uses in several areas of woodworking. The term dowel drill is something of a misnomer, since these tools are used for many purposes. The name does, however, describe the tool's original function of drilling very accurate holes for inserting dowels in furniture construction. Dowel construction allows a close to zero tolerance. These tools may take several variations regarding their tip features. In most instances, the entire tip is made of a solid piece of hard metal, usually tungsten carbide, and the center point, spurs, and knives are ground into it. A few manufacturers use individual segments of hard metals to form the individual features.

The tools depicted are identical, save one feature—how they are held and driven. The drill bit in E has a threaded shank and will be used with a corresponding holding device on a drill machine. This drill configuration represents technology imported from Europe since the early 1960s. Such tools are very effective, accurate, and cost reliant because they can be sharpened repeatedly. The tool in F has a smooth shank and length-adjustment screw feature to modify the length as it is sharpened. (See Illus. 5-20.) These tools are a good example of the close relationship between the machine and tool manufacturers.

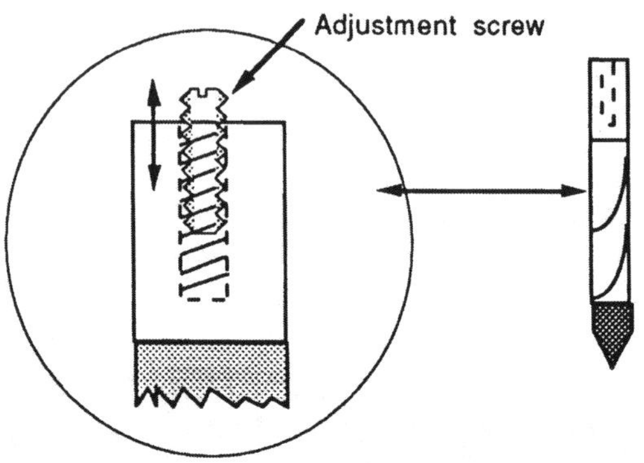

Illus. 5-20—Drill length adjustment

• G and H—*through-hole drills* are similar to dowel drills in most characteristics except tip features. The name through-hole drill describes their intended function. The leading edge angle is 60°, the commonly accepted angle necessary to drill through most wood materials and not leave a torn edge when the tool passes through the back side of the hole. As the tip protrudes through the back side, the angle reduces the contact surface pressure because the material is cut only gradually from the center outward. On the drill in H, the shank has an internal thread for use on a mating machine chuck.

Through-hole drills are used for more processes than their name implies. Clearance holes for screws, hardware, or other mechanical attachments and devices make good use of these compact, well-designed cutting tools. This technology continues to improve as more manufacturers use well designed, automated metal working machinery and the dimensional accuracy of the cutters becomes more precise.

• I—the more standard *fluted drill* is common in all areas of wood and metal working. There are many types of these cutters available for numerous applications. Among standard fluted drills are the following variations:

1. Left- and right-hand rotation versions;
2 Shanks that have external and internal threaded shanks;
3. Drills that are straight and parallel;
4. Drills that are tapered to mate with a particular screw pattern;
5. Cutting diameters measured in decimal, fractional, letter, number, or metric increments;
6. Shanks of the same diameter as the cutting diameter;
7. Shanks all the same size regardless of cutting diameter;
8. Shank diameters in English dimensions, with metric cutting diameters, and the converse;
9. Drill lengths that vary widely, from less than 1" to perhaps 24" or more;
10. Cutting tip features in numerous configurations to suit the material to be cut;
11. Flute inclination angles that vary to suit the nature of the material to be cut and evacuated; and
12. Flutes that vary in number from one to three.

Actually, this list could continue for several pages if all features were precisely described in great detail. The point is that standard fluted drills are manufactured by hundreds of companies worldwide and have a wide range of machine applications. With correct cutting geometry and metal alloys, the drill tools perform very well for a very wide range of materials from softwood to moderately hard steels.

Standard fluted drills are no doubt the most common of all shaft-mounted tools. For woodworking, metal working, hobby work, building trades, light industry, and even heavy industry these are the basic tools that we all need. Specialization of manufacturing processes in recent years has created the tools now prevalent in specific industries. Wood cutting drills, for instance, are very efficient at cutting wood and wood-based materials, but these same tools could not be used to cut metals at all. While general-purpose tools are available, they are falling out of favor in light of how very well the newer tools perform, especially in an industrial environment.

Technological Specialization and Improvement

Today as never before, we are seeking improvements in everything that we do. Market competition here in the U.S. and all over the world is becoming more aggressive. This competition is driving all segments of all industries to make better use of time, costs, energy, and raw materials. Cost effectiveness takes on new and much broader meanings. In terms of cutting tools, if, for instance, we can design a drilling tool that is even 6 or 8 percent faster or uses 5 percent less energy to perform the work, it is well worth the cost of the engineering study that it took to do so. The specialization of tool design that seeks cutter geometry for drilling only a particular man-made panel material may at first seem unworthy of the effort until we realize that millions of these panels are manufactured each day. Then, a drill designed to cut holes in particleboard or medium density fiberboard makes a lot of sense.

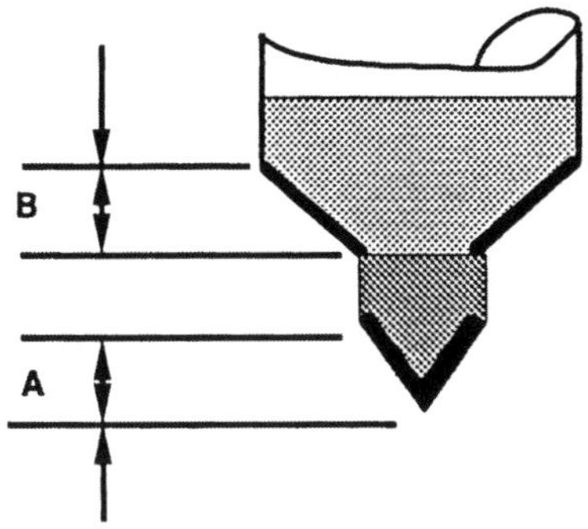

Illus. 5-21—Multi-plane drill tip

The tool in Illustration 5-21 represents a small step forward in tool technology, making use of a mechanical principle to divide the work into two segments, using two cutting surfaces (A and B),thus conserving energy. The work of drilling a hole into or through the workpiece is divided into two individual processes because each segment differs in geometry and surface speed. The smaller diameter has a more aggressive cutting angle and smaller relief angle. The larger diameter, while traveling further with each rotation and removing more material, requires more relief angle and a shallower cutting angle. Though its name, *multi-plane*, is somewhat inaccurate, the tool performs very well in cutting very hard, dense materials.

Improvements of this kind are sought for every machine process and material. Drills and boring bits will continue to be studied and improved, though there are, of course, limits of mechanics and physics that will not be overcome. We will see more specialization of cutting tools as they relate to specific machine processes and the characteristics of materials to be cut. There may be, and in some instances already are, saw blades, drills, and router tools for man-made panel materials and a very different set of tools for solid, natural woods.

Chapter 6

Shaft-Mounted Cutting Tools II: Router Tools

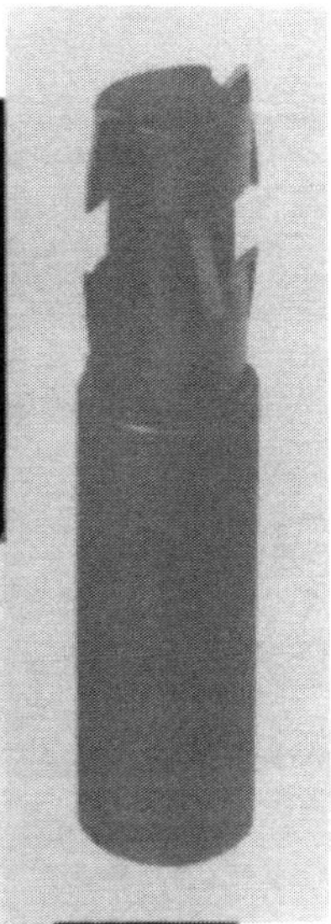

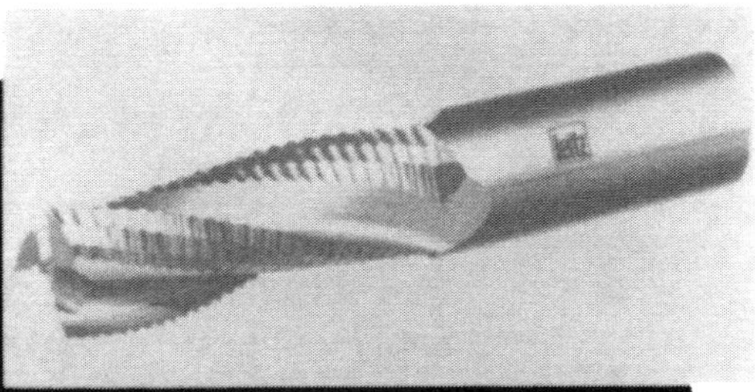

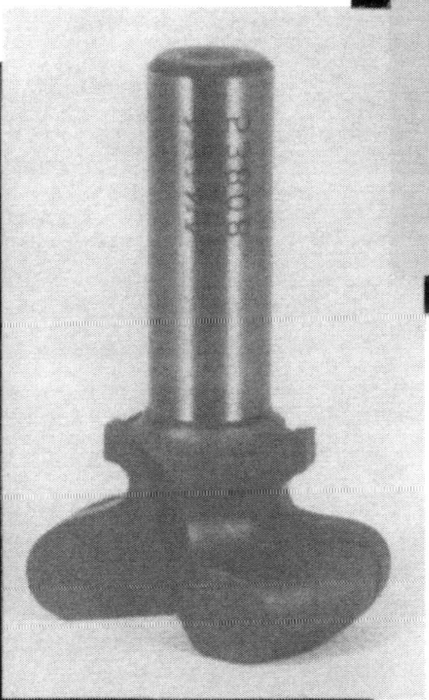

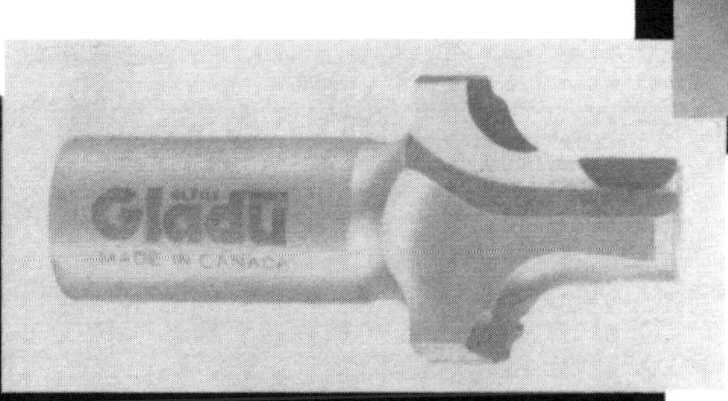

Photos courtesy of Amana Tool Co., Guhdo USA, Inc., Lach Spezial Werkzeuge GmbH, Leitz Tooling Systems, Inc., Leuco Tool Corp., and Gladu Tools, Inc.

Shaft-Mounted Cutting Tools II: Router Tools

Router machines and cutters are in the best sense modern: The tools and processes are such that they have no counterparts among hand-held, hand-driven tools. The unique processes possible with router tools were previously done with a combination of saws, drills, and chisels. The development of precise machine movements and electric motors to provide power changed machining processes not only in woodworking but in all other industries as well. As we've noted before, there has been parallel development of technology in the metal working and woodworking industries.

As machine technology developed worldwide, hundreds of new ideas came forth every year. In what might be called the golden age of mechanical invention, between 1650 and 1900, literally hundreds and thousands of devices and systems were developed. Many of the machine concepts developed in those years are still with us in some form. Perhaps most important about that period of history is the creative effort directed at improving the efficiency, accuracy, and speed of manufacturing.

Craftsmanship, while still important, was giving way to technology. A machine could do much of what the artisan-craftsman could at speeds not even conceived of before. We will not attempt here to resolve the old argument about craftsmanship versus technology. Will the artisan-craftsman be replaced by the engineer-craftsman? In my opinion, there should be no conflict. The artisan-craftsman should understand and make use of all the technology available. Speed, efficiency, and precision, as concepts of machine technology, cannot be seen in a finished product, except perhaps in hand-carved details in a finely done cabinet. If a single very expensive cabinet is finished each week or month, but several dozen can be produced in the same time, with no loss of structural integrity nor appearance, can it matter that power tools were used in the construction? Each craftsman will have to answer the question for himself.

Router machines and cutting tools fill a unique place in woodworking, being capable of such a wide range of tasks. They can cut grooves, rebates, and dadoes, as well as complex profiles. They cut both forward and backward. They can be made to plunge-enter the workpiece and travel laterally, or to enter the work from a sidereal location.

Router machines vary from the size of a 12-oz. soft drink can to that of a four-car garage. Costs of such machines vary proportionally to their size—or nearly so at least—from about $50 to $800,000 and more. Even the smallest hand-held router may have a small computer inside its case that is capable of sensing the load on the tool and varying the energy to perform the work. Further, the machine can adjust the speed of the tool and give a visual display of the rpm to the operator. With this technology and versatility, speed, and efficiency, the increasing popularity of router tools and machines is easily understandable.

Misperceptions about Routers

For the hobby worker, craftsman, and industrial worker, routers have a similar appeal—that they can perform simple mechanical surface processes such as grooves and rebates, as well as complex decorative profiles. And the machines hold further broad appeal—namely, low cost and utility.

Despite this almost universal appeal, there exists a disparity between how workers at the hobby level perceive router cutting processes and how industrial workers perceive them. Oddly, the two different misperceptions bring about nearly the same outcome for the cutting tools.

Cutting tools, as you remember, generate enormous amounts of heat in the course of their work, and the heat is carried away by the waste chips. This is frequently overlooked by hobby and industrial workers alike, with consequent damage to cutting tools. The hobby worker too often cuts too little or too slowly, allowing the cutter to spend too much time in contact with the workpiece, which elevates the temperature dramatically.

In industry, high feed rate in relation to possible rpm creates the real possibility that the tool or workpiece will move forward so far that the area of the relief angle rubs against the workpiece in part of the contact cutting circle, elevating the tool temperature to a dangerous level. Both hobbyist and industrial worker need to understand the possible feed speeds and work loads for router tools.

Another misperception regarding router cutting tools is that nearly all that have no tracer bearing will plunge-enter the workpiece. Router cutters can be made to plunge as a drill does and then travel laterally, but this process requires some unique geometric features. (See Illus. 6-1 and 6-2.)

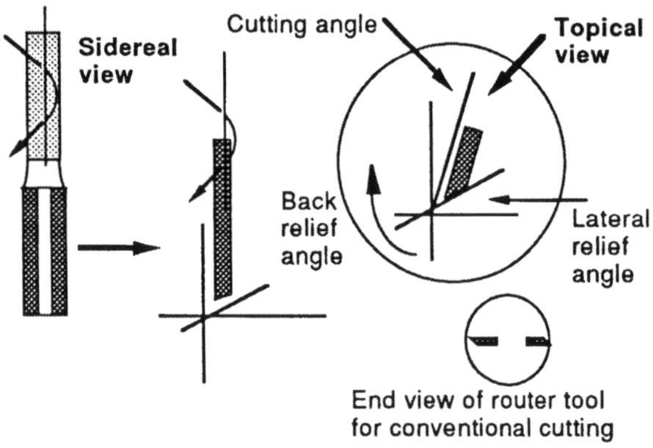

Illus. 6-1—Basic conventional router tool features

Plunging Routers

The router tool traveling laterally behaves as most other cutters that cut on the periphery. A drilling tool has no capability to cut on the periphery but rather only the small cutting circle on the end of a shaft. To expect a router tool to drill and also cut laterally makes three assumptions:

1. There is a cutting knife on the end of the shaft, complete with the correct relief angle to function well.
2. There is a means to evacuate the waste in the direction opposite to feed.
3. The drill knife cuts across the entire diameter.

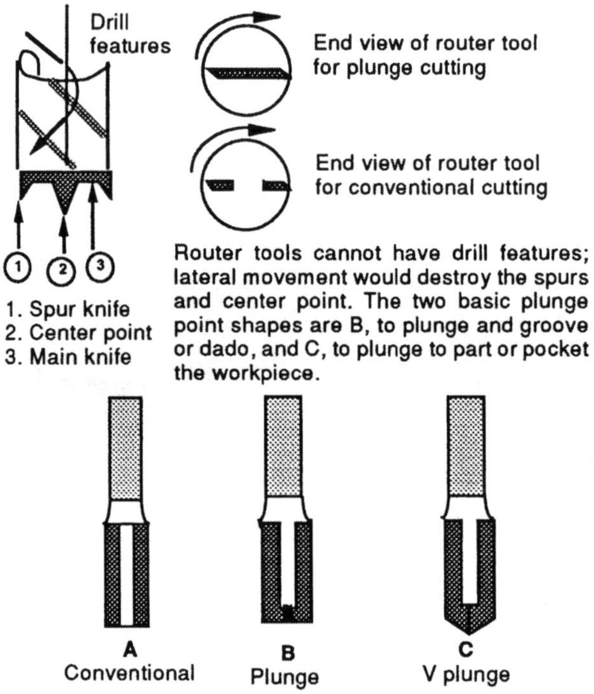

Illus. 6-2—Plunge-entrance features of a router tool

For a router tool to plunge-enter the workpiece without a knife section that has bottom relief and completely crosses the diameter would not be possible because (1) a conventional cutting router does not have sufficient bottom relief to sustain the cutting process with a downward feed direction, though laterally, it will have enough to rebate, groove, or dado, and (2) if there is not a knife section completely across the end of the shaft, as with a drill tool, the knives at the periphery will cut only as far as they may protrude beyond the bottom of the shaft, and they will only be able to rub. (See Illus. 6-1 and 6-2.)

Plunging router tools take on several forms, but all have design features to allow them to perform two distinctly different tasks. Industry, in using automated machinery, has found a different solution to the problem of tool entrance. Whenever possible, router cutters are manufactured having plunge points. Where that is impractical, machine movement allows the cutter to enter the workpiece without damaging the cut-ters or the material being cut. The machine holding and driving the cutter directs the tool's entrance into the workpiece at an inclined progression instead of vertically. This is called a ramp entrance. (See Illus. 6-3.) Router cutters generally have sufficient bottom clearance to allow ramping, though not enough for a vertical entrance. Even hand-held router machines will

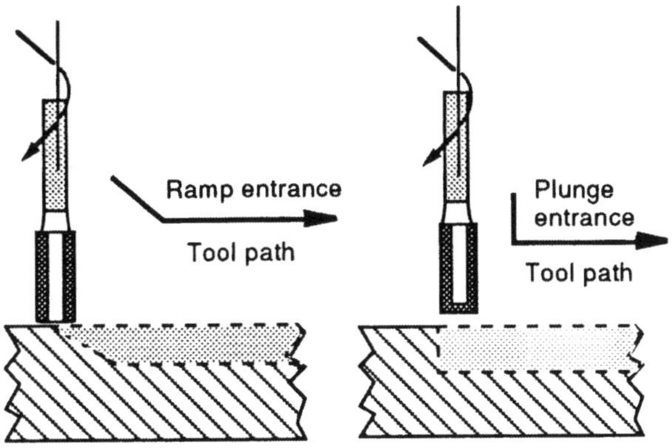

Illus. 6-3—Router entrance, tool path

permit a ramp entrance with use of a special fixture.

Evacuating waste chips when plunge-entering a workpiece is seldom a problem, even though router tools do not usually have a specific means to do so. The velocity of the tool creates considerable air movement, and the progression is brief and relatively shallow. Attempting to plunge a deep hole with a router tool at its usual rotation speed, however, would generate enormous heat. Even with a plunge point, a router tool is only rarely designed to cut anything but shallow entrance holes.

Advantages of Routers

As mentioned earlier, among the virtues of router tools is their versatility. A longer list of advantages includes these:

- Low initial cost compared to other tools;
- Small size for cutting detailed features;
- Capability for either climb- or abrasive-cutting;
- Easy manipulation into any axis;
- Ability to cut straight mechanical surfaces as well as decorative profiles;
- Easy sharpening and maintainance;
- Ease of holding and driving compared to other cutting tools;
- Adaptability to fixtures and ease of control with inexpensive devices;
- Capability of significant work loads relative to tool size;
- Low-cost manufacturing in short periods of time relative to other tools;
- Availability from many sources with increasing improvements in quality and precision; and
- Numerous shapes, sizes, and configurations for many machine adaptations.

Router tools are the most compact and versatile of all cutters, and if used properly, can be successful at many processes. At the end of of this chapter, we will provide guidelines for correct feed speeds, rpm rates relative to feed speed, and an approximate means to measure horsepower consumption of router tools.

Router Capabilities

- *Mechanical joint surfaces*, such as those in Illustration 6-4, are common in several areas of woodworking. The construction of furniture, case goods, cabinetry, seating, and so forth, all make use of mechanical construction joinery. The simplicity of the use of router tools makes machining these surfaces inexpensive and fast, especially for smaller parts. Industry has in recent years developed both portable and small stationary router machines that are very efficient and productive. Holding devices, jigs, and fixtures for these machines are similarly inexpensive.

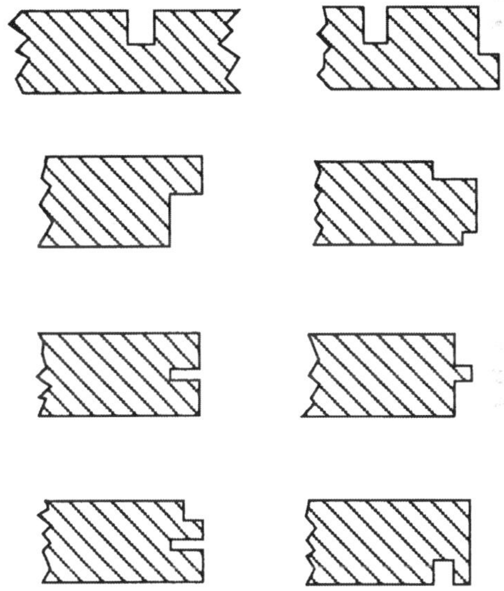

Illus. 6-4—Router-cut mechanical joints

- *Decorative profiles* (Illus. 6-5) can be accomplished easily with router tools and machines. The versatility of the machine processes makes even fairly large profiles possible, if not in a single pass, surely in multiple operations, and still more cost effective than with other machines.

The greatest strength of router tools and machines is in cutting edge-detail profiles. Treatment of edge surfaces of either rail or panel parts in all areas of furniture construction is difficult, regardless of the machining system. Router tools in either portable or stationary machines can more easily accomplish the work, especially for smaller manufacturers and hobby

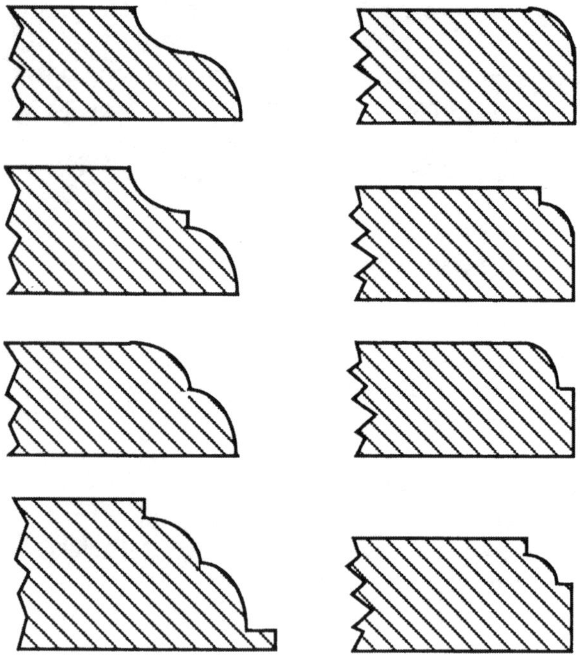

Illus. 6-5—Router-cut decorative profiles

workers. In very large manufacturing settings, routers play an increasingly important role, but not expressly for edge treatment processes as in small companies. Without router tools and machines, the work processes demand very large industrial machining systems or more labor-intense hand methods. Historically in cabinet work, edges were decorated with hand-carved rails, then later with machine-produced, molded rail parts. With an inexpensive, compact cutting tool and machining system, edge treatment could be done even after assembly of the rail or panel part.

Illustration 6-6 shows the same edge treatments either as a shaped feature of a part or as an attachment to the part. Either method holds distinct advantages in specific instances. In both methods, however, router machines and tools play an increasingly important role.

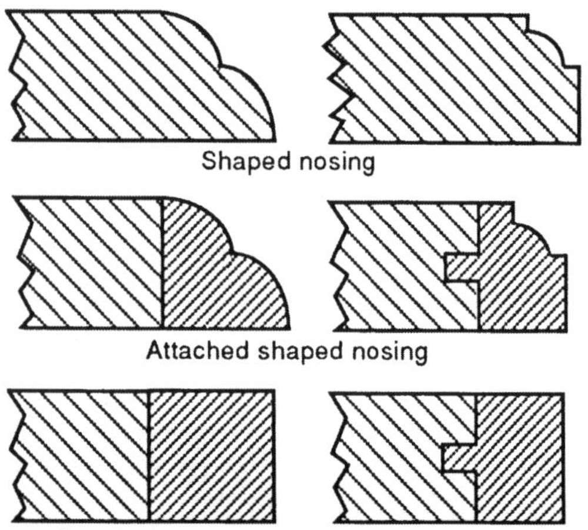

Illus. 6-6—Shaped vs. attached edge treatments

The illustration shows (1) the detail shaped directly into the rail or panel part, (2) the detail cut into a shaped part attached to the rail or panel part, and (3) a rectangular part attached to the rail or panel part for later profile shaping. A router tool and machine system could be used in all three situations, though other machining systems may be more efficient and cost effective.

Router tools and machines can be successful in the assembly of nearly any furniture component. Because they are available in such a wide array of tool shapes, sizes, and profiles, and because they are compatible with numerous machine types produced by manufacturers all over the world, routers are perhaps the most prolific of all cutters. Further, the cutters can be held and brought to the cutting process in any of several methods, with no single method having any distinct advantage.

Router tools' compact size makes them ideal for the machine processes that employ them. The amount of work performed in each of these machine processes and the little mechanical manipulation necessary to accomplish the work favor router tools and machines over alternatives.

With little expense for machines, fixtures, or holding devices, router tools can function with considerable success. Illustration 6-7 depicts two similar router profile cutters, one with what is commonly called a tracer bearing, though it may have other names. The tracer bearing allows a router to cut the radius shown on the upper corner of the workpiece without any other mechanical device. The radius detail can be accurately cut and controlled with even a hand-held router, which can easily produce the necessary horsepower.

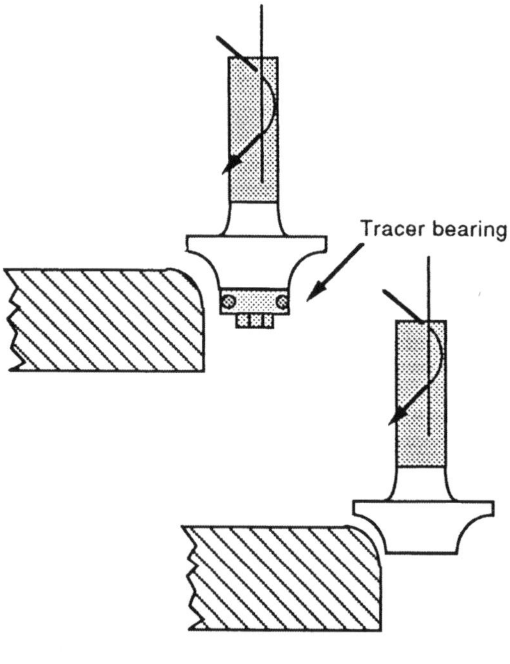

Illus. 6-7—Routers with and without tracer bearing assembly

The second tool in Illustration 6-7 is nearly identical to the first, except that it does not have the tracer bearing. With this tool, a means will have to be provided to control the cutting path. Depending on the machine employed, any of several methods could be used. On a hand-held machine, perhaps it would be a device for the base. For a larger, stationary machine, the choice of fixtures or other holding devices could be almost limitless.

It is often overlooked that the machines and workpieces have to be held with sufficient strength to contain the forces of the cutting process. If the cutting process requires, say, 2 hp to cut the detail features, the same 2 hp has to be contained and controlled. It must be clear that if a router machine is rated at 2 hp, that is the maximum energy that the machine is capable of, and it is not likely that we would need to reach that maximum and stall it to a standstill. Though this may seem redundant and needless to discuss, the point must be made for the sake of safety: Using cutting tools in any process without carefully and completely containing the energy is dangerous.

The use of tracer bearings is popular for both hand-held and stationary machines. Hand-held machines of nearly every imaginable cutting tool size, shape, and profile use tracer bearings. This benefits hobby or craft woodworkers, making even complex work easy to accomplish with little cost.

Tracer bearings are also popular for stationary router machines because the cost of holding devices is less and set-up times are often shorter. In industry however, the use of tracer bearings is not as widespread because the bearings do not have sufficient strength to withstand the rigorous work load placed on them in many hours of operation producing hundreds of parts per set-up.

Again, the most common work for router tools and machines is edge treatment of rail and panel component parts in furniture and cabinetry construction. Illustration 6-8 shows such a process. With a correct means to "steer" and control the cutting process, the work can be

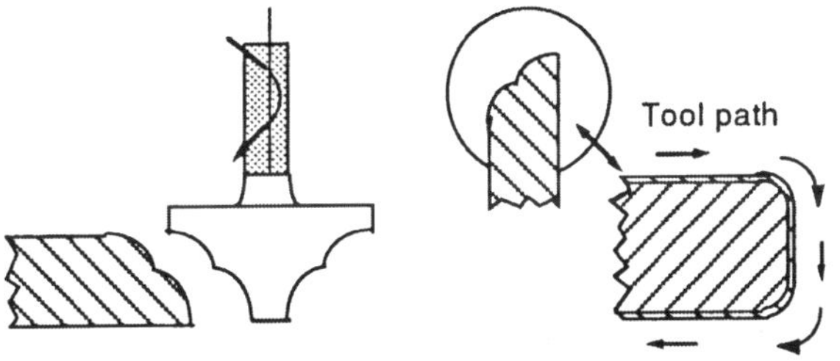

Illus. 6-8—Router tool edge profile cutting process

easily accomplished, and the number of shapes of products that could be created is endless.

The exact edge detail is limited only by the imagination of the designer, the toolmaker, and the worker completing the work. The depth of the edge detail is limited by constraints of the tool's diameter. Router tools have definable limits of diameter that we will address later.

Illustration 6-9 depicts two similar edge detail processes. Simply, one has a square mitered corner, the other a radiused corner. The purpose here is to illustrate further the uniqueness of the router tool machining process and the array of precise and decorative edge treatments that can be accomplished with these small, versatile machines. The drawings could have shown nearly any edge shape, the point being that router tools and machines are very versatile and efficient as well as inexpensive in comparison with any other cutting process or larger machines.

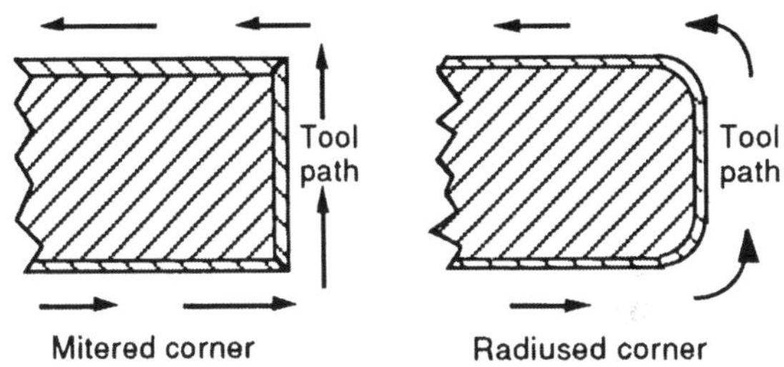

Illus. 6-9—Mitered and radiused corners

Router Performance Characteristics

The first point to be made in examining router-tool cutting processes is that, as with all other tools, there are definable limits to what can be attempted with them. Router tools are small, but their relative size is what makes them so unique and versatile. It is also their greatest handicap.

For instance, the high speed at which router tools turn is necessary to achieve even reasonable feed rates. Other cutting tool and machining systems have much faster feed rates. Shaper machines with 6,000 rpm spindles and 5" diameter cutters can have feed rates of 30' per minute and more; feed-through moulders can easily achieve 80' per minute; and even tenoners commonly operate at 40' per minute. For a 20,000 rpm router machine, a common feed rate would be 140" (11.66') per minute. We express the feed rate of router tool processes in terms of inches per minute rather than feet per minute because it makes the process easier to monitor accurately, and, incidentally, it is the same method of expression used to describe the progression of cutters in metal working. (In Europe and elsewhere, though, feed rates for all machining systems are expressed in meters per minute.)

To understand more fully the abilities of router tools and especially their limitations, we need to review the tool performance formulas (from Chapter 1) again.

Tooth/knife progression and *rim speed* calculations will readily illustrate that while router tools have many advantages, given their physical size, they also have considerable limitations. Incidentally, there is a somewhat sequential order to the calculations. *Tooth/knife progression* (Sz), also called length of cutter mark, is needed to find *feed speed* (U), *mean chip thickness* (hm), and *cutter mark depth* (t).

Rim speed (V), also called peripheral speed or surface speed, is most often the first index we seek when examining the performance of a cutting tool. Rim speed is the relationship of circumference and rpm. It tells us a relative speed of the tool that we can generalize into broad groupings, such as charts and graphs to guide us to correct tool selection and usage. Rim speed is of such importance that often an encountered problem can be identified with only that figure.

As mentioned earlier, because of their small physical dimensions and unusual cutting abilities, router tools can be used for many processes with great success. Their small size, though, creates limitations that must be examined carefully. Router cutters perform much as any others, in that they have teeth or knives on the periphery, but the details are very small. Thus, attention to detail is magnified, and the need for precise concentricity, balance, and correct cutting angles is acute.

For the sake of clarity, we will review the tool calculations and performance characteristics with the same size tool operated at the same speed, i.e., rpm. Tool diameter is 5/8" (0.625"); rpm is 20,000.

If we know that the material we want to cut has a particular strength and density we can select the appropriate rate of tooth/knife progression and feed rate to be used. Or, after we have completed a cutting process, we can work backwards to find what the cutter was doing to give the results achieved—good or bad.

Rim Speed Calculation

Illustration 6-10 shows a 5/8 (0.625)" diameter tool turning at 20,000 rpm. It will have a rim speed of 54.4'/sec.

$$0.625 \times \Pi \times \text{rpm} = 39{,}269.8$$
$$(12 \times 60) = 720$$
$$39{,}269.8 \div 720 = 54.4'/\text{sec.}$$

This means that the outer periphery is traveling at that speed. The use of feet per second to express rim speed is commonly accepted; to convert the number to inches for any reason would yield a figure too cumbersome, especially for larger tools of any type. Rim speed gives us a good generalized index from which we can make other observations and changes if required. In the lower portion of the rim speed calculation, the 12 converts the figure into feet (as 12 inches per foot), and the 60 converts the figure to one second (as 60 seconds per minute).

To seek (or select) correct feed rate and progression

Diameter = 5/8"; RPM = 20,000; # knives = 2

Feed rate (U)—to be selected
Tooth/knife progression (Sz)—
to be selected

(Tooth/knife progression)

$$Sz = \frac{U \times 12}{N \times Z} = \text{progression in 1/1000 inch}$$

$$\text{(Feed speed) } U = \frac{Sz \times Z \times N}{12} = \text{feet/second}$$

$$\text{(Rim speed) } V = \frac{D \times \Pi \times N}{12 \times 60} = \text{feet/second}$$

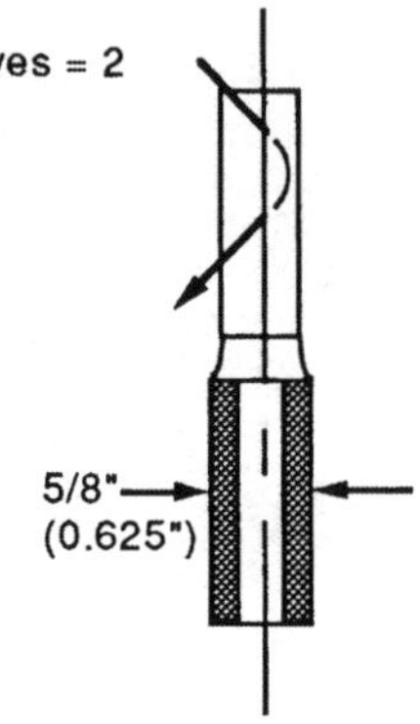

D = diameter in inches
pi = 3.14157
N = rpm
Z = number of teeth/knives
Sz = tooth/knife progression
U = feed rate (ft./min.)
V = rim speed (ft./sec.)

Illus. 6-10—Router tool performance indices

for each tooth/knife for a router cutter, we will either select from a text reference or call on our own experience. As stated earlier, we can, after completing a cutting process, work the formulas backward, "after the fact," to seek the indices and compare our results with what we have, in fact, done. Experience, while perhaps not the best teacher, is surely the most profound, and in this instance fairly safe, in that most of the variables are contained and controlled by either the router machine or the features of the cutting tool purchased from a competent manufacturer.

To find tooth/knife progression, multiply the feed rate (U) times 12, divide that figure by rpm (N), and multiply by the number of teeth/knives. If the feed rate is 140" per minute, the tool has two teeth/knives and is turning at 20,000 rpm. The progression of each tooth/knife (length of cutter mark) is 0.0035". Note that the 140" has to be first converted to feet; that is, 140 divided by 12 = 11.66'. Though we could have selected any particular feed rate, the 140" is a fairly common rate for stationary or computer-controlled automated router machines.

Feed Rate Calculation

To find feed rate, the calculation is tooth/knife progression times the number of teeth/knives present, times the rpm, divided by 12 (to return the figure to feet from inches).

$$0.0035" \times 2 \times 20{,}000 = 140 \div 12 = 11.66'/\text{sec.}$$

This is the rate at which the cutting process is taking place; that is, in one second, 11.66' of material is cut.

The figures given here represent commonly expected performance of cutting tools and processes in industry, and, with the use of very large computer-controlled machines, tooth progression and feed rates could in fact be two or three times faster. It is not likely, though, that we would be able to manipulate a hand-held router machine at any speed approaching those used in industry.

The examination of the feed rate and tooth progression for router tools has several dimensions that have to be understood before all of the possible criteria are clear. Given all of the processes that the tools could be used for, seeking optimum performance could take considerable time if we could not make useful generalizations.

Arc length

Illustrations 6-11 and 6-12 show the same processes viewed from over head and laterally, describing two different perspectives of tool use. In all instances the material cut is the same, as are the depth of the cutting processes, the tool used, the rpm, and other performance indices. Only the width of the cut is different.

Illustration 6-11 examines the arc length during the period in each rotation that the tool is in contact with the material being cut. While this examination is not often critical, as can be seen, the results are dramatic. There are no specific indices nor chart-type instructions that can be drawn from the information, but it is needed to understand some considerations for selecting a particular tool for a given process at a desired feed speed when cutting a material with certain physical characteristics.

In the instance of contact arc length, the advantages sought will be higher feed rates, lower tool temperatures, or other improvements. The advantage when cutting a rebate profile will be found using a tool with a diameter that is considerably larger than the width of the rebate. There will be better clearance angles, larger gullets, and shorter contact arc length. Waste evacuation will present no difficulties, keeping tool temperature lower and allowing higher feed rate without danger to the cutter. Additionally, while the horsepower necessary to cut any given rebate will not change dramatically with a larger diameter cutter, the effect of having lower tool temperatures and better gullet distance and clearance will reduce the energy needed by as much as 10 percent or more.

Last, a portion of the horsepower will be "stored" in the momentum of the heavier cutting tool while it is in motion. With the wide range of physical characteristics found in the materials that we cut in the woodworking industry, any advantage that can be gained will be worthwhile.

Illustration 6-12 again examines rebate and groove cutting processes, but from an end view. To estimate variation in consumption of horsepower from one to another will not be as simple as the illustration might convey. Viewing the drawings, we can make conclusions that are not entirely accurate; that is, the different sizes of rebates do not directly equate to the amount of horsepower needed to perform the work. Actually the horsepower needed relates more to the contact arc length, but not in a directly proportional progression. Cut with the same tool, a rebate two times wider than another may require nearly three times the horsepower in many materials. There are no precise means to fore-

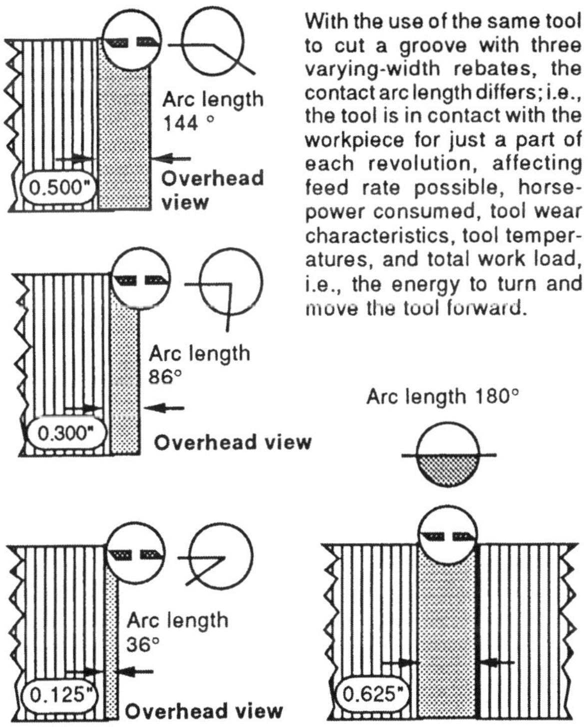

With the use of the same tool to cut a groove with three varying-width rebates, the contact arc length differs; i.e., the tool is in contact with the workpiece for just a part of each revolution, affecting feed rate possible, horsepower consumed, tool wear characteristics, tool temperatures, and total work load, i.e., the energy to turn and move the tool forward.

Illus. 6-11—Cutting circle arc lengths

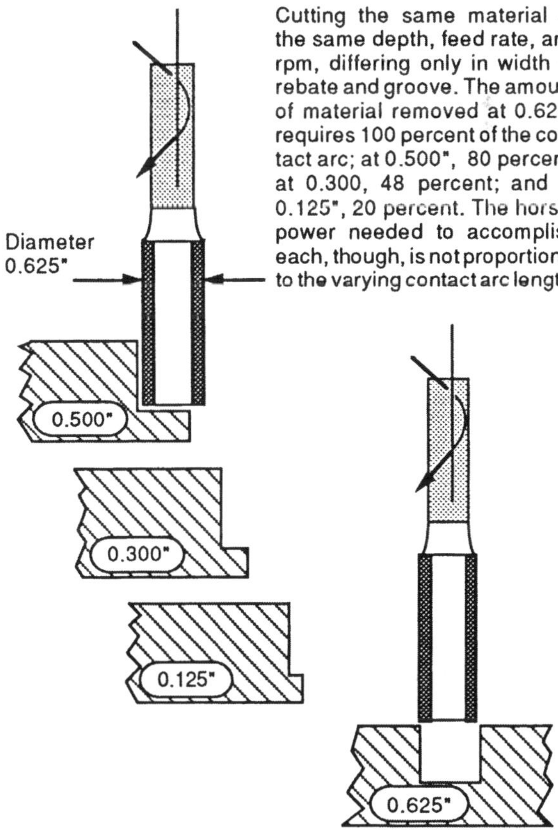

Cutting the same material at the same depth, feed rate, and rpm, differing only in width of rebate and groove. The amount of material removed at 0.625" requires 100 percent of the contact arc; at 0.500", 80 percent; at 0.300, 48 percent; and at 0.125", 20 percent. The horsepower needed to accomplish each, though, is not proportional to the varying contact arc length.

Illus. 6-12—Power consumption examples

cast exactly how much horsepower will be required to perform a given cutting process; the closest figure will only be 80 to 90 percent accurate.

Concentricity and Parallelism

Continuing the examination of router tool processes, we need to recall the tooth/knife progression figures from Illustration 6-10. The cutter in our example had a progression (Sz) of 0.0035", less than the thickness of the paper that this text is written on. During that particular cutting process, each tool knife/tooth removed that amount of material and moved forward that far. The earlier reference to precision and concentricity and the importance of exact details will now become clear.

Router tools, more than any others, need feature details that are completely parallel and concentric. Their small physical size, the rate of speed at which they work, and the amount of work they perform in the cutting processes make every detail of great importance.

Concentricity—what it is and, especially, what could cause a lack of it, i.e., eccentricity—needs to be completely understood. (See Illus. 6-13.) Most commonly, within the body of the cutting tool, either the knives may be placed eccentrically during manufacture or the tool may be ground into eccentricity during sharpening. Remembering that a common tooth progression of a router tool is only from 0.002" to 0.006", any measurable eccentricity could result in the tool cutting at only 50 percent efficiency. For instance, if we have a two-knife router tool with one knife that is eccentric by 0.004", and our desired progression is also 0.004", we will cut with only one knife!

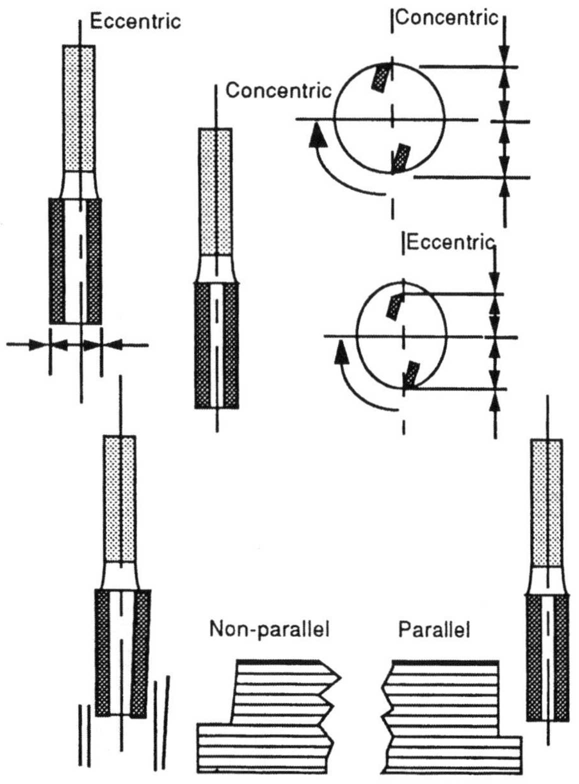

Illus. 6-13—Concentricity and parallelism

While we would hope that this example is a rare exception, the condition can easily occur by any of several oversights. The most common is when, during a sharpening operation, a different amount of material is ground from the individual knives—for instance, to remove 0.006" from a chipped knife and only 0.002" from the opposing knife without chips, then forgetting to balance and check for concentricity. Tool-grinding machines are very precise but still prone to human error. Most manufacturers have elaborate quality control systems in place to avoid most obvious errors, but none is completely foolproof.

Concentricity, or lack of it, can be easily mistaken for other mechanical problems that may or may not relate to the cutting tool itself. There can be an eccentricity in the placement of the cutting tool knives, and there can also be instances where the knives may not be parallel.

Exactly identifying a problem in concentricity will, in many instances, require special measuring devices. For instance, measuring the diameter of a router tool will not give any information regarding concentricity. With a single-knife router tool, measuring its diameter while stationary will not even give its (cutting dimension) diameter because it has to be turned in a fixture device. The point is that, given the complex geometry of all cutting tools, router cutters, with their compact features, are difficult to study and diagnose in the instance of any irregularity.

Parallelism (or lack of it), as mentioned earlier, can be created during the course of manufacturing or altered during subsequent sharpening service. Within the tool body, a lack of parallelism is not difficult to detect and measure. If the cause is mechanical within the router machine, however, the diagnosis may not be as easy.

Router Holding and Driving Systems

Perhaps the most common misconception of router machines is the life span and viability of the collet-type tool-holding system. (See Illus. 6-14 and 6-15.) The perception is that the collet itself does not wear and will last the life of the machine—not true. With the amount of horsepower that the machines produce, and the sidereal forces against the tool shaft, collets have a very definable life expectancy. The base, that is the lower portion of a collet, is positioned so as to act as a fulcrum or pivot for the length of the tool cutter body. As the collet ages through metal fatigue and wear, the opening may become noticeably oval in shape.

The forces acting on the collet are such that there is

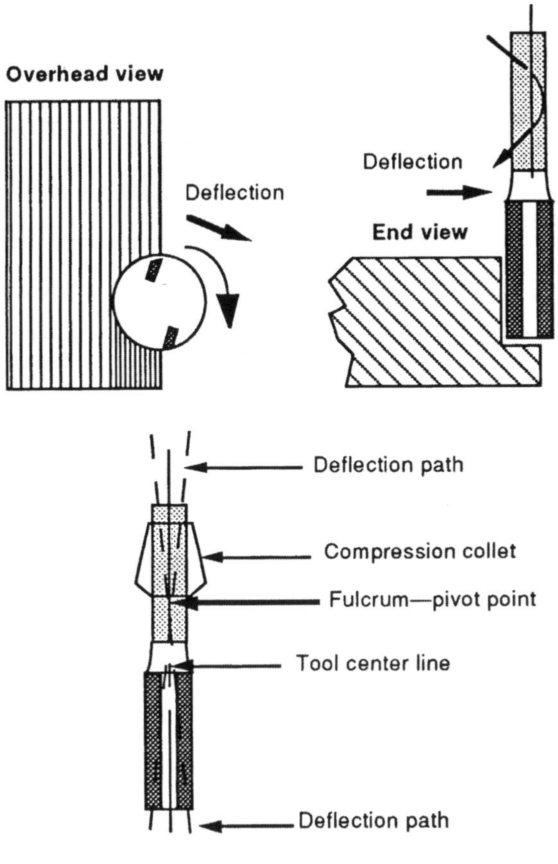

Ilus. 6-14—Router tool deflection

a predictable progression of wear and fatigue. In industry, it is common to inspect and change compression collets at least every 1,000 hours of operation, more often in the instance of heavy work loads. This does not imply that the collets are in any way poorly made or inferior, just that they have a more-or-less normal life span, and replacing them is an accepted practice. Regular inspection of tool-holding devices of any type will ensure not only the life of the device itself but also that of the cutting tool and the accuracy of all of the processes that the machine is used for (to say nothing of saving the cost of the workpiece material, which could be damaged if the collet is worn).

Another overlooked characteristic of router cutting tools is the fact that, especially as the cutting edges become dulled, the tool body itself can flex and bend against the forces turning it and forcing it into the workpiece. This is more pronounced with larger stationary and automated machines than hand-held ones, where such stress would more easily be noticed by the operator. The small cross-sectional size of router tool shafts and the alloy content of the steels that they are made of allow a minor deflection; that is, the tool shafts are not brittle. The amount of deflection is very small and is in part a safety consideration directed at being able to withstand both impact shock and high torque loads without fracturing and becoming a missile in the workplace.

There are several prevalent methods to hold and drive router tools. (See Illus. 6-16.) In the U.S., for most router machines, hand-held, stationary, and even the largest computer-controlled type, the straight, smooth shaft, compression collet method is the most popular, as in drawing A. The sizes vary from 1/4" to 1", with a mix of metric sizes finding growing acceptance. The safety of the smaller sizes is questionable, given the range of tasks given to them and the horsepower available in the newest hand-held machines. The cross-sectional area of a 1/4" shaft is only 1/4 that of a 1/2"

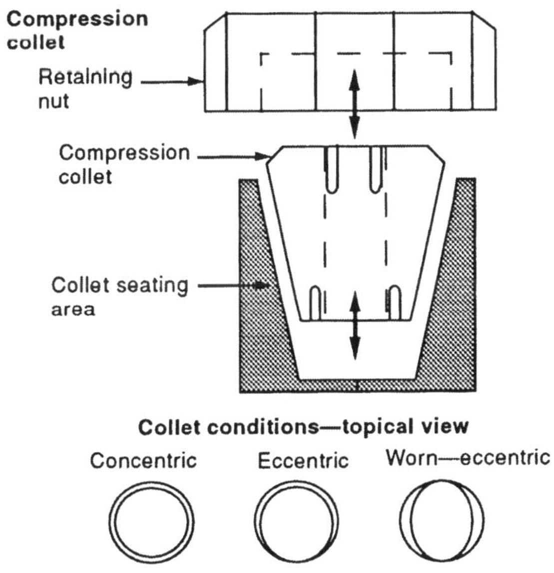

Illus. 6-15—Holding and driving methods

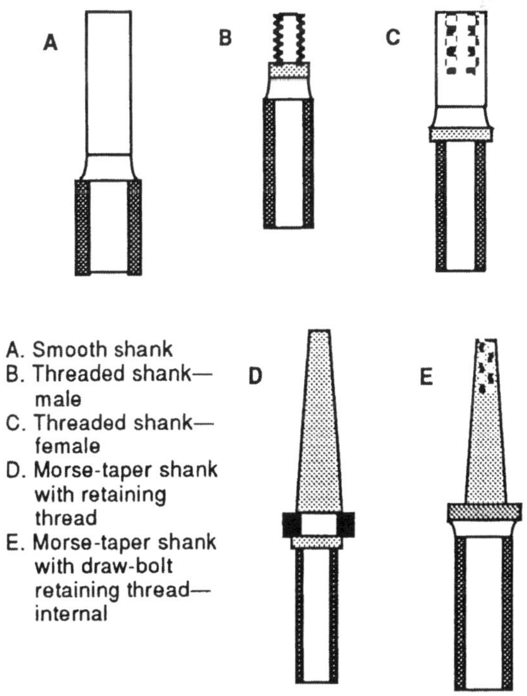

Illus. 6-16—Holding and driving methods

shaft, and driving such a small tool with 3 hp or more does not make for safe practice.

In drawings B and C, the cutters have either a male or female threaded shaft. In Europe and Asia discontent with smooth-shaft collet devices has made these tools very popular. For most hobby and craft applications, the threaded shank tools are both fast and easy to use. Another factor is that this method provides sufficient accuracy and strength for smaller stationary and portable machines.

In drawings D and E, the tool shafts have the same configuration as many of the popular American, European, and Asian metal-cutting tools. In Europe, however there are numerous safety laws that regulate cutting tools for all industries, and, depending on the horsepower of the machine, a unique means to hold and drive the tools may be required. The two types of tools shown have either an external or internal thread to pull the tool into the tapered socket. The tool in D has the external thread, and throughout Europe has the name of the Hess system. The tool in E, with the internal thread, is used on several brand names of woodworking machines, but is more common in metalworking industries.

The Morse-taper tool shank, whether made in the U.S. or elsewhere, has a taper of 3/4" per foot, or a 16 to 1 slope. The advantages of such a system to secure and drive cutting tools are twofold: First, the taper provides a very accurate and stable means to provide concentricity. Second is the great mechanical advantage of the taper's ability to transfer large amounts of horsepower with little vibration and few movable parts.

In industry, router tool performance continues to be pushed to new limits with regard to feed rate, the amount of work that can be attempted, and the material to be cut. The relatively small diameter of router tools places limits on them that will not be easily nor completely overcome. More rpm are not an answer to higher feed rates because the mechanical and heat problems created are more costly and bothersome than any minor gain that might be seen. The mechanical problems that relate to router machines are as difficult to solve as any encountered in the design and manufacturing of cutting tools, except perhaps that the machines themselves are only indirectly involved in the cutting process.

Router Tool Features

The features of a router tool that make cutting possible at all are the *clearance distance*, *cutting angle*, and *flute size*.

Clearance distance is the area within the cutting circle about the body of the tool into which the knife or knives protrude. (See Illus. 6-17.) In drawings A, B, and C, the tools are identical except for the size of clearance distance. The cutting circle that the cutter travels in is as perfectly circular as possible; the knife portion of the tools are not. The three single-knife tools illustrated

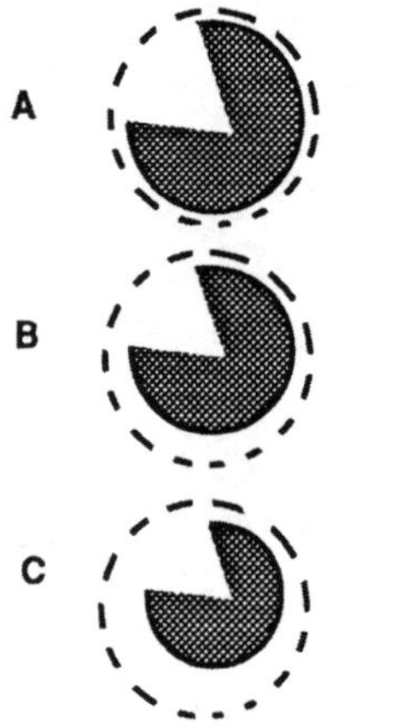

The three tools at left all have the same cutting diameter and cutting angle and different clearance distances. A will withstand the greatest work load and cut at the slowest rate. B is the most versatile, with a moderate feed rate and reasonable strength. C will cut more freely but will not withstand a high work load.

This tool is free cutting with high feed rates, but the edge is frail and prone to breakage

This tool is slow cutting with low feed rates, but it has very strong body edge and body because of the large cross-sectional area.

Illus. 6-17—Clearance distances

could have identical cutting diameters and still look radically different from each other.

As described in the illustration, the functional intent of each tool is different. Clearance distance essentially is a design criterion that seeks a balance between feed speed and strength; that is, a larger clearance distance will require a smaller and consequently weaker tool body not capable of withstanding the rigors of high feed rates. Single-knife router tools can be produced with the largest clearance distance (and most are noticeably eccentric), giving them many advantages over cutters with two knives. Having a single knife will in most instances provide sufficient feed rate for the process, especially when such a tool is so much stronger and cuts more freely than a similar two-knife cutter. Most conventional cutting tools with either one or two knives have sufficient clearance distance to fit the needs of most machining processes, though special cutters can be produced for unique materials and applications. Measuring clearance distance requires that the tool be held and turned, measuring incrementally about the tool, a common ability for a metalworking facility.

The cutting angles on router tools are generally less aggressive than on most other cutters. (See Illus. 6-18.) Given their relatively small diameter, light chip load, high turning speed, and moderate feed rates, routers do not need an aggressive angle to sever most materials successfully. Another element of cutting angle consideration is the heat generated in the cutting process; a

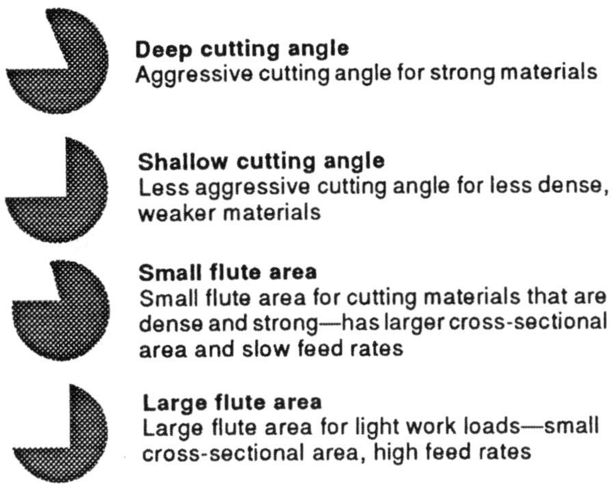

Illus. 6-18—Relative cutting angle and flute areas

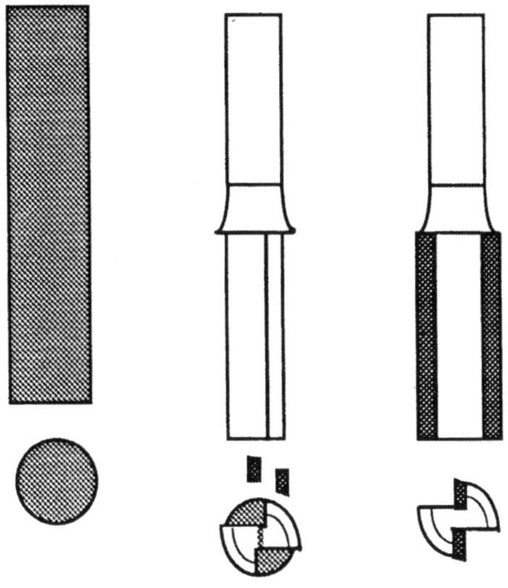

Machining a two-flute router bit begins with a round bar of tool steel. The shaft, knife seats, and flute areas are cut to correct dimensions, and the knives are attached. The knives are ground to dimension and polished, and, finally, the remaining features are polished. Automated systems can produce hundreds of these tools per hour.

Illus. 6-19—Router tool construction

knife with a cutting angle that is too aggressive will have a tip that is more frail and prone to thermal damage and fracture. Also in the selection of cutting angle are the considerations of the knife material wear life and how subject the alloy is to abrasive or thermal-corrosive wear—the difference, for instance, between high-speed steel and tungsten carbide.

The primary element in the selection of cutting angle relates more to the strength of the material to be cut, followed by feed rate needs and other criteria in the process. Cutting angles will vary from -2 to 4 degrees for very dense and strong materials to +12 to 15 degrees for most wood-specie materials.

The shape and size of flute space on a router tool relate to the other features present. It would not, for instance, be possible to provide a large flute space on a tool having a large clearance distance and an aggressive cutting angle. The amount of material on the tool body that will be removed to provide space for all of the needed features takes away from the strength of the tool. The portion of a router tool where the knives are attached can have any shape that can be machined, but the shape of the body and arrangement of the knives have to be such that subsequent sharpening can be easily accomplished with a minimum of time and expense.

The construction of a router tool, as Illustration 6-19 shows, begins with a cylindrical shaft. The needed features are prepared by removing sufficient material to size the shaft dimension, cut the shape of the area(s) for the teeth/knives, and shape the clearance distance. With current machining technology, hundreds of these tools can be produced in an hour. Tool manufacturers attempt to monitor constantly the needs of the marketplace and to be able to produce what is needed.

In the marketplace worldwide, the manufacture of cutting tools has changed dramatically in only the last 20 years. In the finest sense, the world is becoming smaller in that here in the U.S. we can purchase technology from anywhere. Only a few years ago, there was very little technology sharing in the woodworking industries. Today, with improvements in communications and marketing, we could actually seek price quotations and place orders from the screen of the computer that sits on the desk in our office, complete with drawings and specifications.

The features present on any cutting tool have to be such that they will complete the work process as expected, are cost efficient, are readily available, and can be sharpened and maintained without difficulty. While this list of criteria may not seem that difficult, the needs of woodwork manufacturing are in a constant state of change with new materials and machining processes being added.

Router Knife Arrangements

The wide range of materials and machine processes in the woodworking industries have created a considerable collection of cutting tools that are now common. Illustration 6-20 shows the most basic router tool knife arrangements found in the marketplace. With variations, there are at least three times more than the seven shown.

The knife arrangement of a router cutter is directed at a solution to a manufacturing process. The initial cost of the cutting tool is generally not the first consideration, especially if the process is going to be repeated hundreds or thousands of times and is an integral part of producing a particular product. In industry, the criteria of tool selection include:

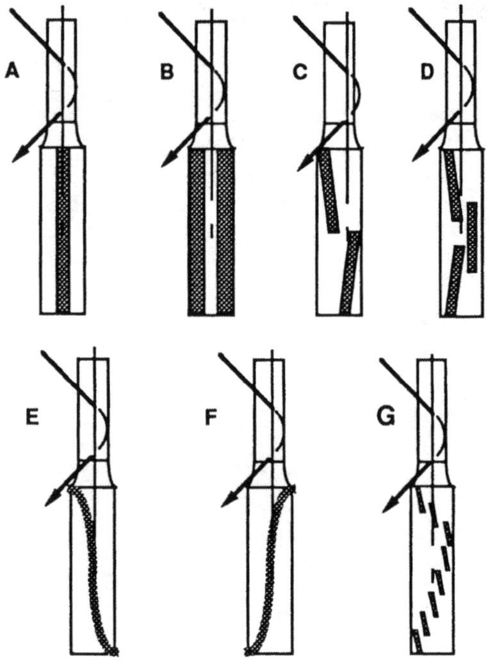

Illus. 6-20—Knife arrangements

1. Effectiveness—will the tool do the work, cut fast enough, leave a needed finish quality, be strong enough?
2. Cost effectiveness—can the tool be well maintained; is the knife material correct for the material and the process; how expensive is the sharpening process per cycle; how many times can the tool be sharpened?
3. Total cost—what is the sum of all of the cost indices: initial cost, service cycles, and service costs?
4. Availability—is the tool readily available; is it mass produced or a special production tool; can we expect difficulty to reorder the tool; will delivery be a problem?

Single knife. Initial cost of a cutting tool may be the last consideration in the purchase of a particular cutter; effectiveness is usually first. With effectiveness, a primary consideration is the arrangement of the cutting knives. In Illustration 6-20, drawing A shows a single-flute/knife router tool, a configuration finding renewed acceptance in industry. Where a single-knife tool is finding a "new" use is with large automated router machines that perform heavy work loads. A single-knife cutter, though slower than a comparable two-knife tool, can have a larger clearance distance and larger gullet and is stronger because less of the body has been removed to provide space for the second knife.

The loss of feed speed is easily compensated by the advantages of being able to perform a larger work load and having longer tool life because a larger knife is possible. There are several possible configurations of single-knife router tools, as will be explained later.

Two knife. In B, a two-knife tool is shown in perhaps the most common configuration of all router cutters. This arrangement gives an excellent feed rate and has sufficient strength to accomplish most work processes. The tool, while not an answer to every process and material, will provide adequate finish quality to serve most needs. The limitations are that very high work loads are not as possible and the total tool costs may be higher with fewer sharpenings possible because the knives are smaller than with single-knife tools. Two-flute router tools are the most prevalent, finding widespread use in all areas of woodworking and being perhaps the least costly to manufacture.

Alternate shear. Drawings C and D illustrate cutting tools that are a product of modern manufacturing need. The tools are essentially single-knife tools, except that the knives perform a bit differently from each other and at different locations in relation to the workpiece. The tools are generally called a 1 + 1 or 1 + 1 + 1 alternate-shear configuration. The need was cutting a panel material covered on both exterior surfaces with a fine wood veneer or a man-made lamination and having the edge surface free of any disfigurations, chips, tears, or irregularities. (See also Illus. 6-21.)

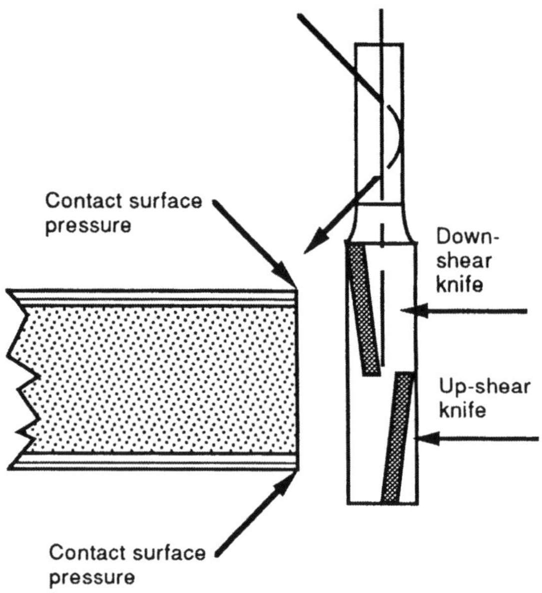

In the tool above, the two knives cut in the same cutting circle and are parallel to the toolshaft. The advantages of the compound shear angles of the knives are the superior edge finish quality and smaller horsepower requirement. At the upper and lower edges of the workpiece, the contact surface pressure is driven against the material, leaving the corners smooth and free of tears and roughness. The same effects will be seen in all similar alternate shear tools.

Illus. 6-21—Compound shear angles

The 1 +1 configuration is very successful in providing adequate strength and feed rate to cut any such panel material. The opposing shear angles function to create contact surface pressure toward the upper and lower surfaces and not parallel to the direction of tool travel. Like other single-knife tools, the alternate shear cutters have large clearance distances and cut freely. Another benefit of this configuration is that the shear-angle tools require less energy than a similar straight-knife cutter. The amount of shear angle present cannot be too large or the tool will cut a concave surface and not a straight-planed surface. Most often, the shear angle varies from 4 to 8 degrees, depending on diameter.

A negative factor of the alternate-shear cutters is that they are more expensive to produce and maintain. But given the great benefit of perhaps eliminating a later machine process, the cost can be appreciated into the product easily.

Single spiral. Drawings E and F (Illus. 6-20) show tools that have knives produced in a single spiral configuration, either up-shear or down-shear. This arrangement is a carryover from the metalworking industries, where it is used for cutting ferrous and non-ferrous alloys such as aluminum, brass, and soft steels. The advantage of this configuration is that fairly high feed rates are possible with contact surface pressure divided in a smooth continuum about the cutting circle, requiring less energy. The obvious disadvantages are high initial cost, a rough surface edge at the upper or lower corner, and higher relative sharpening costs. There are applications where the spiral knife configuration will, together with other considerations, be the best possible solution to the manufacturing process. The spiral-knife tool's most prevalent use is for machining very dense wood-based materials, plastics, or nonferrous metals. As with other tool-knife configurations, the choice has to be what is most reasonable and cost efficent for the material and the machining process in total.

Multiple knife. In G, the cutter is a multiple-knife tool that has several knives in a spiral configuration. The logic of this is correct as far as the idea is pursued; that is, a cutter becomes more efficient by dividing the contact surface pressure about the cutting circle over the length of the body. This configuration is also more efficient in creating small chips that are easier to cut and evacuate. The negative aspects of this arrangement, however, far outweigh the advantages. The initial cost of manufacturing all of the features is very high, especially if there is any shear angle present in the individual knives. There are two more problems—the sharpening cost of a spiral multi-knife tool may equal its initial price, and it would be extremely difficult to keep all of the knives in the same cutting circle during all sharpening service cycles. An excellent idea but, unfortunately, not practical for a router tool. There are numerous examples of spiral multiknife tools for other machine processes, shapers, moulders, and the like that are very practical and successful, but only on larger-diameter tools.

"Chip breaker." A means to improve the performance of two-knife router tools is to divide the work load between the knives, that is, shape them in such a manner that the work done by one is unlike that of the other. (See Illus. 6-22.) The term chip breaker in this instance is a misnomer. The waste chips are not really broken into smaller particles; rather the individual knives cut chips a bit differently from each other. The advantage is that the total work load is divided between the two knives and the chips are smaller, hence easier to evacuate. The greatest advantage of this arrangement is when cutting solid, dense wood, especially along the grain. Chip-breaker router tools are also finding popularity in cutting some man-made materials because the cutters are inexpensive to produce and maintain and do not require anything unusual beyond the staggered cut to be successful.

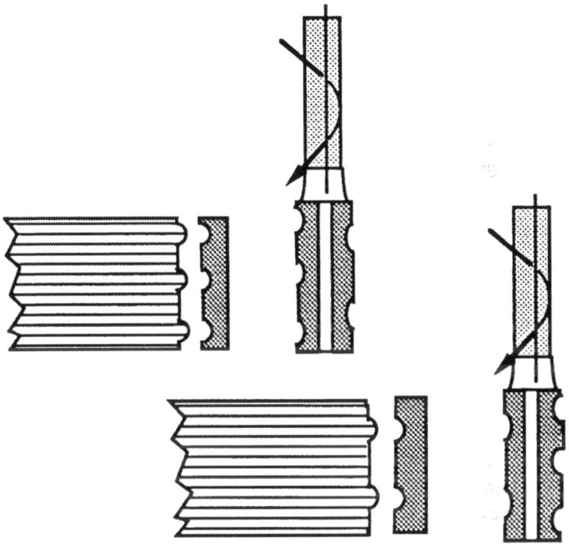

Chip breaker is a minor misnomer because the waste chips are not actually broken into smaller particles. With a router tool, the two knives, as in this example, simply cut different locations along a common plane by virtue of the unique profiles cut into the edge of the knives. This is particularly useful when cutting solid wood along grain, and it produces small, easily discharged waste chips. There are several successful knife shapes produced, from the scallop shape shown here to ones that resemble serrations, and the chip breaker effect works very well with other knife arrangements.

Illus. 6-22—Chip breaker

A disadvantage of the chip-breaker shapes is that, in some instances, there needs to be a second operation to provide a smooth surface (as when the tool knife has the appearance of serrations or teeth). The shaped teeth are particularly useful to reduce the size of the chips, and hence are less prone to damage dense or brittle materials. The router tool in illustration 6-22 has two flutes and a different scalloped profile on each. In the one case, there are three radiused shapes; on the other, there are only two. As one knife completes its contact arc and leaves a particular profile on the workpiece, the following knife removes it and leaves a different one, and so on. The advantage is that, along the entire plane of contact, the work is divided between the two knives, effectively altering the chip mass and contact surface pressure. This is of greatest importance when cutting any material composed of layers that would otherwise be difficult to cut without tearing and chipping, such as solid woods, man-made panels of wood or synthetic materials, and even paper-based products.

Summary

Throughout the woodworking industry, many unusual materials and processes have to be dealt with effectively. Newer materials and coverings for materials come into the realm of woodworking with increasing regularity. Where cutting tools can solve the inherent problems of such materials, they are welcome additions to our abilities. Often in the past several years, cutting tools that were at first manufactured as special productions became standard catalog items after finding widespread use and acceptance. The alternate-shear router tools fall into this category. The chip-breaker profiles are also becoming a standard production item for several manufacturers. Spiral-knife router cutters, orginally used in metal working, are finding more use, especially for non-wood materials such as plastics and soft, decorative metals used in certain furniture components. The single-flute router tool for many years was not widely used because of the belief that greater feed rate was the most desired characteristic to pursue. With improved geometry, however, single-flute routers are proving to be superior in numerous ways, especially for the large computer-controlled systems that are fast becoming the standard of industry.

For all woodworkers, it is increasingly important to know as much as possible about the materials we cut and the tools that we use. Everything is more expensive—our time, our machines, the raw materials we use, and our cutting tools. The most common mistake is not to ask questions. If your experience and knowledge tell you that a certain feature is needed on the cutting tool, find it or have it made. Work as though there are still improvements to be made and you will be correct.

The list of criteria to guide the selection and purchase of router cutters will include:

1. Effectiveness;
2. Finish quality;
3. Strength;
4. Availability;
5. Initial cost;
6. Service/sharpening cost per cycle;
7. Sharpening life, each cycle;
8. Sharpening life, number of possible sharp enings;
9. Possible feed rate;
10. Appropriate to purpose, cost effectiveness;
11. Cost per operation/ cycle, per foot or hour of operation;
12. True cost of entire cutting process.

Outside of industry it is not likely that many of the criteria listed will be pursued, except perhaps initial cost and effectiveness. All twelve are important to most industrial tool users, though, where the cost effectiveness of machine processes are an inherent part of their product cost and profitability.

In Chapter 3, we explored the knife materials for all cutting tools and the means by which they are attached to the cutter bodies. Today there are several choices to be made when selecting the cutting tools needed in any machine process. For instance, we may need four router tools of the same size to cut different materials in different situations:

1. A hobby project to cut solid natural wood without glue lines;
2. In industry, to cut plywood panels with infrequent use;
3. In industry, to cut particleboard panels every day for eight hours;
4. In industry, to cut medium-density fiberboard 16 hours each day.

Solutions:

1. High-speed steel—cost approx. $24.00;
2. Carbide-tipped alternate-shear—cost approx. $80.00;
3. Carbide insert tool—cost approx. $180.00;
4. Diamond-tipped—cost approx. $800.00.

In closing, what must be very clear regarding the success of router cutters is the attention to details. Concentricity, balance, parallelism, correct cutting and relief angles, and clearance distance all have to be precise as with no other wood cutting tools. Routers turn at enormous speeds to achieve even reasonable rim speeds and feed rates, given their relative size, and chip loads are very high. All features need to be as accurate as possible to make routers completely functional. As users of cutting tools, we have to make every effort to comphrend and use the technology available to us.

Chapter 7
Movable Knife-Cutters

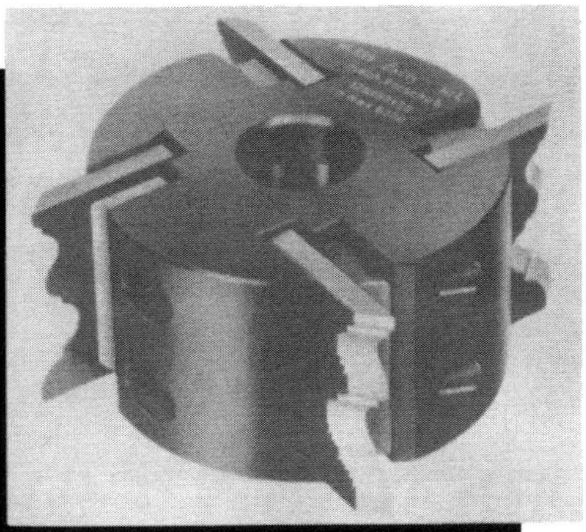

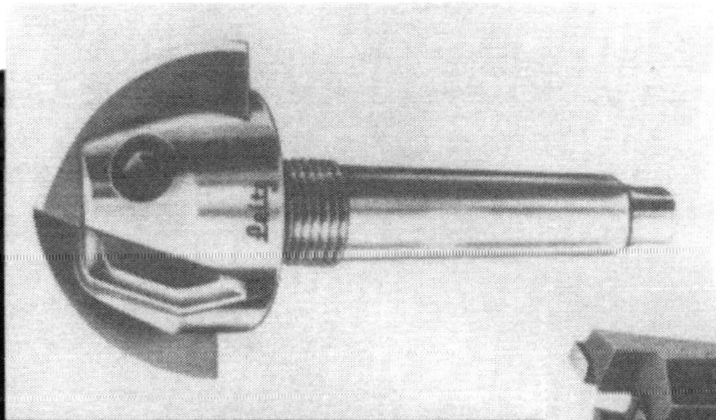

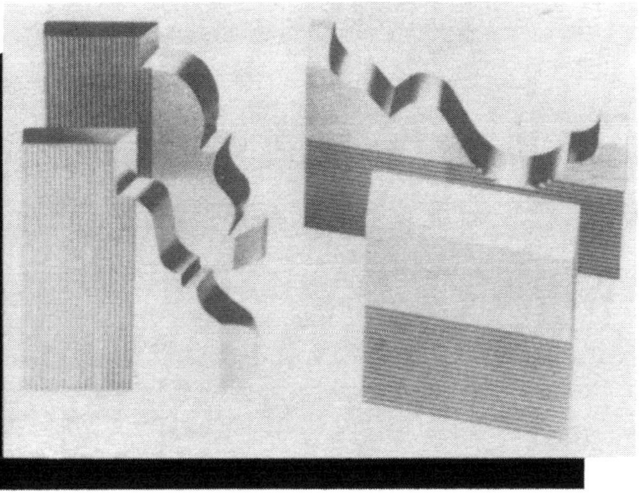

Photos courtesy of Leitz Tooling Systems, Inc., Stehle GmbH & Co., and Leuco Tool Corp.

Movable-Knife Cutters

With the development of the first powered woodworking machines, the need for versatile multipurpose cutting tools became apparent. The first cutters for powered machines were tools that could perform only one function. There were planer heads and individual cutters made for each profile shape and size that was needed. While the tools no doubt functioned well, there were problems to be overcome, including:

1. If two or more cutters were used in conjunction and sharpened repeatedly, keeping track of the size of each became increasingly difficult. (As tools are sharpened, they become smaller.)
2. The cost of manufacturing individual cutting tools for each profile size and shape was huge. (Some woodworking facilities may maintain over 1,000 different profile shapes and sizes.)
3. There was no way to reduce the set-up and operating times to give the versatility of short production runs.

Evolution of Movable-Knife Cutters

There were several stages of development from the earliest cutting heads to the tools of today. The initial ideas were to reduce the cost of individual cutters, the time required to set them up for operation, and the enormous price of maintaining them. When the first powered woodworking machines were developed, toolmakers did not have the understanding of metallurgy we do today. The first tool bodies were steel, but not nearly of the quality that we now take for granted. Recent advances in mechanics and physics make the earliest cutters look very large and clumsy, which they were by present standards.

The inherent problems of placing a cutting tool on a machine spindle are numerous and include difficulties in fit and tolerance among all parts of the cutting tools and machines, concentricity to ensure the most accurate cutting circle possible, and reduction of balance and vibration to as near zero as possible. A fairly imposing order, but we continue to seek technological solutions to the problems, making gains constantly.

The first cutter heads for powered machines were little more than square blocks of steel bored to fit onto spindles. The cutting knives were ground with the needed patterns and simply bolted to the outside of the body. (See Illus. 7-1.)

About this time, a couple of metal working terms came into woodworkers' vocabulary—"grind to pattern" and "mill to pattern." Both relate to machining practices involved in producing cutting tools. The first relates grinding a shape into the cutting knives themselves, and the second to creating a shaped cutter head or head part to which slotted knives could be attached later.

The square-head cutters were successful enough, but required constant attention and were costly to maintain. Another obvious feature of the square heads that used slotted knives was their unfortunate and dangerous tendency to lose the knives. At today's motor speeds, the tools could not function at all; in fact, most safety regulatory agencies have banned their use unless motors and spindles are completely covered and protected.

Development of the first cutter heads with movable and exchangeable knives gave us the ability to maintain

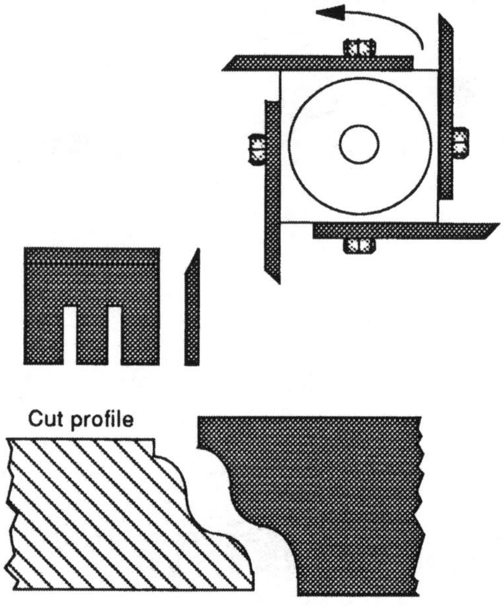

Illus. 7-1—Cutter heads

a large number of profile shapes and sizes at a much lower cost. There were problems, though, keeping the knives the same weight and length. The weight and balance of the knives and cutter heads became increasingly important as higher spindle speeds were attempted.

The size of the cutting circle also became a consideration of tool design because larger cutters have greater centrifugal forces acting upon them, and size, to a certain extent, could be manipulated. Illustration 7-2 shows the effects of centrifugal force acting on cutting

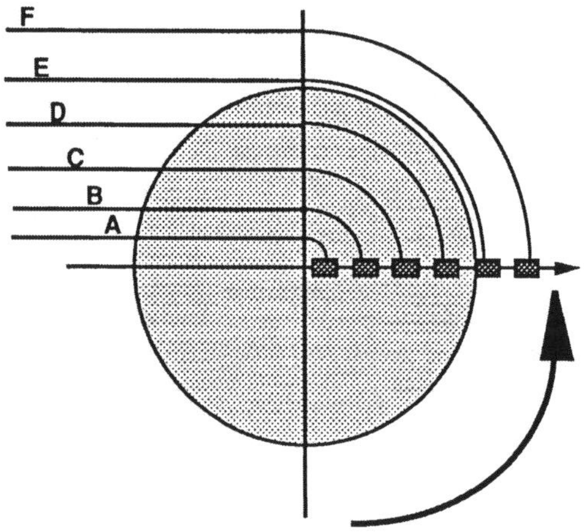

In each of the examples below, the cutting knife weighs two ounces and is turning at 6,000 rpm.

Cutting circle	v (ft./sec.)	Centrifugal force (lbs.)
A. 1-2 inches	52.2	127.8
B. 2-4 inches	104	255.7
C. 3-6 inches	156	383.1
D. 4-8 inches	209	511.5
E. 5-10.5 inches	275	671.3
F. 6-12 inches	314	767.2

The upper limit of tool velocity is 100 M/sec. (328 ft./sec.). An 8" cutting circle would turn at 9,416 rpm and would have a centrifugal force of 1,254.4 pounds for each knife.

Illus. 7-2—Effects of centrifugal force

circles of different diameters. In all the examples, the weight and rpm are constant; only the diameters of the circles vary, from 2" to 12". The 6,000 rpm figure was chosen because most tool spindles in use today turn at that speed, though the range will vary from 1,800 on old or special machines to 15,000 for the newest special machines.

Note that the centrifugal force is enormous even for a very small circle turning only 2 ounces of weight. Most cutting knives are considerably heavier than the examples. For instance, a steel bar measuring only 1/4" × 2" × 3" would weigh 6.79 oz. A tungsten carbide bar of the same dimensions would weigh 16.8 oz.! To calculate centrifugal force, use the formula below:

$$F = \frac{W \times v^2}{g \times R}$$

F = force in pounds
W = weight of object in pounds
v = peripheral speed in feet/sec.
R = radius distance in inches
g = acceleration due to gravity
(32.16 feet per sec./sec.)

Tables and charts are readily available and provide a simple multiplier factor based on rpm to which you simply supply the weight of the knife and radius of the cutter for the calculation. That is, F = W × R × given constant (C). For 6,000 rpm C is 1,022.980.

Another important characteristic is the relationship of the weights of all the knives used in a particular set. The accepted standard for most woodworking applications is that the knives weigh within 0.1 gram of each other (28.35 g. = 1 oz.). To deviate too far from the 0.1 g. standard would greatly increase vibration and balance problems.

The point is that to hold cutting knives mechanically, a system must withstand the tremendous forces exerted on each knife in the cutter. Additionally, the tool body and knives have to withstand the forces of the cutting process encountered during contact with the material being cut, and, equally, interruptions and abrupt contact with the workpiece as cutting begins.

The next steps of cutter-body evolution involved several variations of knives set into shaped pockets. While the T-slot and similar knives (see Illus. 7-3) offered greater safety, they were expensive to produce and maintain. Further, it was difficult to make the assembled cutter balanced and concentric. Grinding the profiles was done as for other tools, and to grind each odd-shaped knife to exactly the same weight was a time consuming effort. It was also difficult to prepare the knives accurately so that all in a set cut in precisely the same circle.

Considering that the advantages sought from movable-knife cutters were to reduce the overall costs and time, the progress, seen from today's perspective, appears to be minimal. While T-slot heads were much safer than the earlier square heads, there was a cost in time for maintaining them. T-slot heads are used today in a few woodworking plants in the U.S., but most have been obsoleted, even though safety laws do not forbid them in most states.

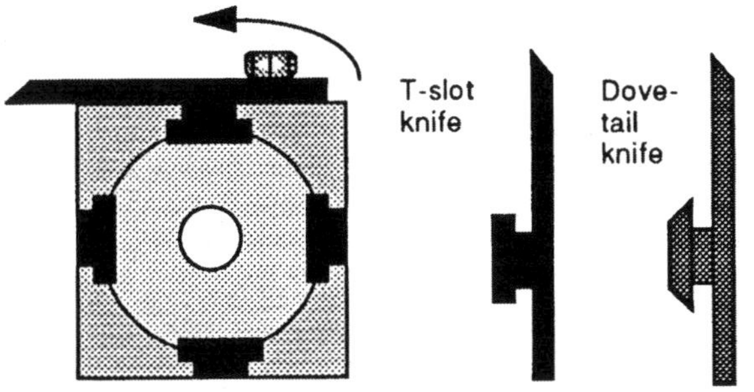

Illus. 7-3—Slot knife cutters

The mechanics of the earliest cutters of all types did not have to deal with the great motor and tool-spindle speeds of today. The first machines did not turn much beyond 1,500 rpm, while most modern cutters—shapers, moulders, and the like—run very easily at 6,000 rpm and more. Advances in several technologies have made this possible:

- The first drive belts were made of leather and, while strong enough, could not withstand the speed even if there were bearings and support mechanisms available to reach higher speeds.
- Bearings capable of high speeds are relatively recent, developing in the last 75 years.
- There were no alloys that would withstand the torque loads and temperatures of higher speeds.
- Metalworking technology and mechanics had not yet advanced to be able to mass produce the machines, tools, or other components.
- Today's machines rely on very recent technology of high-speed electric motors, precision bearings, efficient drive belts, gearing, and unusual metal alloys, some not seen until the past 35 years.

Holding and Driving Methods

The means to hold and drive the cutter bodies with motors or machine spindles have also changed, though not as dramatically as other features of woodworking technology. Illustration 7-4 shows the most prevalent method to hold and drive cutting tools that have an arbor hole. It is important to understand that the cutter bodies are being both held on the shaft and driven by the motor. Thus, the horsepower has to be transmitted from the spindle to the cutter body to perform the work, and the tool has to remain on the spindle, as precisely and concentrically as possible with no deviation of vibration, mechanics, etc.

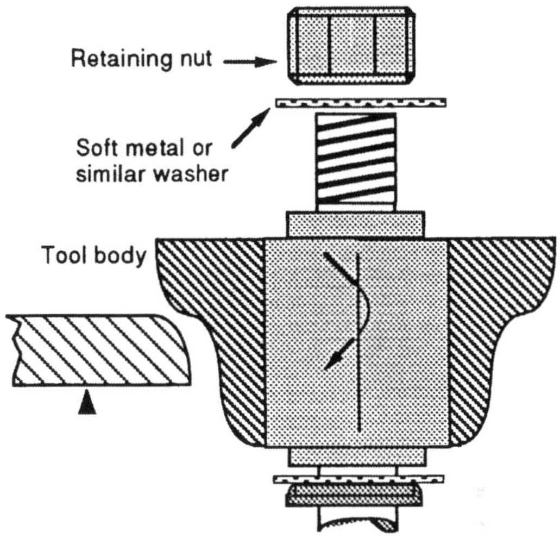

Illus. 7-4—Threaded shaft

Standard Threaded Spindle

The standard threaded spindle may use any of several thread patterns to secure the tool. The torque necessary to completely tension the retaining nut usually results in several thousand pounds of force against the soft metal washers. Greater horsepower, though, calls for different methods.

For several machine types, the standard thread is produced in both left- and right-hand configurations to allow for change of rotation. The only serious problem with the standard spindle is the dimensional difference between the bore of the tool body and the diameter of the spindle itself. Additionally, wear characteristics of both bodies add to the tolerance difference; that is, as the tool-body bore and the spindle each wear through normal use, the bore becomes larger and the spindle smaller.

There are several classes of fit that describe these features in great detail. Ordinarily, the cutter-body bore is larger than the machine spindle by 0.002" to 0.004". But even with minimal difference, we can conclude that the cutter will not turn in a perfect circle but rather will be eccentric by the amount of the difference. And this is but a very simple example; in very large, complex machines, tracing all possible eccentricities can be very involved. Any means of controlling the eccentricity is worth the expense and effort given that, for many wood cutting processes, each knife is doing only a very small amount of work, only a portion of the knives are doing any work at all, and only one or two knives of, for example, a four-knife head are dictating the finish quality if there is any measurable eccentricity at all.

Self-Centering and Clamping System

Among the first attempts to correct eccentricity for standard threaded spindles was called the self-centering and clamping system. (See Illus. 7-5 and 7-6.) Though cumbersome, expensive, and time consuming to use, if carefully implemented, the collection of tapered collars and machined bodies did reduce problems significantly.

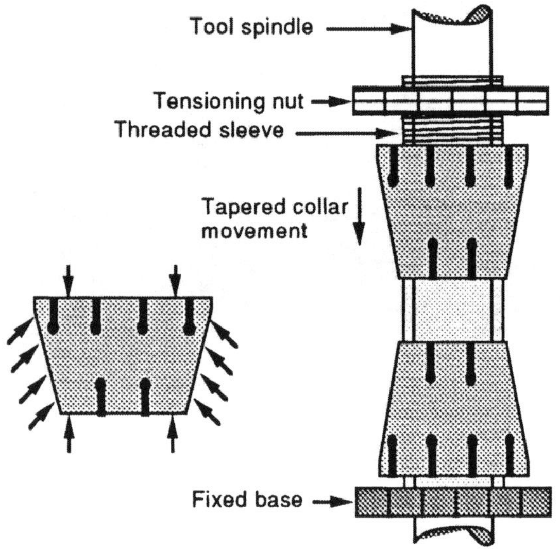

Illus. 7-5—Self-centering and clamping system

The mechanical principle involves a set of tapered collars, similar to those used to hold and drive router tools, though much larger. The inside of the collars are straight, parallel, and the same size as the spindle they are placed on. The tool body is machined to mate

This system assumes that the tapers are perfectly concentric and will remain so. Without careful attention, however, the effects of wear and the ambient environment can cause this tool system to perform poorly. The first consideration is to keep the movement surfaces clean. Problems can involve inadequate gripping power, allowing the tool to slip on the spindle, or lack of concentricity, yielding an irregular finish surface and poor tool performance.

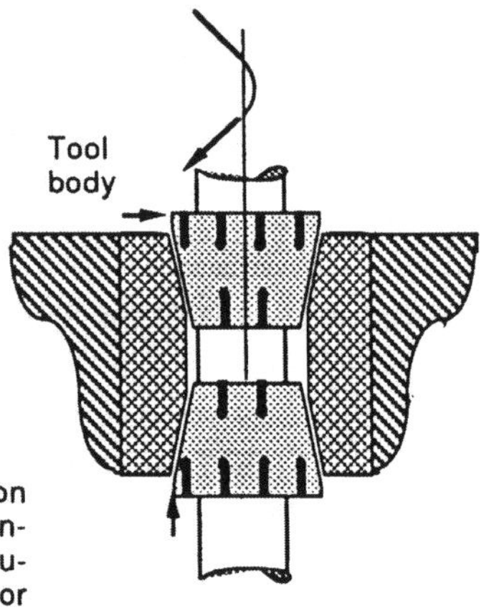

Illus. 7-6—Self-centering and clamping function

exactly with the taper of the collars. Theoretically, tightening the retaining nut on the sleeve will compress the collars to grasp the spindle and force them against the tapered inner walls of the cutter body. The greatest problem with this system is the predictable wear of all of the movable parts, compounded somewhat by elements of their environment. Dirt and dust are a problem if there is not careful consideration given to their effects; that is, the compression collars will not seat properly, creating a dangerous condition by not securing the tool firmly to the spindle. There is a normal wear life of any metal part that is expected to flex as the compression does. The collars should be changed according to the manufacturer's suggested schedule.

The self-centering and clamping system was in part developed to reduce the tolerance differences between tool bore dimensions and machine spindles. At the time of their development they were so successful that many machines were produced with spindles that were perfectly smooth and without any other means to secure and drive the cutting tools. The two features that make this system somewhat obsolete is the relative high cost compared to current technology and the difficulty of use.

The Keyway System

As woodworking machines become larger, faster, and more complex, there is need to transmit more horsepower to the cutting tools. Another part of the increasing complexity of machine processes is having motors that, for the benefit of particular processes, can be reversible. Until recently, a change of spindle direction required a change of the spindle, otherwise the torque movement would loosen the retaining system and possibly cause loss of the tool. The keyway spindle configuration (Illus. 7-7) solves at least two problems of more recent machine and tool technology:

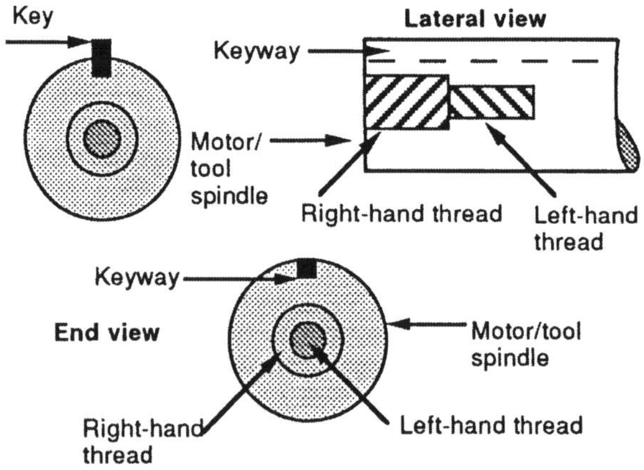

Illus. 7-7—Keyway system

- A key and keyway can help transmit greater horsepower to the cutting tool and provide a safety measure when the tool strikes a solid object, because the key will shear off, protecting the machine and the machine operator.
- With the shaft internally threaded with both left- and right-hand configurations, the rotation can be reversed simply by changing

the retaining bolt, providing the ability to either climb cut or abrasive cut, a considerable benefit.

This configuration would be useful even on simple spindle moulders—what we in the U.S. call shapers. The advantages would be the same as for the larger machines—the safe transmission of large horsepower and the opportunity to change direction easily.

Concentricity with this shaft configuration remains a concern, but manufacturing standards have improved so that this issue is not of the magnitude that it was even 15 years ago. Additionally, the metallurgy and heat treatment of machine parts of this type have improved dramatically.

Hydraulic Clamping Systems

Where great precision of concentricity is demanded for very high feed rates, in machining solid woods for instance, a relatively new method of holding and driving cutting tools using hydraulics has developed. The system is now over ten years old, but as propagation of technology goes, ten years is a relatively short time.

Clamping a cutting tool to a machine spindle with hydraulic pressure requires a level of machining technology and precision that was not known even 30 years ago. The other technologies involved in the construction of this tool system are metallurgy, heat treatment, and welding.

There are two types of hydraulic clamping systems marketed today. (See Illus. 7-8 and 7-9.) One uses a sleeve body with which we would use an existing cutter or any number of cutters. The other is a complete system built into a specially constructed cutter body. The difference between them is that the sleeve can flex in both directions, inward to grasp the machine spindle and outward to grasp the tool body.

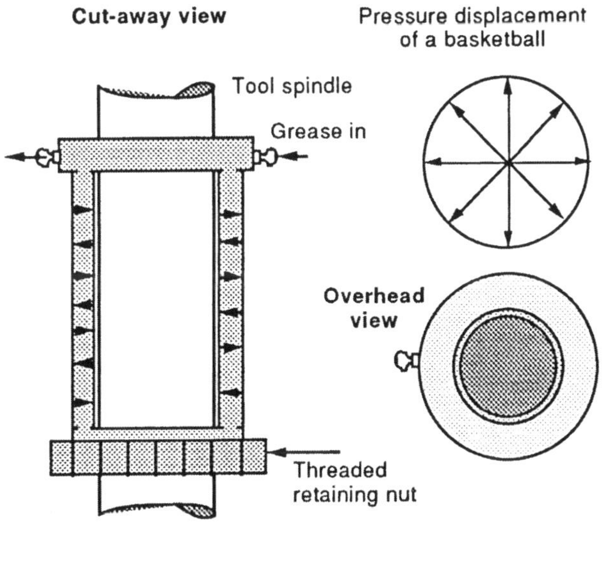

Illus. 7-8—Hydraulic clamping/driving

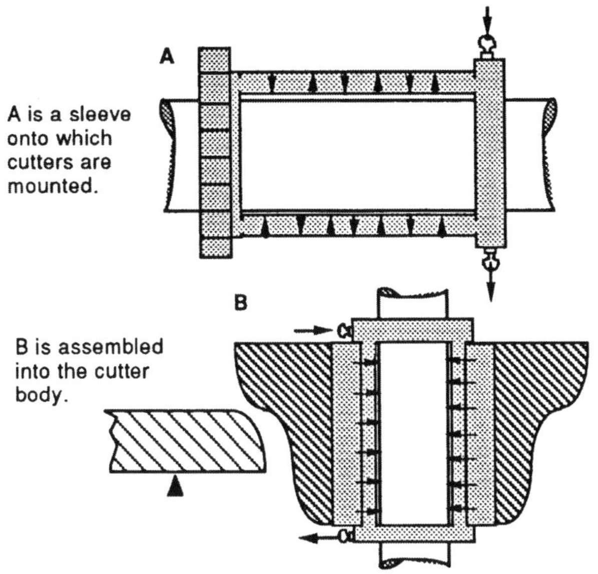

Illus. 7-9—Hydraulic clamping devices

Great precision of concentricity is possible because of a very basic law of physics, i.e., Bernoulli's principle: Gases and liquids displace pressure equally in all directions. As the cavities of the sleeve or body device fill and become pressurized, the movable walls flex equally in all directions. No matter what the difference between the inside diameter of the sleeve or tool body and the diameter of the machine spindle, the spindle will be held exactly in the center.

The advantage of having such precision of concentricity begins with feed rate. When we estimate what our feed rate is or needs to be, we are looking for both finish quality and the amount of work involved in the process.

It is likely that more horsepower is available than is needed, and the strength of the cutting tool is adequate. Finish quality is identified or predicted based on a theoretical number of knife marks per inch, assuming that all knives on the cutting tools are in exactly the same cutting circle. The theory of being able to predict finish quality for most woodworking processes worked well before elevated feed rates were demanded and machine tool spindle speeds were increased to where they are today.

A point about work and finish quality needs to be made clear: On a cutter having four knives, if all protrude from the head into different cutting circles, and the difference between each is only 0.002", the work would not pose any mechanical problem, though only a single knife, the longest one, would be controlling finish quality. Consequently, the finish quality would be the same if you removed the three shorter knives. Or stated differently, if you could get all fourknives in the same cutting circle, your feed rate as regards finish quality could be, theoretically, four times faster.

Gaining the benefits of the hydraulic clamping system begins with preparing the cutting knives—grinding the profile to be cut. The grinding machine is fitted with spindles identical to those on the woodworking machine, and the cutter is held exactly as it would be when in the cutting process. After grinding the knives, the tool is placed on the machine and put into its intended process.

The hydraulic clamping system can be adjusted to provide concentricity within 0.0001" repeatedly and in so doing increase feed rates and finish quality dramatically. The original machine processes that used this holding system were in the manufacture of solid wood trim, though now other machines and processes, even single-spindle shapers, use hydraulic holding and driving systems. The only drawback to the hydraulic clamping system is the initial high cost, but, given the advantages, the price will be justified in the volume of production, though, it is not likely to become a popular hobby woodworker's tool.

We must take care in every detail of securing cutting tools to the machine spindle and transmitting energy to the process. The greatest concerns are tolerance and conditions created by deviations of tolerance, i.e., vibration, irregular finish quality, balance, and concentricity.

Cutter Body and Knife Configurations

Development of various types of cutter heads sought advantages in reducing overall costs, making them faster and easier to use, adding to their versatility, increasing production, and making them more accurate. Some designs were significant and continue in use today; others proved to be less so and ceased to be produced.

Split-Collar Cutters

Progression of cutting tool development reflects parallel advances in other technologies, namely metallurgy and manufacture of steels and other alloys. A cutter head called a split collar utilizes straight, flat bars of steel, onto which the needed profile is ground. (See Illus. 7-10 and 7-11.) As originally designed this cutter was effective, but dangerous; with modification, it is still in use today.

The original split-collar cutter was designed for use on a spindle moulder, or shaper, and consisted simply of two thick, mating discs with a machined groove profile identical to the shape and size of the steel bar used as knife stock. Initial problems were keeping the knives at the same weight so balance was not a serious problem and ensuring that the retaining-bolt assemblies would keep the tool body from coming apart, which would lead to obvious consequences. Too often, the two-knife sets were not identical, and, even with correctly tensioned bolts, the tool assembly could not be held together.

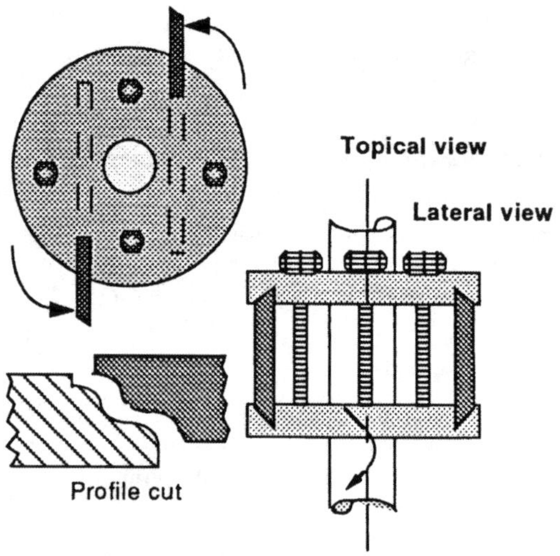

Illus. 7-10—Split-collar shaper head

Knife stock for this type of tool and others is manufactured in 25" bars from which small lengths are cut to make cutter knives. The details within each bar are very accurate but may vary slightly from one bar to another; hence, the problem. The toolmaker too often attempts to use small pieces of bar stock from different bars. The differences might be very small but sufficient, with the centrifugal force present and horsepower used, to make the tools fly apart.

An improvement in this type of tool was addition of a toothed profile on one of the knife-edge surfaces and better tolerance details in the bar stock from which the individual segments are taken. A couple of tool manufacturers provide a screw assembly with the cutter, allowing easier set up and balance of the cutting circle. The split-collar cutter head, as it is produced today, is

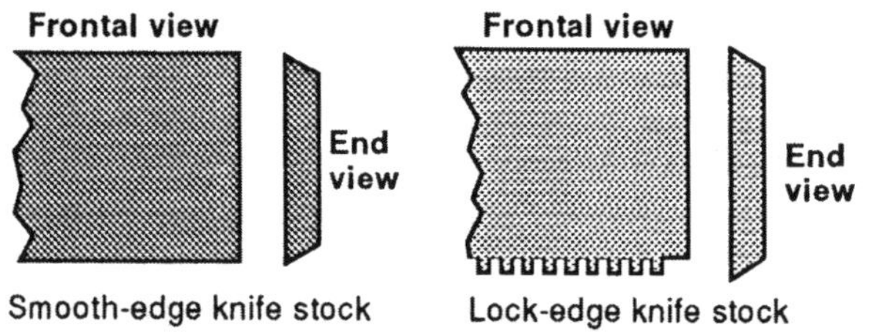

For the several types of split-collar shaper cutters that are produced, there are two basic knife shapes utilized: Smooth edge knives should be handled very carefully, observing all safety precautions, especially taking the knife stock from the same bar to ensure exact dimensions. Lock-edge knives provide a measure of safety, having teeth that fit into a mating profile to facilitate adjustment of the knives and better secure the knives in the mating collars. The knife stock for both types are available in numerous sizes, madeof different materials, or tipped.

Illus. 7-11—Split-collar shaper knife stock

popular for many wood-product manufacturers because it is inexpensive and a large number of profile shapes can be easily maintained.

Another feature of this tool that greatly increases its versatility is a tracer bearing on either the top or bottom segment of the cutter body, which makes production of irregular shapes easily possible with holding fixtures. Last, the tools have for many years been produced for either left- or right-hand rotation.

Historically, developments of cutting tools and those of machines were largely parallel, though occasionally it seemed that they were not completely so. That is, often developments in the machine technology did not seem to correlate with those in cutting tools, and vice versa. Putting obsolete cutting tools on more modern machines or placing very good cutters on old, inaccurate machines happened more often than finding any improvement for both simultaneously. The past 150 years of woodworking technology have been characterized by such repeated changes, first in one area, then in the other—a new cutting tool leading to a new machine process leading to a new cutting tool, and so on.

Moulders

The machine development that dramatically changed most of our concepts of shaping wood was the feed-through moulder. Basically, the machine can be thought of as several connected shapers, reflecting the earlier name for a shaper—"spindle moulder." This machine process produces long, profile-shaped wood parts, generally called trim moldings. These long stick parts are used as decorative trim rails or structural members of cabinets, case goods, or other furniture pieces. We'll describe the machine and its arrangement later in this chapter.

The first cutters for the feed-through moulders were versions of the square heads. These proved costly, cumbersome, and dangerous, and the search continued for solutions to these problems. The moulder cutter head used today retains the form it was given at the time of its development, though there have been significant improvements in the means to secure the knives.

Illustration 7-12 shows the basic features of modern moulder cutter heads. The differences reflect how the manufacturer elects to hold the knife stock in place, using several sizes and shapes of wedges and screws. All use knife stock that has a serrated profile either ground or milled into its back side. The knife pocket, where the knife is to be seated, has a mating profile cut into it. The wedge shape of the knife pocket further strengthens the tool, using centrifugal force to push the profiled surfaces more tightly together. The advantages of the modern moulder cutter head include:

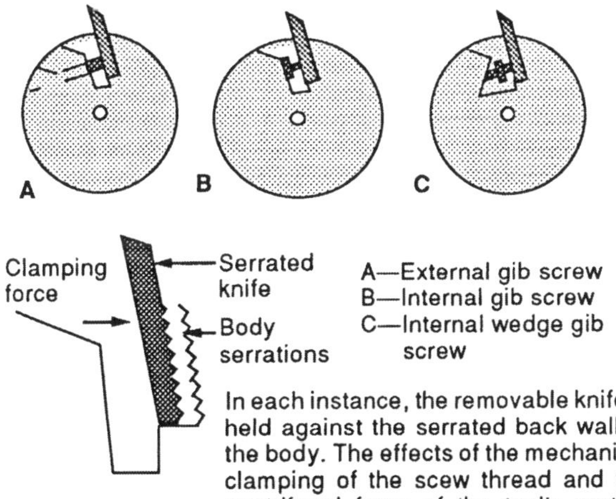

In each instance, the removable knife is held against the serrated back wall of the body. The effects of the mechanical clamping of the scew thread and the centrifugal force of the tool's motion keep the knife firmly in place.

Illus. 7-12—Moulder-shaper head configuration

- Numerous sizes and shapes of needed profiles can be maintained with only a few heads.
- Knives can be easily and quickly set up in the heads.
- Knife stock is readily available in several sizes and in different alloys or tipped with hard metals such as tungsten carbide.
- Tool bodies can be easily balanced by carefully weighing the knife bar stock as it is cut and ground.
- Knives can be sharpened several times and moved outward into the original cutting circle.

There are several manufacturers of moulder heads around the world, and each offers a slightly different version. The cutter bodies are now used in different forms and sizes in many different machine processes, but the basic mechanical principals remain largely unchanged.

Just as parallel developments take place in different technologies for woodworking, they occur in different parts of the world at the same time, which may spark some heated discussion regarding who introduced them. An example of this is the knife stock used in modern moulder head cutters. Apparently, American and European developments in the steel knife bar stock were parallel:

- Both European and American manufacturers use a serration profile on the back side of the knives to retain them in the knife pockets;
- Both space the serrations at 16 per inch, or 16 in each 1.5875 mm;
- The angle of American serrations is 60°;
- The angle of European serrations is 90°.

Safety Factors. The problem here is that having two readily available knife stocks for nearly identical cutter heads may pose a danger. (See Illus. 7-13.) There exists the possibility of mixing the knife stock used to prepare the cutter knives. While the difference of angularity in the serration profile may seem minor, especially since both are arranged at 16 to the inch, the

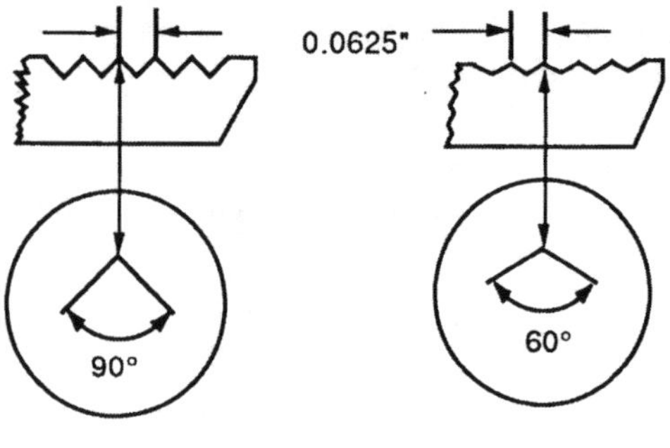

Illus. 7-13—Back serrated knife stock

problem should not be ignored. The danger of mixing unlike knife stock on cutter heads has three elements;

- The possibility of the knives being thrown from the head,
- Permanent damage to the cutter head, and
- Damage to the cutter knives. (See Illus.7-14.)

Given the amount of force exerted by tensioning the gib screws, the likelihood of no damage at all is remote. The softer material, either the knife stock or the cutter head, will be deformed by the harder material. Most likely, the knife material will damage the cutter head because cutter bodies are usually only surface hardened, while cutting knives are completely hardened to protect against abrasive wear. As in all other woodcutting processes, there are exact details to keep track of, and not mixing unlike knives and cutters is a good example.

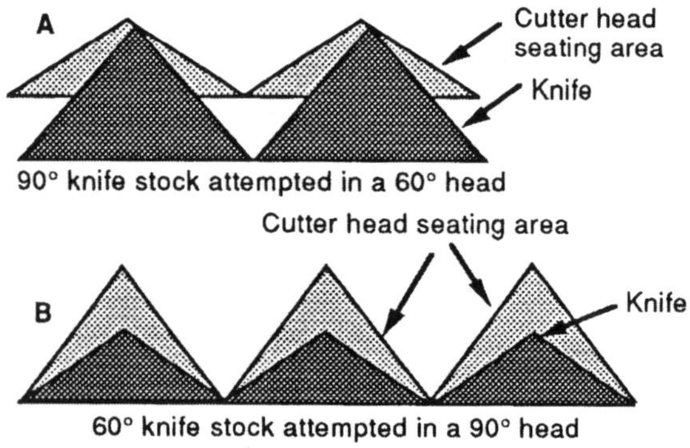

Confusing knife stock can be dangerous because the knife will not be well seated in the cutter body. In A the damage will be most likely to be to the knife. In B, the damage can be to the tool body. The best advice is to always to use a gauge to identify the angularity of the serrations.

Illus. 7-14—Angularity of serrations

As mentioned earlier, with all cutting tools that have cutting knives secured to the tool body by some mechanical means, how we tension and how much we tension the retaining screws is important for several reasons. There is most often a definite sequence for tensioning and a specific amount of force that should be applied to the screws to hold all the movable parts of a cutter body.

Not following the manufacturer's prescribed method of tensioning might lead to using force sufficient to distort the cutter head. If the screws are overtightened before placing the cutter on the spindle, it may not go on the spindle at all. If the screws are overtightened after the tool has been installed on the spindle, it may not be possible to remove it without special tools. If the sequence of tensioning the screws is incorrect and the head distorted, the knives may not seat correctly.

The mechanical advantage of a screw thread pattern is not something that we commit to memory, nor is it something that we use in our everyday work experience. But the force exerted by an inclined wedge—what a thread pattern really is—is tremendous.

In the example in Illustration 7-15, a 7/16" bolt having 14 thread turns per inch is being turned with 15 lbs. of force on a 10" wrench. The resulting force on the end of the bolt is 13,200 lbs., omitting friction loss. If we assume that each knife pocket has two screws, and the head has four knife pockets, the head, totally, has to withstand 105,600 lbs. of force just from the screws being tightened about the periphery at four locations 90° apart. The cutter further has to withstand forces created by the cutting process itself—the force of the energy turning it, the momentum of the weight of the cutter, the centrifugal forces acting on the knives, and

The formula to calculate the force (Q) exerted by tensioning a screw (in pounds of force):

$$Q = \frac{F \times 6.2832 \times R}{p}$$

Q = pounds force
F = force exerted on the lever to turn the screw
6.2832 = the mechanical advantage multiplier of an inclined plane
R = radius of the lever
p = thread pattern pitch (1 divided by the number of threads per inch). In the example, 1 divided by 14 = 0.0714".
The force in the example is 13,200 pounds.

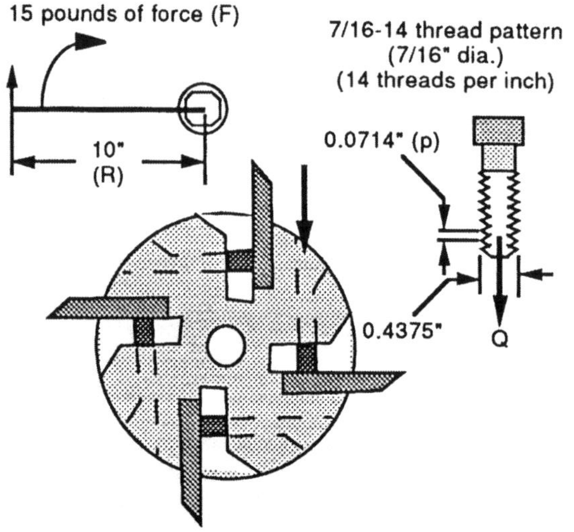

Illus. 7-15— Tensioning force calculation

the impact or interrupted contact with the material being cut.

The best methods to secure movable parts of any cutter body are those recommended by the manufacturer. There are no practical, safe shortcuts to these procedures. There cannot be enough emphasis placed on safety. (See Illus.7-16 and 7-17.)

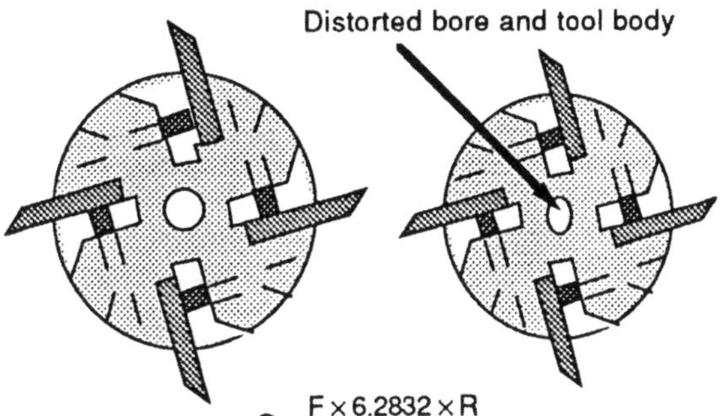

$$Q = \frac{F \times 6.2832 \times R}{P}$$

There is tremendous mechanical advantage in the tensioning of a common screw thread. With a 7/16-14 thread, a 10" wrench, and only 15 pounds of force, we can exert 13,200 pounds on the gib screws and wedges that hold knives in a moulder/shaper head, neglecting friction. Further, if there are two screws for each knife pocket and four such pockets in the cutter body, we can exert a total of 105,600 pounds of force on the cutter body itself. If the force on the wrench were doubled to 30 pounds, there is possibility of distorting the body of the cutter itself, and, too often, that is what happens. There has to be considerable care given to both the sequence and amount of force applied to the screws that hold any cutter body part into an assembly.

Illus. 7-16—Gib screw tensioning

On most larger industrial woodworking machines, the cutting tools are completely enclosed inside elaborate covers and safety devices. We still have to be mindful, however, that there is no completely safe cutting tool. We cannot ignore the fact that we are working with very large horsepowers and mechanical forces that have to be carefully controlled.

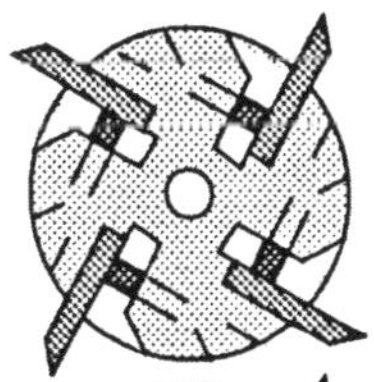

A correctly set-up moulder-shaper head will have knives of exactly the same weight. The screws will have been sequentially tightened to the correct tension. All knives will have been ground to cut in the same cutting circle.

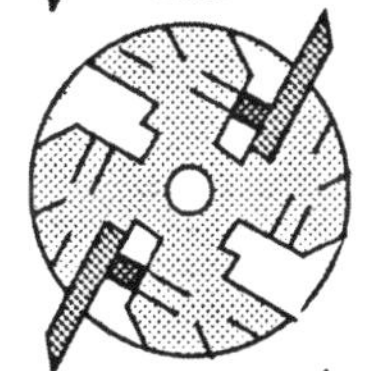

To run a cutter head with some knife pockets left empty invites the possibility of deforming the tool, both by the tension of the gib screws and impact during the work it performs.

If there is the need to run the tool with only two knives for a short time for limited production, always fill the knife pockets with blind knives to protect the cutter body from deformation. The blind knives have to be exactly the same weight.

Illus. 7-17—Moulder head set-up with gib screws

For smaller woodworking machines, the dangers are even more pronounced because the tool spindles are often exposed to the operator. A well designed, safe cutting tool will afford no protection if set-up and use procedures are not followed completely. My own examination of injuries in the woodworking industry over the past twenty years, shows that, too often, an error is the cause, with equipment failure second, and the cutting tools themselves a distant third or fourth as causes.

Cutting circle and cutting angle. The cutting circle and cutting angle on both movable- and fixed-knife profile cutter tools are unlike those on other tools. For simple, straight cutting knives like those on planer or jointing cutters, the cutting angle is fairly straightforward. For a planer head or any cutter body that uses movable or exchangeable knives, the cutting angle is dictated by the angle of the knife pocket and measured at the periphery of the tool body. Planer and jointer cutter knives do not protrude very far from the tool body, so the cutting angle does not change measurably within the knife itself.

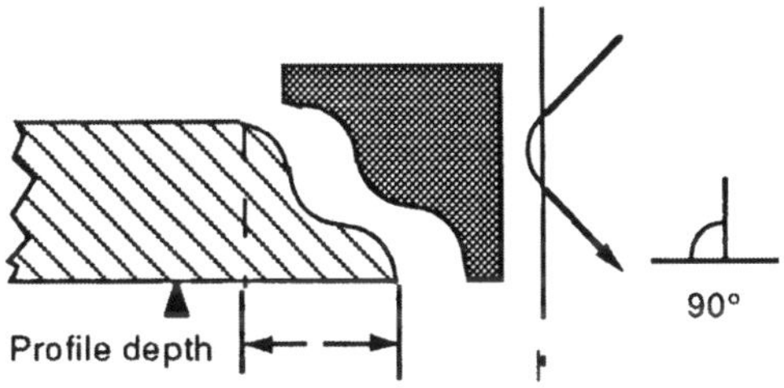

For most cutters with adjustable or exchangeable knives, the cutting angle is identified as a particular angle (between 12° and 20°) measured at the periphery of the tool body, not at the farthest diameter of the cutter knives. As the knives protrude farther from the tool body, the cutting angle becomes more acute. This is true even in the same knife that cuts a very deep profile.

Illus. 7-18—Profile cutting with adjustable/exchangeable knives

But for tools that cut profiles, whether with fixed or movable knives, the cutting angle changes all along the knife as it protrudes from the body of the cutter. (See Illus. 7-18 and 7-19.) Remembering that cutting angle is measured from the center of the tool body to the periphery, as the knife extends farther from the tool body, the cutting angle becomes more acute. For most materials this is not a problem, except for very hard, dense

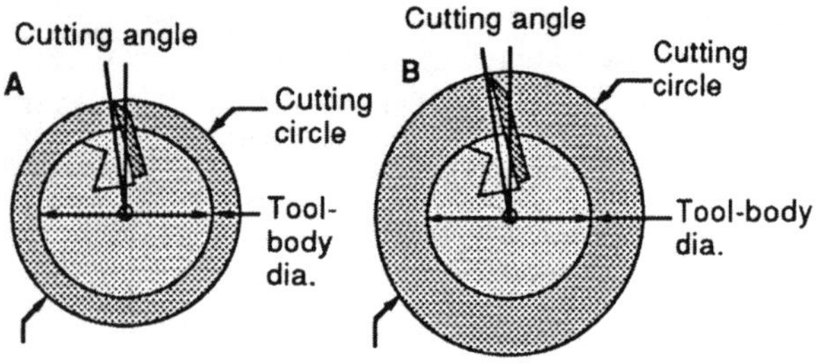

Illus. 7-19—Cutting angle and knife projection

solid woods that cut better with shallow angles—oak, pecan, hickory—and some man-made panel materials.

Another point that is not often made clear is the difference between body diameter and cutting circle, or as many cutting circles as there may be on a particular tool that have to be identified and managed. In setting up and cutting complex, deep profiles, there may be several individual circles to measure and control. (See Illus. 7-20 and 7-21.)

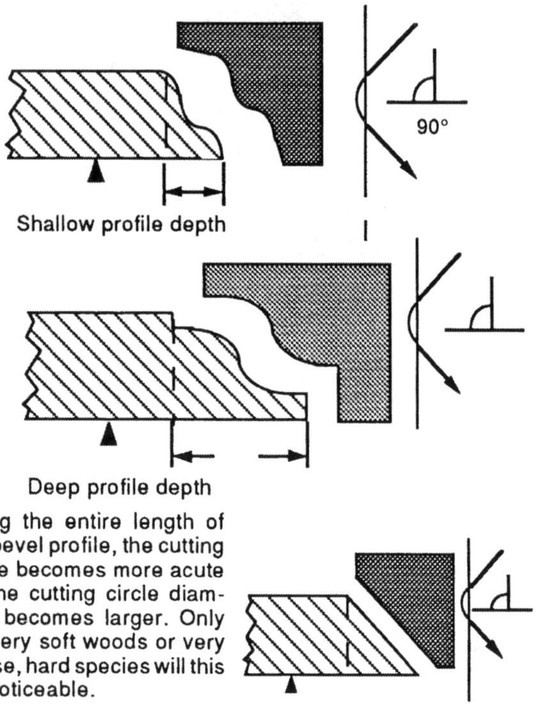

Along the entire length of the bevel profile, the cutting angle becomes more acute as the cutting circle diameter becomes larger. Only for very soft woods or very dense, hard species will this be noticeable.

Illus. 7-20—Profile depth and cutting angle

In Illustration 7-21, a simple extended ogee shape is such that we have to know both the major and minor cutting diameters to control the size of the finished part. For other sizes and shapes of profiles, there are numerous set-up and operating procedures that require that different portions of the tools and machines be carefully defined. Tool-body diameter is just that. Cutting circle has several definitions that always include at least the smallest and largest diameters. We may additionally need to define particular portions of it along the profile, such as steps or other prominent features that are an integral part of the larger construction.

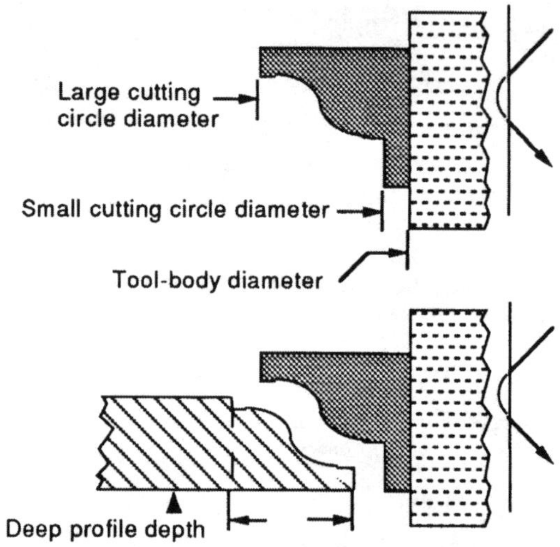

Illus. 7-21—Major and minor diameters

Tool Arbor Manipulation. The next point is not exclusively related to profile cutting, but should be illustrated and explained for several reasons. When deep profiles are cut on any wooden part we most often view the shape in a particular way, based on a preconceived idea of how the cut can be performed or how it has been done in the past. We also view tasks or portions of tasks based on our experiences with tools and machines that are familiar to us, and we pursue the work with only what we have seen or done before, that is, not exploring the possibility of finding a newer or faster or better way of completing a project.

Of all woodworking machining processes, cutting deep, wide profiles is among the most difficult. Often, parts have to be machined several times to complete all of the feature details, especially on smaller machines that do not have several tool spindles, such as a shaper. But if we could manipulate the tool spindle and alter the cutting tool configuration, the work could be completed easily in a single operation. (See Illus. 7-22.)

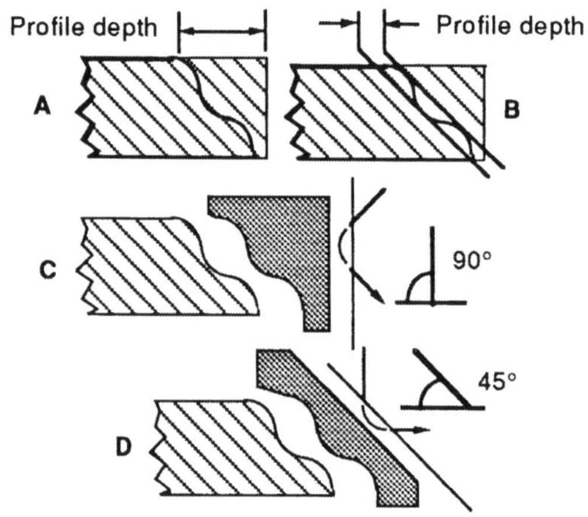

The drawings above illustrate the fact that much deeper profile details can be cut, even with cutting tools having the same diameter, by tilting the machine arbor. The advantage is both in cost and in the safety of the machining process.

Illus. 7-22—Tool arbor manipulation

In drawings A and B, the profile is identical and the only difference is that we have measured the depth of each, from either 90° (A) or 45° (B). In drawings C and D, we have the same profiles as they would be cut with the machine tool spindle at two different orientations. As can be seen, there could be tremendous advantage to tilting the machine tool spindle to 45°:

- Tool diameter could be much smaller, exerting less centrifugal force.
- The smaller tool will most likely cost less.
- There is a measure of safety in turning a smaller diameter cutting tool.
- This profile and similar, even deeper profiles can be accomplished in a single operation rather than several.
- Smaller diameter would allow higher rpm if that would be an advantage to the particular process.
- This manipulation of the machine tool spindle does not adversely affect any part of cutting the profile shown.

While tilting the machine spindle is a means to accomplish a difficult process, we can find other improvements in any woodcutting process if we do not bind ourselves to doing everything the same way all of the time.

Feed-Through Moulders

As mentioned earlier in this chapter, the woodworking industry is becoming more machine intense and less labor intense, in part by relegating more processes to complex machines. Of the machines that have become more or less industry standards, the feed-through moulder is probably the most prevalent, followed by the tenoner and the combination edge bander/tenoner—machines to be explained later.

The feed-through moulder is a system in which the wood stock is fed completely through the machine to accomplish several machine processes. Essentially, the machine takes the place of several single-spindle shapers and operators, complete with all the holding, clamping, guiding, and driving systems.

The wood stock begins as a long rectangular part and, after machining, may be any of perhaps thousands of unique profiled shapes used for almost countless purposes. Moulded stock is that trim found on cabinets, furniture pieces, and case goods; around doors and windows; or in the construction of any of the products mentioned.

Feed-through moulders, together with the cutting tools designed for them, have increased the productivity of the woodworking industry many times over. A skilled worker with a single-spindle shaper could in a day produce perhaps 50 moulding parts—after several set ups using as many cutting tools and holding devices. During a similar period a feed-through moulder could produce, after set up and tooling, the same moulding at a rate of 20' to 60' per minute on average. The advantages of such technology are not difficult to understand.

Illustrations 7-23 and 7-24 show sequentially what may take place during the processes inside a feed-through moulder. Initially, it must be said that there are other ways to cut the profile in the example—using fewer

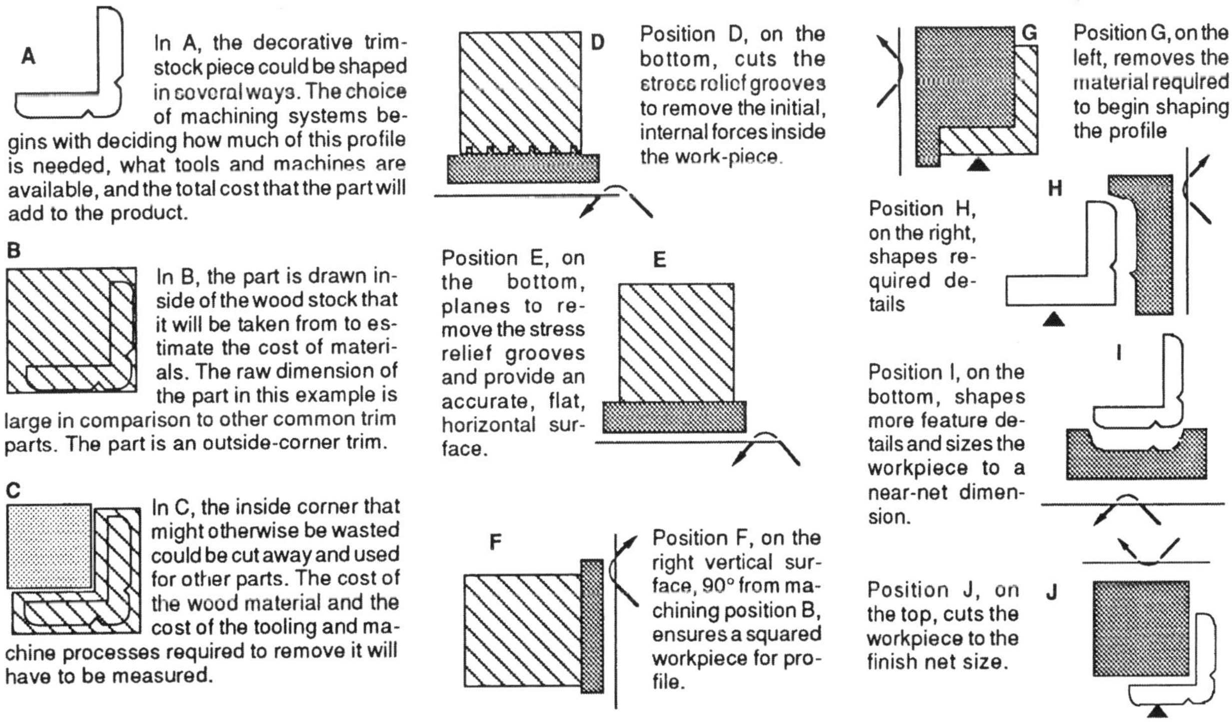

Illus. 7-23— Feed-through moulder machine processes

World wide there are more than 40 manufacturers of feed-through moulders that produce several models for specific tasks. The particular arrangement of tool and motor spindles will vary somewhat, though current technology often places a bottom cutting tool first, followed by the right-hand then left-hand cutting tools. Not all manufacturers produce the grooove bed and groove cutting systems for straightening and stress relieving. On the machine at right:

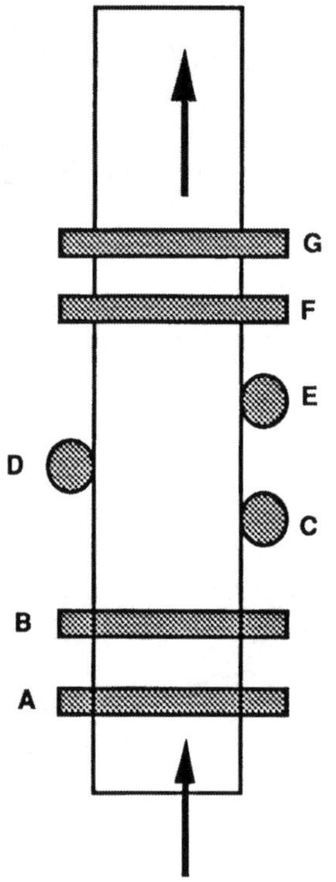

A—Groove bed cutter
B—Bottom planing/surfacing
C—Right-hand profiling/jointing
D—Left-hand profiling/jointing
E—Right-hand profile cutting
F—Bottom profile cutting
G—Top profile cutting

Additional spindles could be used for more complex profiles or multiple parts. Feed-through moulders can have as few as four cutter spindles and as many as 10 or more. Total horsepower will vary from as little as 20 to over 150 on very large machines.

Illus. 7-24—Feed through moulder assembly

cutting tools perhaps in a different orientation and sequence. The illustrations, however, show both the possible size of the piece and complexity of the processes.

The reason for showing such a machine process in this text is to illustrate the importance of cutting tools in such a complex environment. The cutters have to be as precise and trouble free as possible to take advantage of the machine's abilities. If, for instance, the cutting tools used on a feed-through moulder required extensive maintenance or were inaccurate, the machine processes would be hampered considerably. All of the technology has to work in unison.

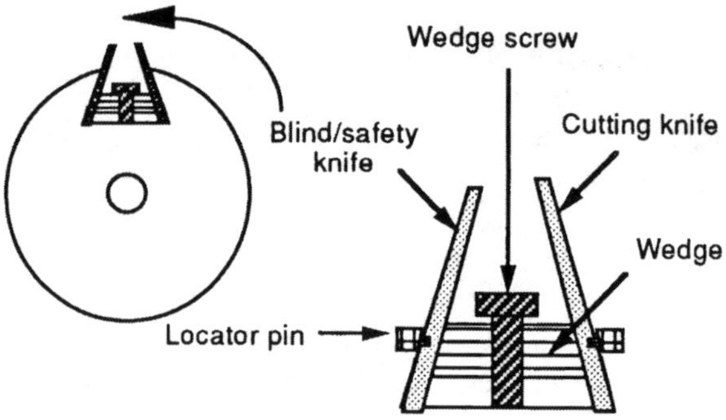

There are several sizes and shapes of cutters called universal cutters. Their features make them applicable to several types of woodworking machines: shapers, moulders, profilers, and others.

Illus. 7-25—Wedge and knife assemblies

Universal Cutters

The next family of movable knife cutters that is gaining favor has unique features especially for smaller manufacturers who do not wish or have the necessary facilities to produce their own cutting knives. These cutters are generally called *universal cutters.* (See Illus. 7-25 and 7-26.)

The knives are held in the body by a unique inverted-wedge system. As the wedge screw is tensioned, the wedge is lifted into the tool pocket, securing it firmly, with further assistance from centrifugal force when the tool is in motion. The knives are placed in the tool pocket and set vertically on locator pins in the tool body, ensuring the accuracy of the cutting circle.

The choice of profile shapes varies from one manufacturer to another, with perhaps as many as 50 standard shapes offered. There is additionally the possibility of having special shapes to suit the particular processes or products. Custom shapes are limited only by the size and shape constraints of the cutter body.

Among universal cutters, there are two basic types manufactured for cutting shallow or deep profiles, though both may have equally complex shapes. In one case, the tool body is shaped simply as a cylinder, with straight, parallel peripheral walls. The cutting knives

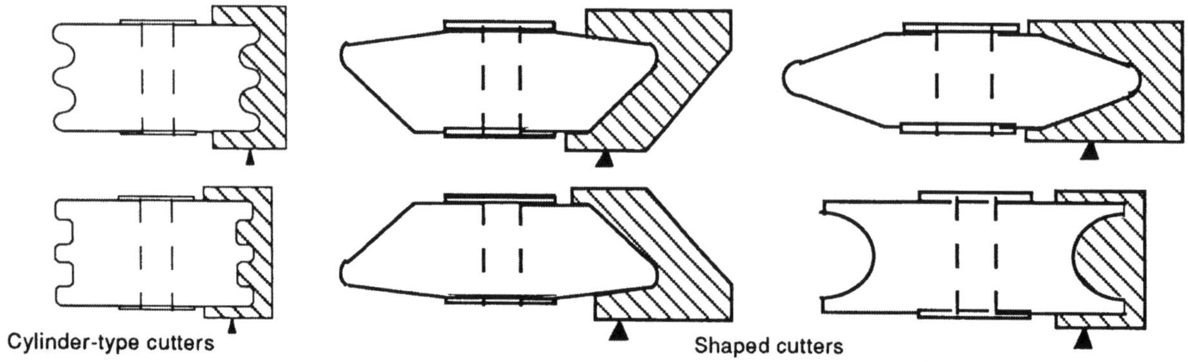

Shaded areas are the possible cutting zones for each tool

Illus. 7-26—Universal cutter heads

are shaped for the profiles and held mechanically. Without support for the knives beyond the periphery of the body, this tool configuration cannot accomplish very deep profiled shapes.

In the second type, the body is shaped to the form that the cutting knives will eventually cut. With this, the knives can be given more support and perform deeper profiles and cut on both the periphery and edges of the tool body. (See Illus.7-26.) The shaded areas in the drawings represent the profile shapes that can be cut by the particular tool. The shaped bodies give greater support to the cutting knives and, hence, can produce deep complex profiles without difficulty.

Profile/Counter-Profile Cutters

There are two complementing cutting tools that do not fit any earlier described movable-knife tool category. These cutting tool systems were developed to satisfy the needs of mill work producers, manufacturers who cut complex window, door, and decorative linear and end profiled parts, especially those that make up what are profile and counter-profile details.

The greatest concerns in this type of manufacturing are that the tools are costly and difficult to set up and keep in a correct size relationship to each other, especially important because many of the tools cut mating profiles such as window and door parts. (See Illus. 7-27.) It is not possible to say which type of wood-profile cutting is most complex, but profiles that have to fit into a similar mirrored image, i.e., mating profiles, are all difficult.

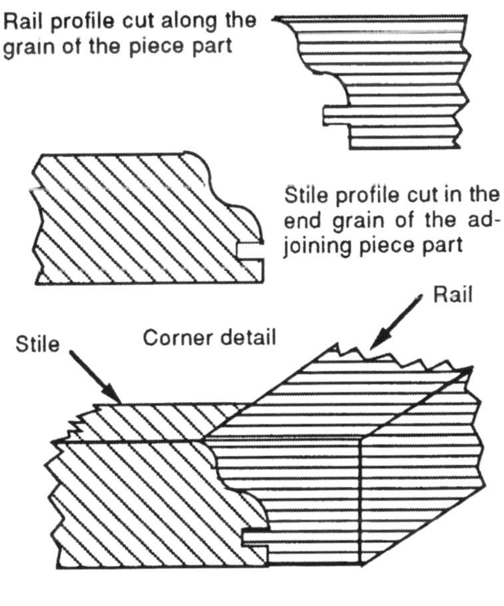

Illus. 7-27—Profile and counter profile

The complexity of cutting profile/counter-profile details is more than just keeping all of the features correct so they will fit together. Another critical need is the size details of this cutting procedure; that is, the fit of the feature details would be lost if there is no way to tell, for instance, how long each part of the assembly is. The size of the wood part produced determines where the feature details begin and end, and can in part be controlled by maintaining cutting tools having exactly the same diameter.

With other movable-knife tools, the cutting circle diameter is adjustable and can be duplicated with each sharpening, though there is a fairly costly time element involved. This is especially true if the knives have to be removed from the cutter head, creating two planes to locate and adjust.

There have been many ingenious machining systems and cutting tools devised to accomplish the very complex details in the construction of windows, doors, and cabinets. Most noteworthy of the features in these products are the details of corner-joint assemblies and the wide array of profiles and counter profiles used in their manufacture. The corner joints are made up of complex profiles and deep tenons that are difficult to cut by any means and present problems that are not easily solved. (See Illus. 7-28 and 7-29.)

The two major problems, as mentioned earlier, are the actual size of the parts produced and the mating of the complex profiles. In sizing the parts, all the cutting circle diameters of the tools have to be controlled in relation to the intended feature details. As with all complex cutting tools, there are probably several diameters to keep track of as well as mechanisms that hold the wood as it is being cut.

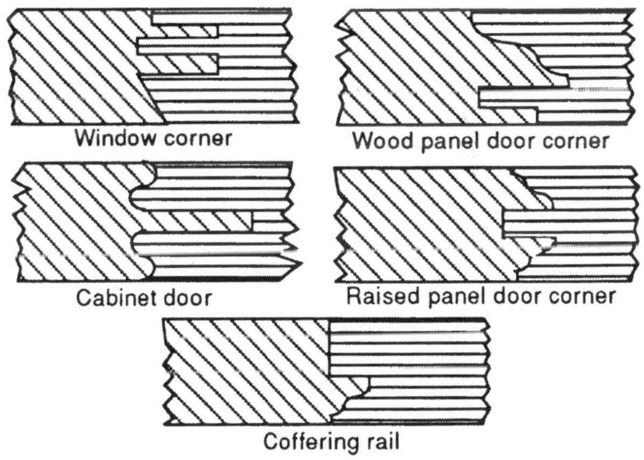

Illus. 7-28—Door and window detail features

Illustration 7-29 shows a portion of the details that have to be maintained for the two mating parts to be produced on an ongoing basis. In some areas of manufacturing, the details of products may not change for many years, and the profiles will have to be duplicated accurately hundreds of thousands of times. For the mating profile and counter profile in the illustration there may be as many as 50 individual details to keep track of on the machine, on the cutting tools, and within the material being cut if more than one species of wood is being used.

Keeping track of a large number of details can be simplified first by reducing the number of variables

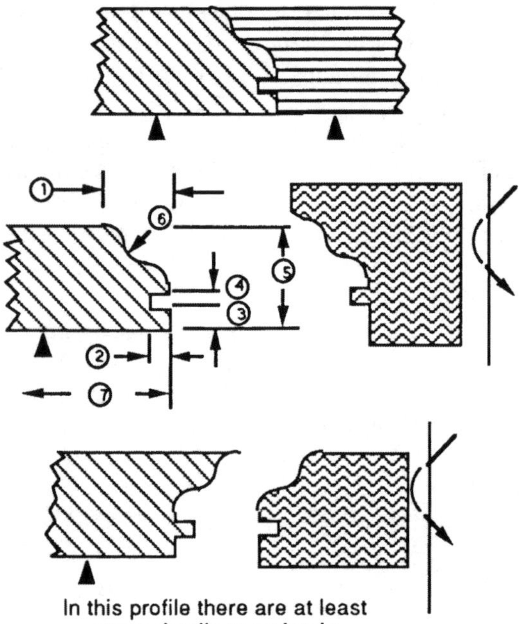

Illus. 7-29—Details to maintain

present, then by grouping the details into manageable units—something that is easier said than done without careful engineering. Manufacturers of complex wood product parts do, in fact, maintain all of the details with a combination of dedicated machining practices and unique cutting tools.

Of the cutting tools developed in the past 50 years, there are two that address the most difficult problems encountered in production of windows, doors, and cabinets, i.e., complex profile/counter-profile mating parts and deep mortise and tenon details. These are the face-located stop and the back-located stop systems.

Face-located stop. As mentioned earlier, to simplify a collection of details, it may be possible to gather them into a manageable group. (See Illus. 7-30.) The tool illustrated gathers the cutting details of a given profile onto a single knife (or if very large, a number of knives), and, as it is sharpened, the knife segment can be adjusted along a machined track to return to the exact cutting circle as when new. The tool knives are only face sharpened, and, throughout their entire sharpening life, do not lose any feature detail of size, shape, or location. There is, then, no change of any characteristic at all: Tool height remains constant; all cutting circles remain fixed; and all features remain as when new.

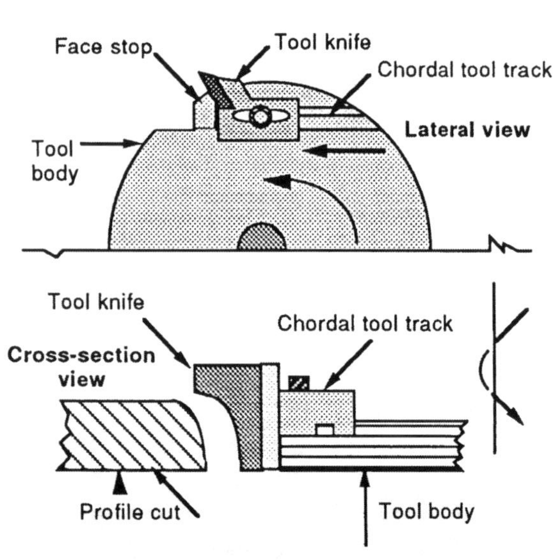

Illus. 7-30—Face-located stop

There is a significant cost to such a tooling system because each cutter is very costly to produce and is unique. The cost, though, can be easily justified in long production runs that do not often call for changes in details, as with doors, windows, and cabinets.

Having all of the fixed details of cutting circle, profile features, and sizes reduces not only machine set-up times but time for other manufacturing processes as well. There is, for instance, enormous time lost in inspection, assembly, and finish preparation of parts that are not manufactured to precise sizes.

The movable-knife face-located stop tooling system works very well for any profile feature, from simple radius details to the most complex blending of several detail features together. The knife segments are manufactured for a particular profile and are attached to a carrier body. The angle of the chordal track on which the knives are located is the result of the required cutting angle, back relief angle, and strength of the material to be cut.

What makes all of this possible is production of the knife segments with all necessary relief angles present to cut whatever complex detail is needed. The individual knives are essentially like those for any other profile cutter, except that they can be moved outward as sharpened.

Back-located stop. In Illustration 7-31, the cutting tool principle is similar, but, instead of having the knife segment rest against a face-located stop, the cutting circle is controlled by a stop on the back of the tool knife. The purpose of this is to allow very deep mortise and tenon details to be cut while maintaining all the necessary details of cutting circle diameter and cutting depths and widths. As with the face-located stop tool, the individual knives are arranged and held to the tool body mechanically. Unlike the face-stop system, where the knives are on a grooved track, the knives are carried in a groove cut into the tool body itself to reduce the thickness of the knife and the entire width of the assembly. Simply stated, the face-located stop cutting tools cut essentially on their periphery, as with radiused profiles; the back-located stop cutters cut both on the periphery and edges, as with deep grooves, tenons, and mortises.

The cost of the back-located stop tools is their only disadvantage. They are custom-made, one-of-a-kind cutting tools that, though costly, can be easily justified

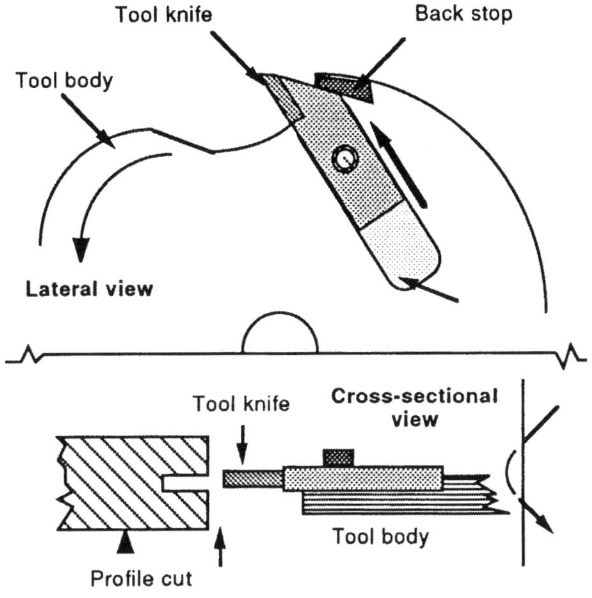

Illus. 7-31—Back-located stop

in the labor and material savings they offer to the large manufacturer.

Illustration 7-32 shows how very deep tenons could be cut with the back-located stop tool. Other knife arrangements are possible that will perform the same processes. The knives could be brought to the cutting circle from different angles, or they could have features that would not require the lateral approach shown.

The virtue of this cutter system, like the face-

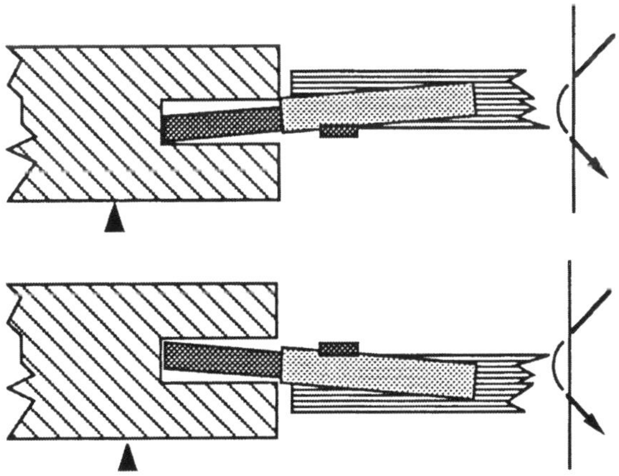

Inside corner details can be preserved accurately by having individual knives on the same cutter do a different portion of the work. These drawings illustrate individual knives on the same cutter in the same process.

Illus. 7-32—Knives on same cutter

located stop tool, is that all the feature details remain constant. Further, after the knives are completely worn from repeated sharpening, only the knife segments have to be replaced and not the carrier body.

The face-located and back-located stop cutting tools have a number of commonalities:

1. Both are custom, one-of-a-kind tools produced for the manufacturer, and one-of-a-kind anything is expensive.
2. Both are easily cost justifiable at a certain level of manufacturing. The cost containment lies in large-volume production, not small or custom processes.
3. Cutting knives for both systems are easily sharpened, though the procedure requires unique fixtures or holding devices.
4. Spare knives eliminate any loss of production time.
5. Only the knife segments need ever to be replaced; there is no wear in the tool carrier body itself.
6. All the required details remain constant, greatly reducing the costs of set-up and tool preparation and manufacturing.
7. There is no need for back-up or spare cutting tools, further reducing the total cost of this particular system of production.
8. Within a narrow range, the tool carrier bodies can be made to perform more than one cutting process, though this would have to be negotiated with the manufacturer.

In this list, there is a point that should be pursued. Item 2 states that this group of cutting tools is cost justifiable at a certain level of manufacturing and perhaps not at others. What this means is that, in industry, many machining systems and cutting tools can only be justified within a large manufacturing facility. A hobbyist or craft woodworker could not justify a cutting tool that costs several thousand dollars and is capable of cutting hundreds of thousands of feet of production. Corresponding to many cutting tools used in industry, there are smaller, less complex, and less costly ones available for smaller woodworking facilities, though movable-knife, fixed-cutting-circle tools do not have such a counterpart on the market as yet.

Where movable-knife tools have their greatest use is in production of edge-detail profiles on several machine types. The virtues of these cutters are that they allow woodworkers at all levels of manufacturing to use, control, and maintain a large number of profiles for their products. From the earliest cutter bodies, there have been improvements to every facet of their use. The result is that woodworkers can better afford short production runs and irregular work schedules with less labor cost and better machine time utilization.

As the true cost of labor in the manufacture of a wood product rises, there will be more changes to the cutting tools and machines used. With rising costs of the materials, cutting tools, and machine processes, the pressure of market competition will require more efficiency yet.

Selection and Use of Movable-Knife Cutters

Suggesting a check list of tool-selection criteria for woodworkers assumes that we know the technical and manufacturing level of all our readers, and that is not true. Allowing then that a certain number of you are students, another portion are hobbyists, another yet are craft workers, and the remainder are employed in industry, here is a list that should be of use to all of you.

Movable-knife cutting tools are among the most versatile and cost effective of all cutting tools in manufacturing processes, but there are basic guidelines that need to be observed in their selection and subsequent use:

1. Be aware of all the mechanical features, including:
   a. how the knives are secured in the tool body;
   b. that the safety features are clear, precise, and effective;
   c. that the means to align and hold the knives accurately in the cutting circle is clear, precise, and quickly accomplished;
   d. that the gauges or other set-up devices are readily available, fast, and easy to use;
   e. that the necessary spare parts such as screws, wedges, gib bars, and the like are readily available;
   f. that the knife stock in needed sizes, shapes, and quality is also readily available;
   g. that factory-made profiles, if you purchase them, are stocked for your use.
2. Be certain that the tool selection is guided by:
   a. the tool's ability to perform the needed work, that is, be aware of the limitations of the tool in question;
   b. the tool's fit with all the physical characteristics of the machine on which the process will be done;
   c. the cost justification of the process and product within the limits under control;
   d. the operating life expectancy of the cutting tool.
3. Understand all of the characteristics of servicing the cutting tool, whether serviced "in-house" or at the vendor. All times and costs have to be carefully examined.
4. Be certain that the machine process and cutting tool are correct for each other. Very small machines will not accomplish large cutting processes. Limits of space in this text do not allow us to list and identify every type and style of movable-knife cutting tool available. We have illustrated the features of the most common, and, beyond certain precise features, most will approximate what we have shown. Features not readily available for examination that we have to be aware of when purchasing cutting tools are characteristics of
   a. concentricity,
   b. balance, and
   c. the metallurgical character of the materials that make up the tools.

The best advice is to deal with a large, well established tool manufacturing company. In selection of a versatile tool system such as a movable-knife cutter, our level of manufacturing has to be a primary guideline. That is, we have to identify, qualify, and quantify our needs. If the first concern is for a cutting tool that will be set up and run for long periods of time, then that should drive the selection process.

But if our need is for a cutting tool that will be set up and run several times each day, we have to pursue that criterion. If, for another example, we have several dozen unique profile shapes that are run each week or month, we have to select cutting tools that will allow this within the cost-containment guideline of the product that we manufacture. Movable-knife cutting tools will address all of these examples easily, with the assumption that we carefully pursue our selection and use the best cutting tools.

There are other guidelines that will help us to choose the correct tool for the process; the best value in terms of function to cost; the best value in terms of versatility, precision, and effectiveness; and the best technology available at the time of the purchase:

- Investigate all of the suppliers available in your area. There are several hundred tool manufacturing companies worldwide, ranging in size from two or three employees to over 4,000. Take the time to understand your needs and find the tool manufacturer who can provide the tooling to accomplish your work.
- Purchase from knowledgeable suppliers who can provide technical support for the products they sell.
- Do not rush the selection of cutting tools and machine processes. There may be alternatives that will simplify or reduce the cost of the production process.

Cutting tools are an expensive, integral part of your ability to manufacture anything. You have the right to expect excellent service. Too often, woodworkers are sold a cutting tool that makes dramatic promises that never come true. There are thousands of examples on tool room shelves in shops everywhere. A part of this lies in a basic premise of salesmanship. The salesperson's primary objective is to make a sale; the woodworker's objective is to make improvements in his work. If both can be satisfied, all the better; but the salesman usually has the upper hand and is probably not a wood-worker.

Get as close to the manufacturer of your cutting tools as possible, through either a representative or an authorized sales agent.

Chapter 8
Cutting Profiles

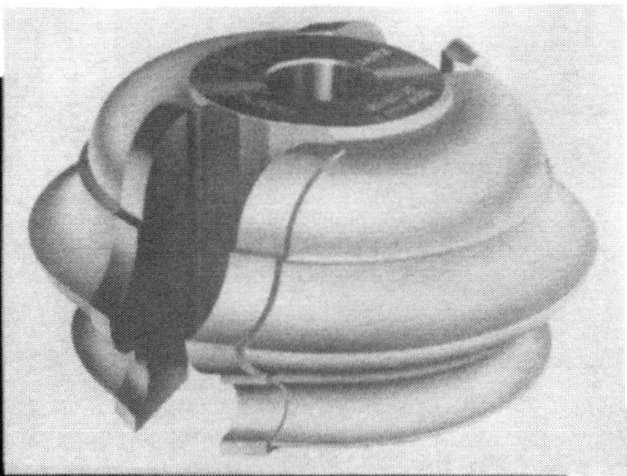

Photos courtesy of Leitz Tooling Systems, Inc., Guhdo USA, Inc., and Stehle GmbH & Co.

Cutting Profiles

Shaping wood into decorative profiles is the most complex and expensive of all machining processes. The cutters are more costly to produce and maintain, and there are more types of cutting tools and machining systems available to perform the tasks. In this chapter, we will examine the features of cutting tools that are used to produce radius shapes and the problems that must be overcome to accomplish the work as an ongoing manufacturing process.

Design elements of case goods, cabinets, seating, and all types of furniture call for an almost endless collection of unusual profile shapes. While toolmakers produce numerous standard cutting tools, over 30 percent of cutters in many furniture manufacturing facilities are unique. The implication of this is the critical need for accurate communication between toolmakers and the woodworking industry.

Furniture designers, woodworkers, machinery manufacturers, and toolmakers have complex, busy, and never completely satisfactory relationships. Each has a perspective on what is needed, reasonable with regard to performance characteristics, mechanically possible for the machinery or cutting tools, and so on. Only rarely are all in complete agreement. The following information must be communicated between toolmakers and woodworkers:

- Woodworkers need to know the capabilities, performance characteristics, maintenance requirements, overall costs, and benefits of a given tool type over another.
- To produce what is needed, toolmakers need to know the nature of the machining process, the machine details, the characteristics of the material to be cut, how much of the material will be cut in a particular time frame, and other similar requirements.

The decorative edges and surfaces of furniture pieces are what place the particular product at its position in the market and give it identity. The preservation and duplication of classical shapes is almost an industry unto itself. Each year, manufacturers introduce small and subtle changes to their products that may affect their success for many years, making the shaping of decorative details the single most important manufacturing process that they pursue.

We cut wood into useful, functional, and attractive products by any of five basic processes:

1. We rip cut material into long flat parts such as planks or panels.
2. We cross cut planks or panels into specific lengths.
3. We shape angular mechanical joint surfaces:
   a. rebates,
   b. grooves across or along the length of the part,
   c. planing and jointing,
   d. mortise and tenon shapes, and
   e. mitered shapes on the ends or sections of parts.
4. We drill holes for such things as mechanical fasteners.
5. We cut decorative, complex shapes on the ends, edges, faces, and other locations on parts of products.

Of all five, the last is by far the most complex, costly, and difficult. Additionally, there is a greater variety of tools for cutting decorative shapes to choose from, as regards cost and utility. There are more types of tool construction for such tasks, more mechanical and geometric considerations to be made, and more cost involved than for any of the others. Given the complexity and cost of cutting profiles and decorative shapes, a close examination of tools and machines is necessary. To be successful, cutting tools and machine processes have to meet criteria regarding the following:

1. Cost added to the final product;
2. Cost and yield of the material being cut;
3. Limitations of the available machine process;
4. Expenditure of time in the machine process;
5. Initial cost, operating/service cost, and total cost of the cutting tools;
6. Availability of resources.

Selection of the correct cutting tool for a specific process is critical, often with little room for error. For woodworkers, choosing cutting tools is getting more involved. Earlier, they may have asked simply, "How do I accomplish this?" Today several questions are necessary: "What machine process will accomplish the task? How much of this needs to be done, for how long, and how often? What is the exact nature of the material to be cut? And, What are the cost and service constraints of any particular cutting tool?"

Toolmakers are critical to woodworkers' effort in every way. They must have complete understanding of what the woodworkers' needs are. As an extreme example, we would not want to spend several thousand dollars for a cutting tool that will be used only occasionally. Conversely, we would not select a frail, inexpensive cutting tool for turning out thousands of feet of production every day with material that is difficult to cut.

Needs of the Toolmaker

Toolmakers need three distinct groups of information from woodworkers to provide them with appropriate cutters. They need descriptions of:

- The machine and machine process,
- The tool, and
- The material to be cut.

The Machine and Machine Process

Given the wide range of machines available, woodworkers must provide toolmakers a complete description of the following when selecting tools for their manufacturing processes:

- Machine type—shaper, table saw, router, tenoner, feed through moulder, etc.
- Machine brand name—the toolmaker may already be familiar with it.
- Work station (where tool will be used)—if the machine has more than one tool spindle/shaft.
- Tool holding/driving method:
  - spindle diameter—if more that one, list all.
  - spindle characteristics—external threaded, internal threaded, smooth, with or without keyway, size of keyway, clamping length, etc.
  - shaft size(s) for drills, routers, and similar machines—list all that are possible on the machine, e.g., 1/2", 1/4", 16 mm, etc.
- Performance characteristics:
  - horsepower available with the tool spindle or shaft to be used.
  - spindle/shaft rpm—if variable, list all.
  - feed rate and method.
  - tool shaft/spindle attitude—is it tiltable or moveable?
  - number of tool shafts/spindles on the machine.
  - means of holding the material to be cut.

It is important that the toolmaker completely understand the machine and its abilities, especially for the manufacture of profile cutting tools. Most often, the type, brand name, and age of the machine are the primary needs of the toolmakers, but there are many special machines that may need further identification.

The Tool

As regards the description of the needed tool, the suggestions below are intended only as a general guideline, and more information may be necessary. Too much information will never be a problem:

- Tool type—router, drill, shaper, saw, groove cutter, etc.
- Tool function—shape, joint, profile, or other.
- Physical dimensions:
  - maximum diameter of cutting circle.
  - minimum diameter of cutting circle.
  - for shapers or similar tool: cutting width and cutting depth.
  - for shaft cutters: shaft length, minimum and maximum, and cutting length.
  - tool body diameter, if applicable to tool type and process.
  - bore/shaft diameter.
  - number of teeth needed for feed rate, etc.
  - knife material—high-speed steel, tungsten carbide, or other.
- Performance characteristics:
  - rotation—left- or right-hand.

feed direction—against or with cutting direction.
feed method—manual, semi-automatic, or fully mechanical.

- Tool construction:
  solid steel, tungsten carbide, or other.
  tipped with high-speed steel, tungsten carbide, or other.
  mechanically attached knives—of what material?

The Material

A complete description of the material to be cut is critical to the toolmaker, as it affects the construction, knife material, and function of the tool in the machine process:

- Material type:
  panel—plywood, fiberboard, particleboard, or other.
  solid wood—species.
  other material—plastic, metal, etc.
  brand names, if applicable.
- Panel construction:
  laminated or veneered— material used.
- Lamination:
  name of any plastic material used.
  species of decorative wood veneer.
  number of laminations if applicable.
  type of adhesive used—brand name if available.
- Wood solids:
  orientation of cut—along the grain, across the grain, or end grain.
  presence of glue lines.
- Dimensions of the workpieces.
- Workpiece sample—drawing of needed features and description of the process may be necessary.
- The amount of the material to be cut in a given period of time.
- Orientation of the workpiece in the machine process—which side is up or down in relation to machine feed?

Knowing the nature of the material to be cut will dictate many characteristics of the cutting tools for any machine process. Most mass-produced cutting tools attempt to address the needs of a broad range of materials, but may not be completely satisfactory for all, as when the woodworker has altered the material in the manufacture of a product.

Tool Selection Criteria

How we accomplish a particular machining process and the selection of the cutting tools we use is directed by the following seven considerations:

1. The machine type available—shaper, router, etc.;
2. The properties of the material to be cut;
3. The size and frequency of the task;
4. The constraints of the cutting tool—accuracy, cost, tool life, and service;
5. The cost of machining and tooling the product with regard to profit;
6. The configuration of the tool—disposable knives; movable, sharpenable knives; brazed-on knives; or some other;
7. The details of the specific processes.

This list could include several more entries, but the point is that cutting-tool selection today is more involved than simply asking, "How do I accomplish this?" The factors that guide our selections relate mostly to the basic premises of *true cost* and *accuracy*.

True cost in this context is the initial cost of the cutting tool, all of the service costs during its entire operating life, and an approximation of how much material can be cut after each subsequent sharpening. The question of accuracy in tool selection is "Will the cutter perform all the details of the process with relation to the machine, its ancillary parts, and the material to be cut?"

The point here is that the woodworker has to provide the toolmaker with accurate and complete information, if for no other reason than the toolmaker may be in a different city, state, or even country and cannot examine the woodworker's needs firsthand.

A few years ago, cutting tools were classified simply as those having a "hole" in them and those without. That may have have been sufficient then, but, considering what is available today, we need much more information even to begin. There are cutting tool manufacturers that publish catalogs of 300 to 400 pages just to describe their standard products, not including the "special" products that they produce every day or the tools they may produce under a different brand name for another marketing outlet or machine manufacturer.

There are still tools with and without holes, but a more sensible way to categorize cutters is by their construction. The construction of cutting tools better identifies their cost effectiveness, function, and applicability to a given machining process. Three general types of construction identify all cutting tools:

- Fixed knife,
- Movable knife,
- Disposable/insert knife.

Within the three are still tools that have a hole in them and those that do not. But if we are attempting to make the best selection of a cutting tool for a particular process, there is much more information that has to be examined. Following are descriptions of the characteristics of these three groups of tools:

Fixed-Knife Tools

For these, either the entire body is made of the same material as the teeth—tool steel, high-speed steel, etc.—or the teeth are permanently attached by brazing or other means and made of tungsten carbide, high-speed steel, or even polycrystalline diamond. Examples include saws blades, router bits, shaper cutters, drill bits, and others. (See Photos 1 and 2.)

- The tools are sharpened on the face as with a shaper, or profile cutter, or on the face and back relief, as with a saw blade, causing the cutting area of the tool to become smaller in diameter, as with router tools, or shorter, as with drills.

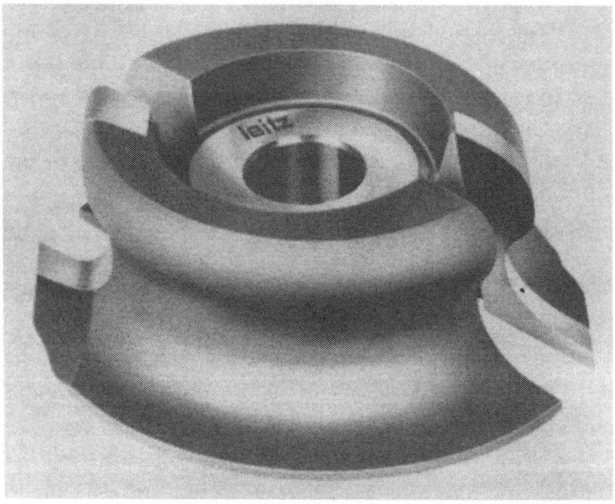

Photo 1—Handrail cutter

Photo 2—Adjustable groove cutter

- These tools can cut the widest array of shapes of the three groups because the necessary cutting features can be built into the body.
- They are the least versatile in terms of performing multiple tasks when compared with other types that can change the cutting knives.
- These cutters can be produced with the safest features of knife support, clearance distances, and protrusion.
- The tools of this group are overall the strongest of all cutters made.

Movable-Knife Tools

In this group, the tool knives are held to the body mechanically by any of several methods. Examples are planers, shaper cutters, moulder heads, and tenoner heads. (See Photo 3.)

- Sharpening a profile cutter involves recutting the entire detail.
- Most knives for tools in this group are made to be sharpened.
- Most knives are sharpened only on the back edge.
- When the knives are made to be sharpened, they can often be extended to the original cutting circle as they are serviced.

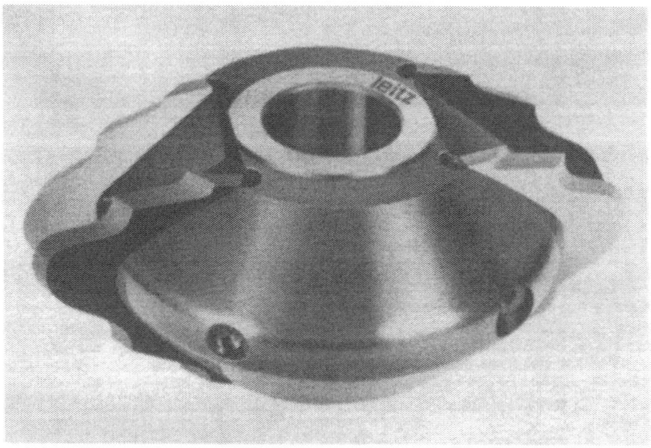

Photo 3—Profile cutter

- A single tool body can hold and drive numerous profile knives.
- In this group are tool bodies that have been manufactured to the shape and size of the profile to be cut, giving strength and support to the process.
- In terms of versatility, these are the most efficient of cutting tools made, capable of a wide range of tasks, from simple surface planing to deep, complex shaped profiles.
- Considerable care has to be given to the set up and maintenance of these tools.

- Knives of this tool group can be either of high-speed steel or tipped with any of several hard metals.

Disposable/Insert-Knife Tools

These are the newest of all tool groups, using the most recent technology of mechanics and metallurgy. (See Photos 4 and 5.)

Photo 4—Variable-angle cutter

- The knives are held to the tool body mechanically, using a wide range of shaped seats, clamping devices, wedges, and screws.
- They are most often made of solid tungsten carbide, though other materials are also used.
- The tools are extremely accurate in all cutting features and profile details.

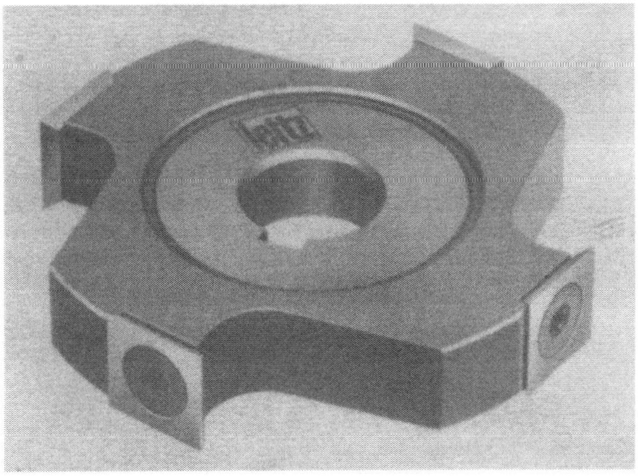

Photo 5—Edgeband trimmer

- The cutting circle is not altered in any measure since the knives are changed as the cutting edges become dulled.
- Nearly any profile is possible as advances in the technology continue to be made.
- The most prolific and easily cost justifiable are those for straight knife shapes, such as for planing, shaping, and jointing.
- There are numerous standard shapes and sizes available, with more being developed, including radiused shapes.
- The knife materials and geometric features will address most materials in the woodworking industry, especially the more abrasive man-made panel materials.

The woodworking industry, like all others, is in an ongoing state of change and refinement because of new machine technology, a flood of new materials to deal with, and new products to develop and market. A part of the effort to manage change is directed at how cutting tools are examined, evaluated, and placed into production.

In industry, selecting, purchasing, and dedicating cutting tools for a particular machine process may take several months and require many evolutionary steps before the final decision is reached. But many of the tools selected survive years of production, simply being replaced by identical copies at the end of their useful working lives. Many cutting tools have been developed just to support and accelerate the decisions of product engineers and designers. There are tools that with a single cutter body can cut numerous complex shapes and sizes of shapes from which product models can be produced.

The tools in Illustrations 8-1 and 8-2 are commonly used to produce prototype furniture pieces in develop-

These cutting knife profiles of four multipurpose tools can produce a wide varity of unique shapes in single- or multiple-pass machining processes. They are intended for only limited work.

Illus. 8-1—Fixed-knife multiple profile cutters

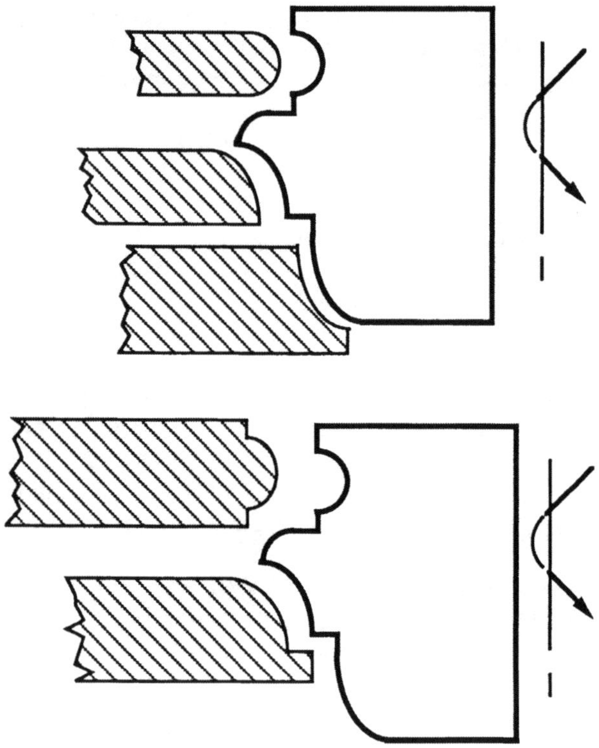

This illustration shows how the same cutting tool can produce five different profile shapes. If the workpiece is machined a second or third time using different portions of the tool, many more are possible. Such tools make production of many shapes possible quickly at low cost.

Illus. 8-2—Profile cutter function

ment or for limited production. The virtue of such cutters is their great versatility to produce extravagant shapes at a low cost, though most often in limited numbers.

The development and introduction of any product is often a long, well documented procedure. And even products that have been manufactured for many years are subject to scrutiny for cost reductions and other improvements. Hence, the search for better cutting tools and machining methods is not just for product development, but is a constant process for the most successful of manufacturers.

This topic may seem to apply only to those in large industry, but that is not true. How industry pursues solutions to the problems of tool selection and machine use can apply to all of us. In the U.S., of all woodworking shops that are registered as businesses, over 80 percent have fewer than 20 employees. Smaller companies may not be staffed with several different engineering groups—production supervision, product inventory control, and others—but they must make a concerted effort to get the best results from the cutting tools and machines that are available to them. Within most areas of the woodworking industry selection of cutting tools is directed by several criteria simultaneously:

1. Volume—the number of pieces that will be needed in a given time frame;
2. Tool expense of the process, both initial and overall operating costs;
3. Labor cost of the machining process that must be included in the cost of the final product;
4. The machine process that can best accomplish the work with the least overall cost;
5. Fixed overhead costs of the manufacturing facility—utilities, taxes, maintenance, etc.

Woodworkers at every level have to examine their manufacturing abilities with regard to these factors when making cutting tool selections. There are correct, appropriate selections, but they have to be carefully and thoughtfully pursued.

Options in Tool Selection

Illustrations 8-3 and 8-3a show that a simple tenon profile could be accomplished by at least five different cutters or cutter sets. Most often, more than a single choice of cutting tool exists for even the most simple profile or machining process. The cutter that is ultimately put into production will be the one that satisfies most of the criteria described previously.

Five different cutting tools could be used to shape a single tenon profile. The most appropriate choice will be dictated by the machine system available, the volume of production, and material to be cut.

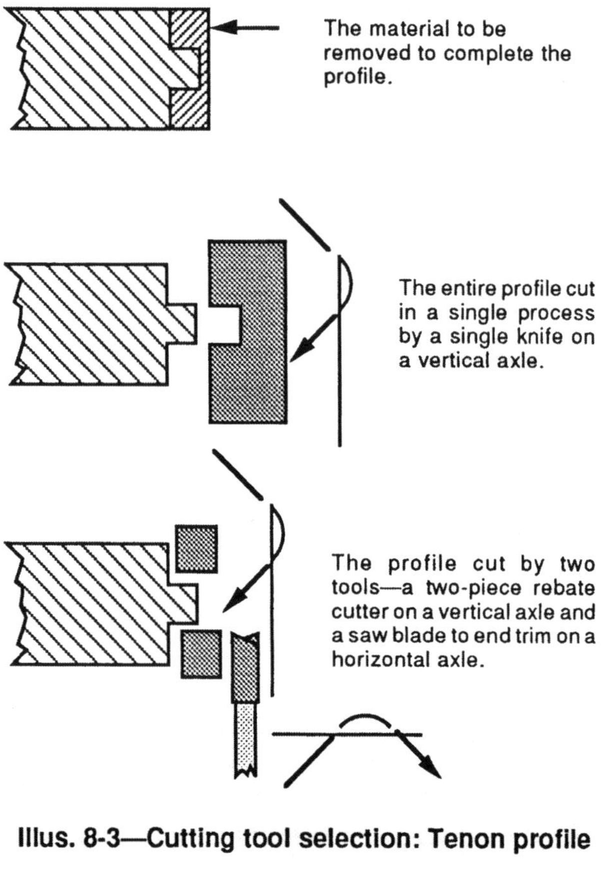

Illus. 8-3—Cutting tool selection: Tenon profile

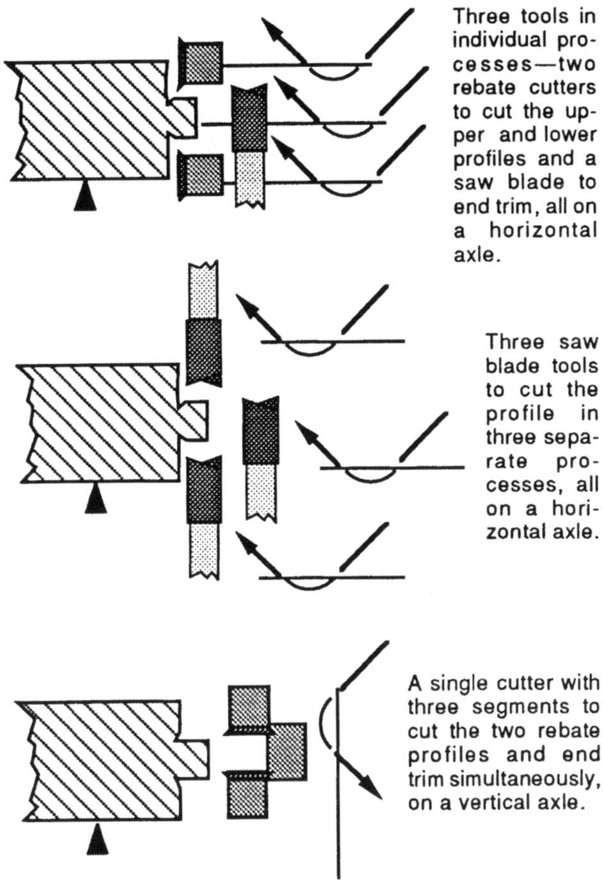

Illus. 8-3a—Cutting tool selection: Tenon profile

The selection of the cutting tools having either fixed, movable, or disposable knives is dictated by several factors including:

1. The type of machine available in the manufacturing processes—manual, single-function process or automated, multiple process;
2. The level of production needed—i.e., a large number of small-quantity production runs or long production runs of a very few products;
3. The nature and character of the material to be cut—natural woods, man-made panels, or other;
4. The "value-added" cost indices of the manufacturing process—i.e., how valuable the process is to the end product;
5. Improvements in cost, quality, accuracy, etc., that the tool may provide.

Even with the best information, the choice of the best possible cutting tool or manufacturing process is often difficult. Illustration 8-4 shows a simple rebate profile that could be accomplished by any of four methods:

In A, there are two cutting tools on a common spindle, operating on a horizontal axle, as might be on a table saw or tenoner. The advantages of this tool set are that both features of the rebate are cut at the same time; the dimensions will remain accurate after repeated service; and the finish quality will be adequate for a relatively low cost.

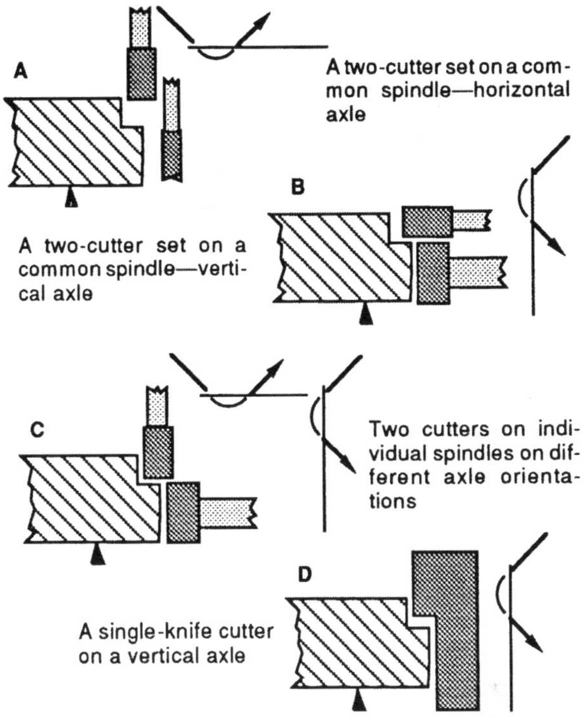

Illus. 8-4—Cutting tool selection: Rebate profile

In B, the cutter set is similar, but the tool spindle is vertical. The machine process could be that of a moulder, shaper, or tenoner. The advantages will be the same, though the cost will be somewhat higher, due to the different tool-knife sizes and orientation.

In C, the machine set up of individual processes will elevate the overall cost dramatically if it is done on different machines, but all costs will be lower if it is accomplished on different tool spindles on the same machine. The finish quality will be no problem, though dimensions may be.

In D, the initial tool cost will be highest, but overall cost may be lowest of the four, depending on other criteria that include the nature of material cut.

Profile Cutter Performance Characteristics

As we examine selection criteria for a particular cutting tool or machining process, we must understand the tools in terms of their geometric, mechanical, and maintenance limitations. This is especially true for tools that cut deep, complex profiles. There are differences among the fixed-, movable-, and disposable-knife tool groups in their ability to cut such profiles.

Several tool characteristics can contribute to cutting problems that are amplified as profiles become deeper or more complex:

1. The diameter of the cutting tool in regard to:
   a. tool body and knife strength,
   b. rim speed of the major and minor diameters of the cutting circles,
   c. finish quality of the cut surface.
2. The relationship of the profile to the other adjacent surfaces.
3. Whether designed to cut an open or closed profile shape.
4. Placement and nature of feature details in the profile shape.

Among the problems of cutting profiles and decorative shapes, the most dramatic are the effects of tool diameter and rim speed. (See Illus. 8-5.) The deep profile in the example shows how the diameter of the major and minor cutting circles will have rim speeds that can alter the performance characteristics of the tool and the finish quality of the material cut. In the drawing, the total work load is considerable because a large amount of material must be removed to complete the shape, though the finish quality at the largest diameter would be the primary consideration in this example. The difference in rim speed between the smallest and largest cutting circles is over 250 percent—27'/sec. at the minor diameter and 69'/sec. at the major diameter. This great difference of rim speed is not common, though it is easily possible in many procedures such as the manufacture of elaborate doors and case goods.

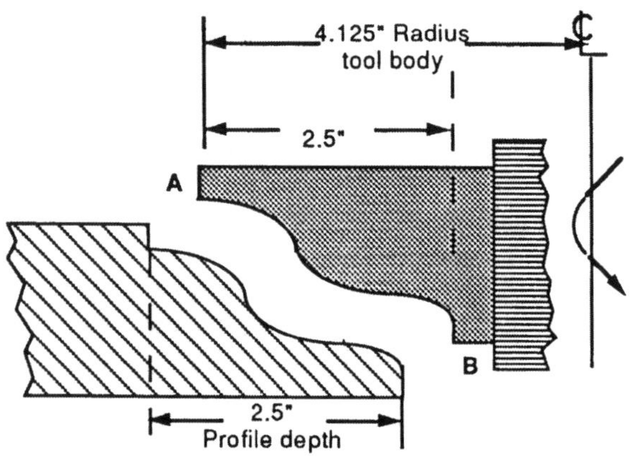

Profile depth affects surface speed of the tool knives at the major and minor cutting diameters. The machine spindle is turning at 6,000 rpm, and the profile depth is 2.5".

A—at the major diameter, the surface speed is 69'/sec.
B—at the minor diameter, the surface speed is 27'/sec.

There will be differences in finish quality all along the cut surface between the two cutting circles. The most common visual evidence is that the wood can be glazed and scorched at the outer peripheral cutting circle where the speed is 255 percent higher.

Illus. 8-5—Tool selection: Profile depth and rim speed

Several measures can be taken to reduce work load on the cutting tools and improve finish quality. (See Illus. 8-6. The profile in Illus. 8-5 and 8-6 is the same, and there are essentially three elements to be cut for the profile.) The first way to reduce the work load is to remove part of the material with another cutting tool, which would be very effective because no consideration need be given to the finish quality since the same area will be cut again later. It may be more costly to use a separate process, but the objective in this instance is to accomplish the unusual and difficult procedure as efficiently as possible.

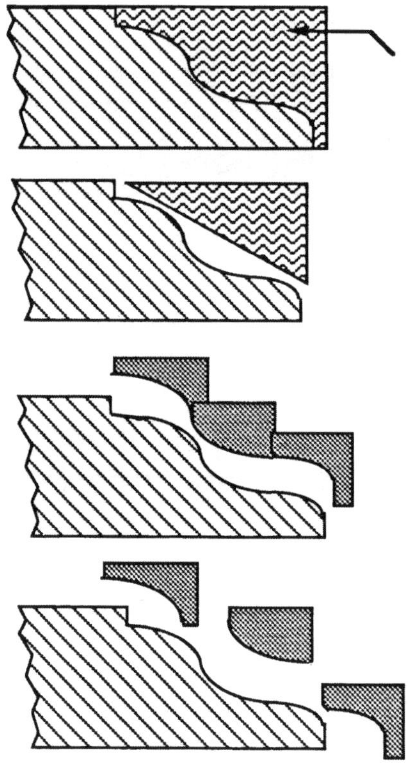

A—The amount of material to be removed

B—The removal of stock to reduce the total load placed on the cutting tool

C—A three-part cutting tool on a common spindle and on the same axis of operation

D—Three individual cutting tools that could be on different spindles, axles, or even different machines

Illus. 8-6—Shaping profiles

Another means of reducing the work load is to divide the profile into three segments that could be cut in any of several ways. (See Illus. 8-6 C and D.) This could easily be accomplished on a complex machine with several tool spindles, and the total of three simple tools may cost less than a single complex cutter. Many decorative shapes are cut with a single tool in a single process, while others may require multiple cutters, procedures, and even machines.

A consideration that relates directly to the selection of cutting tool features is the relationship of tool diameter to profile depth and the subsequent effects on rim speed and other parameters.

If a particular profile shape is 1" deep, that distance represents an inch added to the radius of the cutting tool; that is, the tool diameter has to be 2" larger. Illustration 8-7 shows the angular outlines of profiles to describe how the areas of different profiles are calculated for tool diameter, design, and use. Note the dramatic progression of lengths in the differently angled profile cutting areas. The use of profile cutters requires that all such details be carefully examined. In each example, the material that would be shaped is 1" thick, and the outline area varies in depth only.

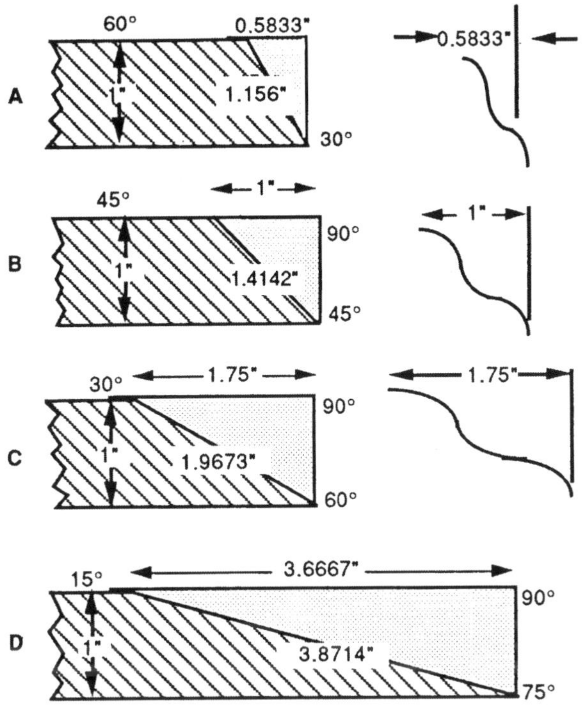

Illus. 8-7—Profile depth and tool diameter

In Illustration 8-8, the profile depth is contained within a 45° angle. For all profile cutting procedures, there has to be clear understanding of the relationships of profile depth to tool dimensions. The effects of rim speed in the different cutting circles and the strength requirements of the cutters are but two of several considerations.

Open and Closed Profiles

So far, we have described cutting *open* profile shapes , as opposed to *closed*, or trapped, profiles. (See Illus. 8-9.) Most commonly, in cutting unusual decorative shapes, woodworkers have problems that affect how well the tools perform the necessary functions and the extent to which the profiles can be successfully duplicated. The two problems may not seem connected, but both are related to one of the geometric features that make all cutting tool processes function—the back relief, or back clearance, angle of the cutting knives.

With cutting tools that cut angular or planar surfaces, such as saw blades, planers, and jointers, the relief angle does not pose a problem. The clearance must simply be sufficient to allow the leading edge of

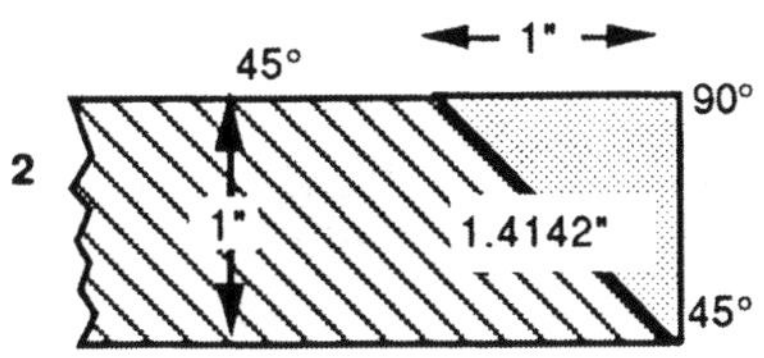

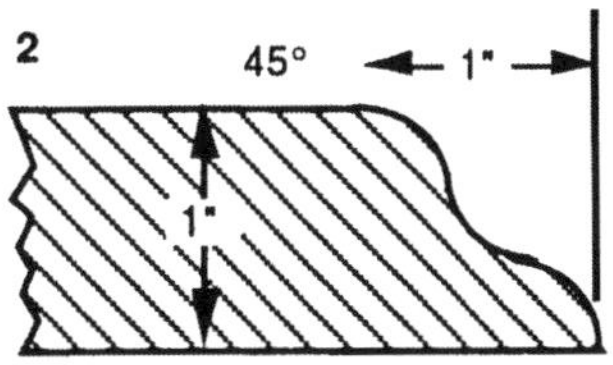

The ogee profile cut is within the 45° angle shown, and with that information we know that:

1. The profile depth is 1"
2. The profile depth has added 2" to the tool diameter.
3. The major diameter is 1" larger than the minor diameter.
4. The length of the knife is at least 1.4142".
5. There will be significant differences in rim speed between the two cutting circles.
6. The design criteria will have to include the geometric details to achieve the needed finish quality, the strength to remove the material, necessary rpm as it relates to rim speed, and appropriate feed rate, etc.

Illus. 8-8—Profile depth and cutting circles

the tool to complete the cutting arc without making contact with the material cut. For a curved surface, however, the problem is entirely different, and with a number of complex features present, the size, definition, placement, and arrangement of such details identify how the ultimate profile can be accomplished. Even for a simple shape like an ogee—a blended profile of three radiuses having the appearance of an inverted, inclined letter "S" (as in Illus. 8-8)—the back clearance angles move away from the leading edge of the cutting knife in a smooth continuum about all three curved details.

Illustration 8-9 shows the basic differences between open and closed profiles. They relate directly to the selection of the most appropriate cutting tool for a process and to the extent that a particular profile shape can be duplicated.

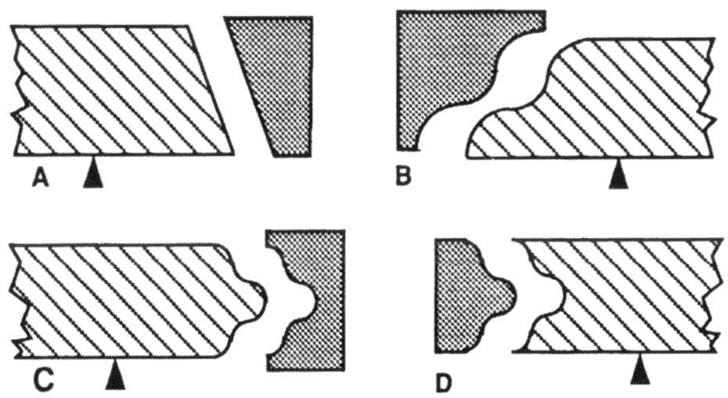

Illus. 8-9—Open and closed profiles

Open profile shapes can be cut and duplicated with fixed-, movable-, or disposable-knife cutting tools, as in Illustration 8-9 A and B. For a fixed-knife cutting tool, there will be no loss of profile shape or size, though the tool diameter will become smaller with sharpening. An open profile poses no problem for a movable-knife cutting tool because sharpening completely recuts the profile details by grinding on the back relief angle, and the knife is relocated to the original cutting circle. A disposable-knife tool is not sharpened at all but discarded, so there is no measurable change to any feature detail.

A closed, or trapped, profile can be either closed outside or closed inside, as in Illustration 8-9 C and D. These shapes pose problems that are mirror images of each other, and both relate again to the back relief angle.

With a fixed-knife cutting tool, an inside- or outside-trapped profile will change dimensionally with each sharpening. For an outside-trapped profile (Illus. 8-9 C), the relief angles are moved away from the center of the profile. As the tool would be normally sharpened on the face of the tool knife, all details and dimensions will become larger, and the tool diameter would become smaller. In the inside-trapped profile (Illus. 8-9 D), the relief angles have to be moved toward the center of the profile. As the knife is sharpened, the feature details would become smaller as the tool diameter does. In most instances, neither a inside-closed nor outside-closed profile would be attempted with a fixed-knife cutting tool with a single knife. More details on relief angles of open and closed profiles are shown in Illustrations 8-10 and 8-11.

Closed profiles do not pose serious problems for either movable- or disposable-knife tools. Sharpening movable-knife tools involves completely recutting the profile. This uses more knife material than sharpening other tools but allows cutting most shapes. For disposable-knife cutters, the shapes are either mass-produced or manufactured specifically for the tool user; they require no sharpening that would change the details. It must not be concluded that all shapes can be cut with any single cutting tool, or even that seemingly simple shapes will be easily cut at all. Ilustrations 8-12 through 8-15 detail methods of cutting closed profiles, and there are a number of other elements of profile shapes that have to be closely examined, which we will address later in this chapter.

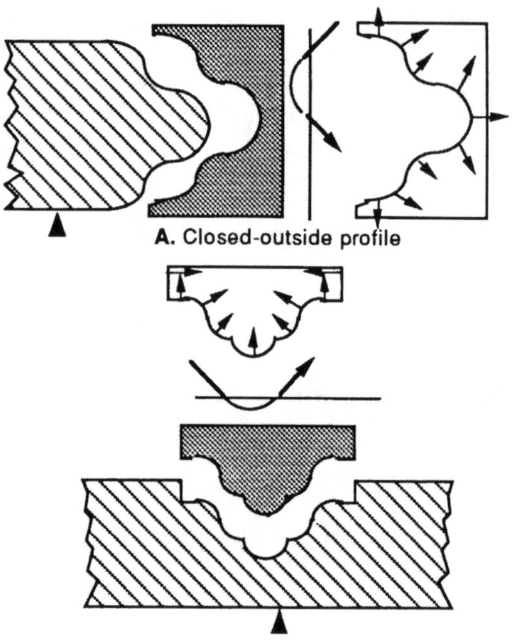

Illus. 8-11—Relief angles of closed profiles

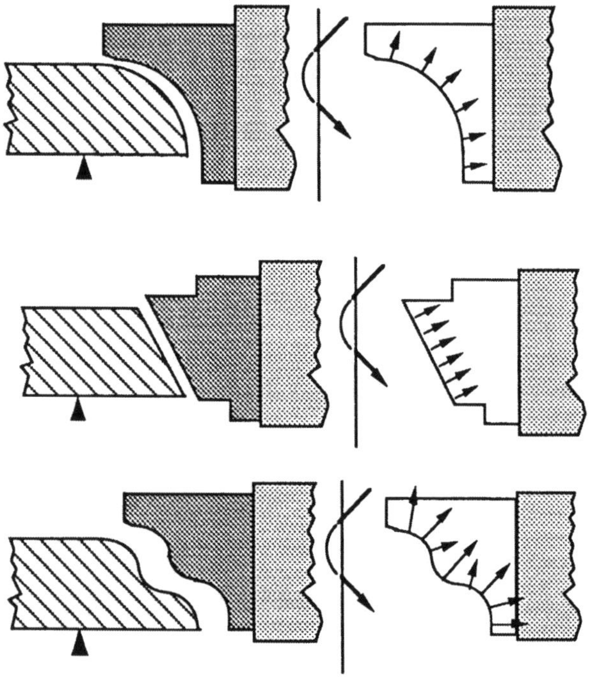

Illus. 8-10—Relief angles of open profiles

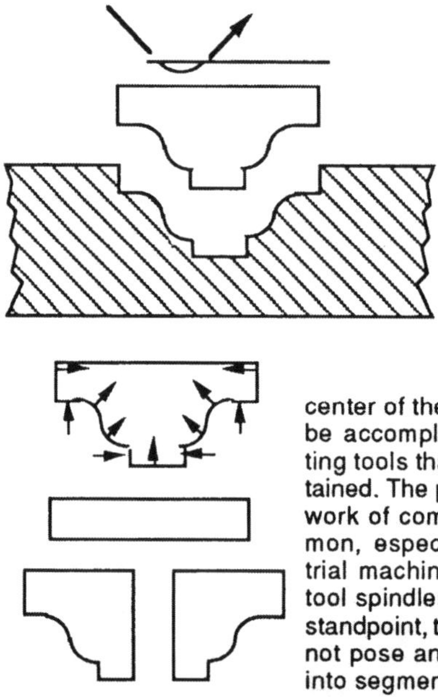

This shape would not be possible to cut with a single fixed-knife cutting tool, as all details would become smaller as it was sharpened; the inside corners do not allow for sufficient back nor lateral clearance; and necessary clearance angles would have to be directed toward the center of the tool. The profile can be accomplished with three cutting tools that will be easily maintained. The practice of dividing the work of complex profiles is common, especially on large industrial machines that have several tool spindles. From a tool design standpoint, this profile shape could not pose any problems if divided into segments.

llus. 8-12—Cutting closed-inside profiles

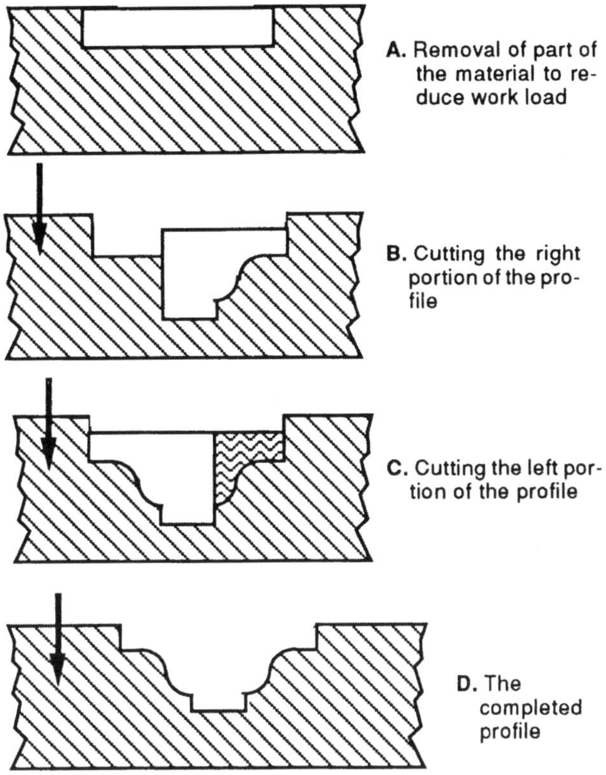

Illus. 8-13—Cutting sequence, closed-inside profile

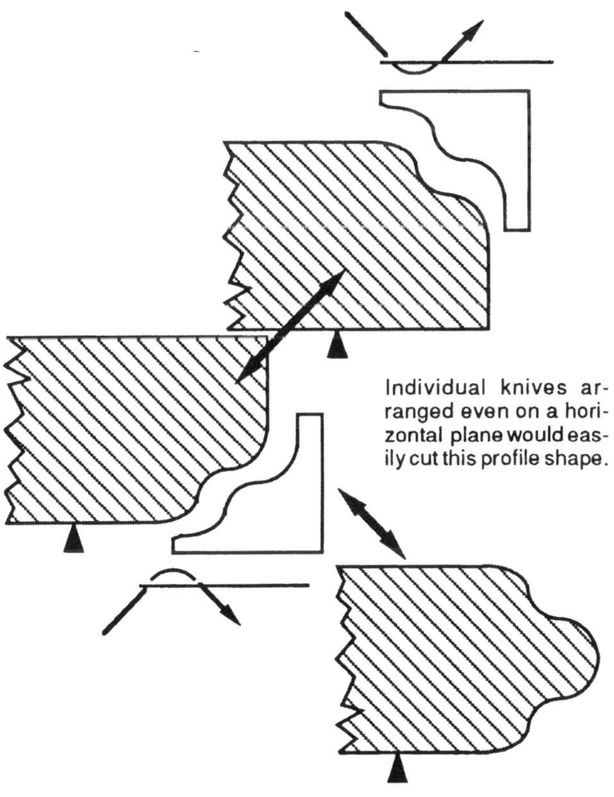

Illus. 8-15—Another method for cutting closed-outside profiles

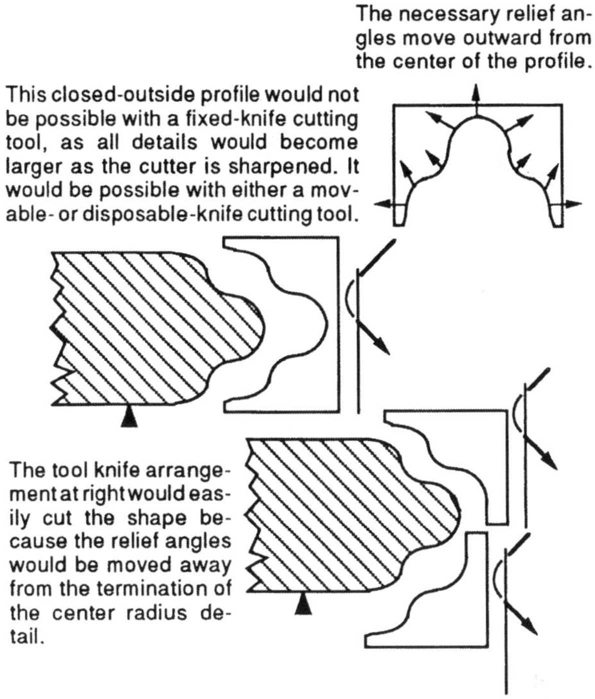

Illus. 8-14—Cutting closed-outside profiles

Tool Sharpening Sequences

Illustrations 8-16 and 8-17 show the sharpening sequence of fixed- and movable-knife cutting tools. A primary difference between the two kinds of tools is in the strength of the knives. Movable-knife tools are not sharpened on the face of the knife because any reduction in thickness would affect the strength of the cutter,

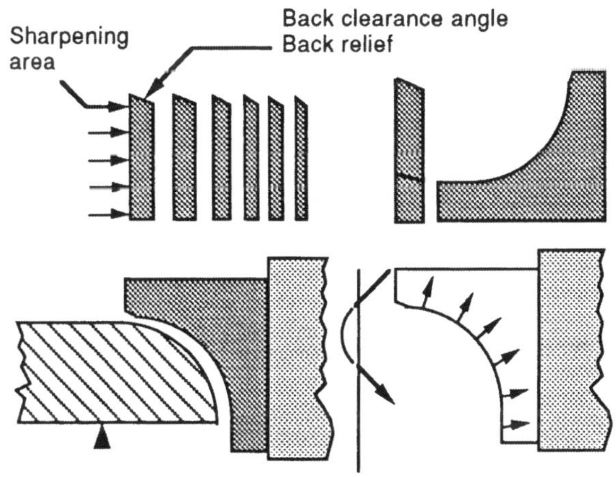

The sharpening sequence of open profile fixed-knife cutters allows the shape to be perfectly preserved throughout the tool's useful life by sharpening on the knife face only. Neither the profile detail nor back relief are involved in any subsequent service operations. The back relief angle and profile detail do not change. The cutting circle diameter does change as the tool size is reduced by the amount of material removed during each sharpening. The total change of diameter is less than the thickness of the knife material.

llus. 8-16—Fixed-knife sharpening sequence

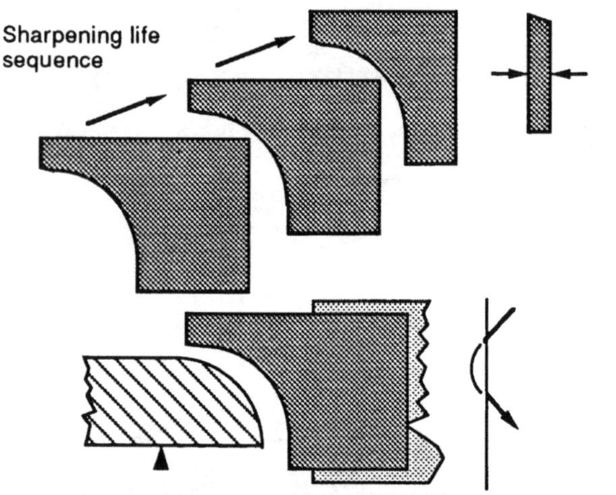

As movable knives are sharpened, the entire profile detail is recut. The knife thickness remains unchanged throughout the tool life. It is critical that an exact amount of material be removed from each knife segment during each sharpening service to maintain balance. The cutting circle is maintained by moving the knives outward to the original diameter.

Illus. 8-17—Movable-knife sharpening sequence

and there is most often no support given to the knife. The advantage of movable-knife tools is that the profiles can be interchanged and the cutting circle maintained without making any machine adjustments.

Fixed-knife tools have bodies that are shaped as the knives are, so reduction in knife thickness does not affect the strength of the tool. Cutters with fixed knives are less costly to sharpen, though they will change in diameter, requiring machine adjustments after each service.

Disposable-knife cutters are generally more costly to purchase and operate than either fixed- or movable-knife tools, but solve many of the problems of both. Since the knives are never sharpened, but rather discarded, cutting circle size is never affected.

Understanding Design Features

In all areas of the woodworking industry, elements of product design continue to be expensive and an often heated topic of discussion. That is, the manufacture of a given product is driven by pressures and realities of the marketplace that all relate to the design features. Before the fact: What features would customers find attractive in a newly developed product? After the fact: How can we better contain the cost of a developed product?

Industrial manufacture of wood products is unlike others in several ways. Most other industries develop and market products in a steady progression, incorporating new design features, materials, and technology. In the woodworking industry, there are similar developments of ever new designs, but there is also a very large effort directed at manufacturing products with accepted design features that identify them as part of a particular era or national origin. For instance, there are Danish, early American, and American Quaker furniture and hundreds of others. In each group of products are accepted design shapes and features. As these are developed into complete product lines, the cost of producing them has to be carefully evaluated. Some profile shapes are easily produced at low cost with inexpensive machines and cutting tools, while others require complex, expensive procedures.

Recognizing the relative ease or difficulty of producing a particular shape or detail requires understanding what the cutting tools and machines are capable of. The most commonly misunderstood process is the machining of any detail that has an inside corner. (See Illus. 8-18 through 8-20.)

The example in the illustrations shows what would appear to be a simple open-profile shape, but the two 90° inside corners do not provide adequate relief angles away from the intersection of the straight surfaces. There are too many constraints against cutting this shape with a single knife in a single process. The tips of the knives at the inside corners would be frail, incapable of withstanding heavy work loads, and relatively short lived. The finish quality of the material cut would be diminished at the inside corners with such reduced relief angles. The machining procedure would have to be slower to avoid elevated temperatures and heavy work loads.

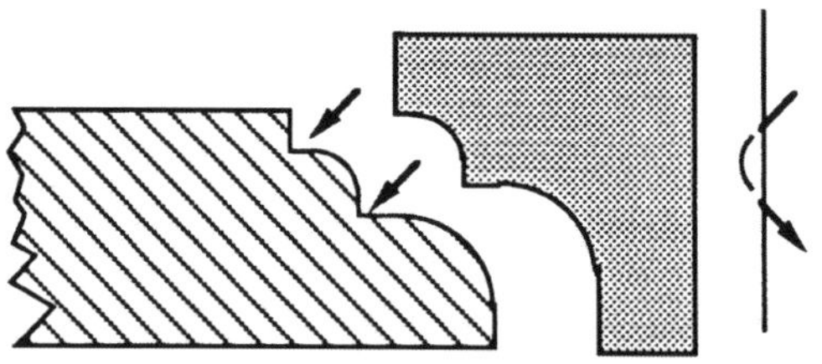

A profile shape that incorporates perfectly square inside corners poses the problem of trapping the relief angles, which would call for very frail knife sections at the intersections. In this instance, the problem cannot be solved with a fixed-, movable-, or disposable-knife cutting tool. There is no reasonable way to provide the needed relief angles at inside corners in a single knife. The work will have to be divided or the shape altered.

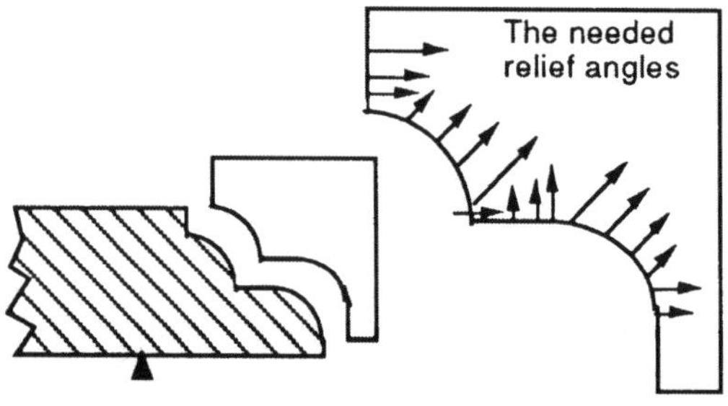

Illus. 8-18—Problems created by inside corners

The machining of nearly any inside corner detail is difficult, and, when placed in the middle of a larger complex profile shape, such details make it necessary to divide the profile into smaller segments for manufacturing. Design features have to be evaluated as they relate to the cost and difficulty of producing them.

The examples in Illustration 8-19, indicate that cutting the inside corner with the multiple-segment cutting tool or even approaching the cutting process from a different axis does not completely solve the relief clearance problem. Simply, it is not possible to divide the back relief in two directions simultaneously.

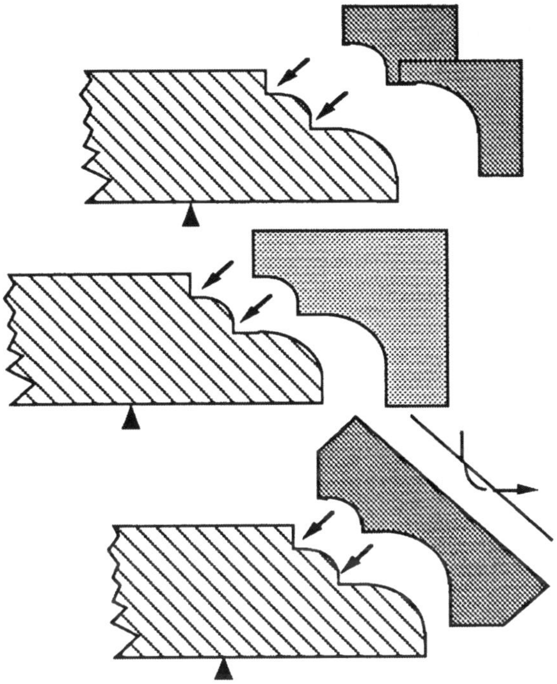

Illus. 8-19—Cutting inside corners

In Illustration 8-20 the two profile shapes B and C will largely resemble A, but the results from the standpoint of cutting tool performance will be dramatically different. The placement of a small radius in the corner intersection or an 8° to 10° bevel on one of the plane surfaces leading away from or into the intersection makes the profile much easier to cut. Providing a profile that allows the needed cutter geometry makes the cutting tool easy to produce and maintain without loss of profile shape and size detail.

There are cutting tools that we know will change dimensionally over their normal operating lives: Drills become shorter; saw blades become narrower and smaller in diameter; and router tools and all others that are sharpened to maintain their cutting ability will undergo such changes. Profile shapes on cutters, however, must be preserved and duplicated accurately to protect the viability of the end products. For most wood products, the ability to produce and repeatedly dupli-

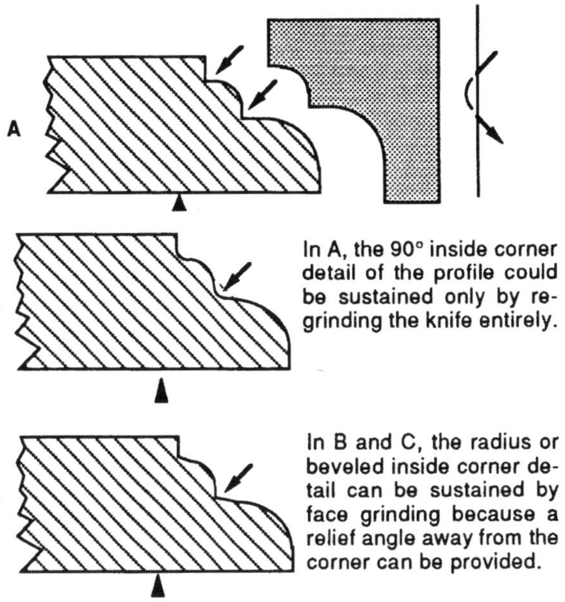

Illus. 8-20—Beveling inside corners

cate profile shapes is the single most important ongoing manufacturing procedure to be accomplished.

Summary

The processes involved in manufacturing any portion of any product have to be carefully examined. Larger manufacturing companies have specific divisions of expertise to bring a product to the marketplace:

1. Corporate management,
2. Sales and marketing,
3. Accounting and data services,
4. Human resources,
5. Research and development,
6. Product designers,
7. Product engineers,
8. Manufacturing engineers,
9. Industrial engineers,
10. Quality assurance engineers,
11. Purchasing,
12. Product inventory control,
13. Production scheduling,
14. Maintenance engineering and staff,
15. Stock handling,
16. Packaging and shipping.

That makes for 16 distinct industrial-professional groups or functions, of which none directly contact the shaping of raw materials into the finished product. A complete list of job classifications in a large, modern woodworking facility would require perhaps several pages. Regardless of the number of employees in a manufacturing facility, however, all these functions are performed, in addition to all of the hands-on procedures needed to complete the finished product, even in a small cabinet shop. Of importance to this discussion are numbers 5 through 10 in the previous list:

5. *Research and development* may have several tasks that would include investigating new materials and procedures for manufacturing of the products. With input from all engineering and management areas, the effort is continually to improve the products and all areas of manufacturing.
6. *Product designers* deal primarily with the visual effects, shapes, and sizes of products. Working with management and sales and marketing, they attempt to satisfy demands of the marketplace within the limitations of their facility.
7. *Product engineers* interpret product design into a manufacturable form. With input from product designers and manufacturing engineers, they define all of the parts in a product that must be understood and dealt with in the course of manufacturing procedures and processes.
8. *Manufacturing engineers* select and dedicate the machines and tooling to produce all the parts for a product. Working with product engineers and industrial engineers, they direct their effort at providing the best machine processes and cutting tools for the process.
9. *Industrial engineers* measure and evaluate costs of producing the parts of the product. With input from product engineers, manufacturing engineers, and all areas of manufacturing, they work for cost containment and efficiency.
10. *Quality assurance engineers* interpret and enforce all manufacturing standards. With input from corporate management, sales and marketing, and product engineering, they provide a prescribed level of product quality.

This chapter has emphasized that selection and use of cutting tools to produce complex profile shapes is a major concern at all levels of woodworking, from the smallest cabinet shop to the largest industrial facility. In the earlier list of tasks involving design criteria and manufacturing of a product, we can see that at least six groups have a responsibility, directly or indirectly, to understand and participate in the performance of cutting-tool processes:

- The research and development group should work as a factory laboratory where the problems of manufacturing are rehearsed and solved. All pertinent manufacturing processes and materials should be studied and evaluated.
- Product designers and engineers must have at least a basic comprehension of machine and cutting tool processes.
- Manufacturing and industrial engineers must have a specific understanding of machine and cutting tool processes.
- Quality assurance engineers must have a clear understanding of what the machine and cutting tool processes are to accomplish.

The size of a manufacturing company or its number of employees is not as important as its understanding of and ability to accomplish the necessary tasks. The objective of all manufacturing is to produce the best possible products for the least total cost. To that end, a significant benefit will be found in carefully examining every detail of the manufacturing processes.

Dramatic technological changes are taking place in woodworking. As advances in manufacturing methods and systems are implemented, it is clear that woodworking is becoming more machine intense and less labor intense. This means that there is a need to find and make use of all of the technical information available to us. As production standards become more stringent, there will not be the luxury of earlier trial-and-error methods, but we will see benefit in the careful evaluation of what the machining processes and cutting tools can do.

Cutting profile shapes requires careful examination of details as well as an understanding of the abilities and characteristics of the cutting tools used. There are profile shapes that, given their geometric features, can be easily shaped by the revolving motion of cutting tools. There are other details that cannot be completed except by segmenting or altering the shapes.

Of all wood cutting procedures, none is more complex and costly than shaping decorative profiles, those unique shapes that give our products their identity, beauty, and character. To create successful products, we must clearly understand all the details of the desired profile shapes and what our cutting tools can do. As woodworkers, we need to understand our needs, and we need to be able to communicate those needs to our partners, the toolmakers.

Chapter 9

Special-Purpose Cutting Tools

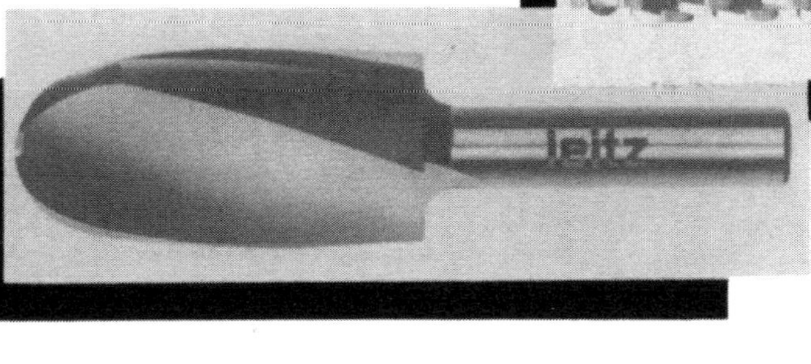

Photos courtesy of Stehle GmbH & Co., and Leitz Tooling Systems, Inc.,

Special-Purpose Cutting Tools

A number of cutting tools are not common to all woodworkers, and, in this chapter, we'll answer some of the "I've always wondered how they did that" questions about some of these special-purpose tools.

After several hundred years of woodworking experience and development of technology, the likelihood that a woodworker will have a totally unique problem to solve is remote. There are hundreds of different cutting tools and machine adaptations, but, too often, information regarding their advantages and uses is not widespread. In fact, there are cutting tools that have as yet been undiscovered by most woodworkers. If the only tool available to us is a hammer, every problem will no doubt be perceived as a nail, but the range of tools available allows us to see things differently.

Within the many areas of woodworking, cutting tools have been developed to solve problems as they arose. Some have moved laterally into other areas, while occasionally unique cutting tools have moved into entirely different industries. For the past 25 years, many of the cutting tool developments in woodworking have been modifications of metal working tools.

Unique Cutters

Furniture manufacturers who produce a particular group of products will develop or purchase cutting tools that are unique to their use. All manufacturers of, for instance, ornate solid wood tables and chairs will most likely have similar cutting tools and machines. Manufacturers of other products may have tools that are specific to their use and seemingly quite distant from the basic four—saw blades, drills, router bits, and shaping tools. This chapter will illustrate some of the more unique cutters and suggest that all of us continually examine the tasks that we perform for improvements, with the overarching idea that there is always more to learn.

The Lamello Spline System

A recent innovation in cutting tool use and furniture assembly involves use of a manufactured spline shaped like a football. (See Illus. 9- 1 through 9-3.) The lamello system was originally patented in Europe and developed to simplify the assembly of case-good furniture pieces with an easily operated and maintained machine process. The basic element of this system is the cutting tool used to create the pocket for the manufactured spline. The cutter has a diameter of 100 mm, or 3.937", and a cutting kerf of 4 mm, or 0.157". The spline is manufactured of three or more layers of compressed hardwood, similar to a small piece of plywood. The assembly process, involving a water-based adhesive, causes the spline to swell to its original thickness, creating a remarkably strong joint. Different widths and lengths of splines can be easily accommodated by cutting a slightly deeper pocket with the same machine and cutting tool. All have the same radius on the periphery.

The small diameter of the cutting tool makes the machine compact and portable, though large stationary machines are produced for industrial use. The concept of making a functional and versatile wood joining system that simply uses a chordal section of a circle is a welcome innovation.

Since the development of the lamello spline system, there have been several additions to the idea, including the hanger bracket shown in Illustration 9-3. In the U.S., we take our readily available supply of decorative hardwoods for granted, and, while we attempt to

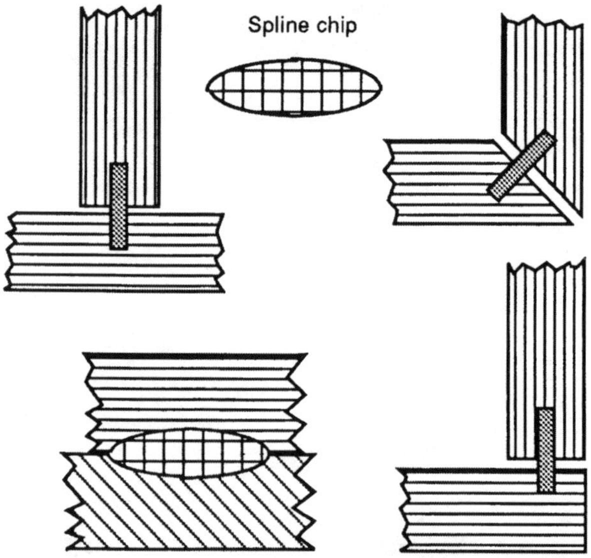

Illus. 9-1—Lamello chip

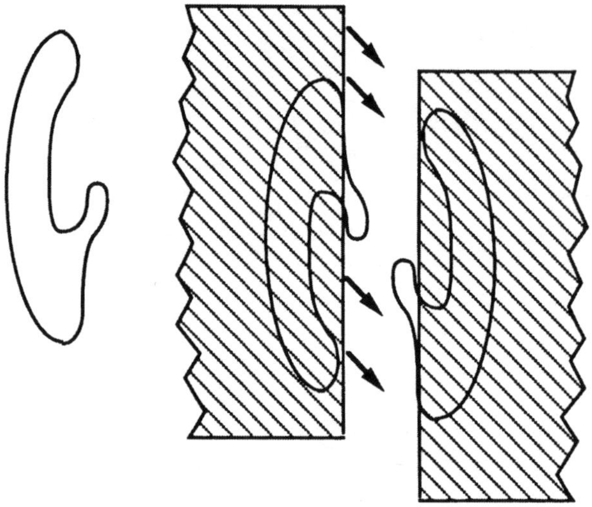

Illus. 9-3—Lamello take-down assembly bracket

gain the best yield from what we cut, there is seldom any specific attention to repairing any wood stock pieces. In other parts of the world, however, there is often an effort to make the best use of what is available, including repairing flaws or, in some instances, completely manufacturing man-made wood materials.

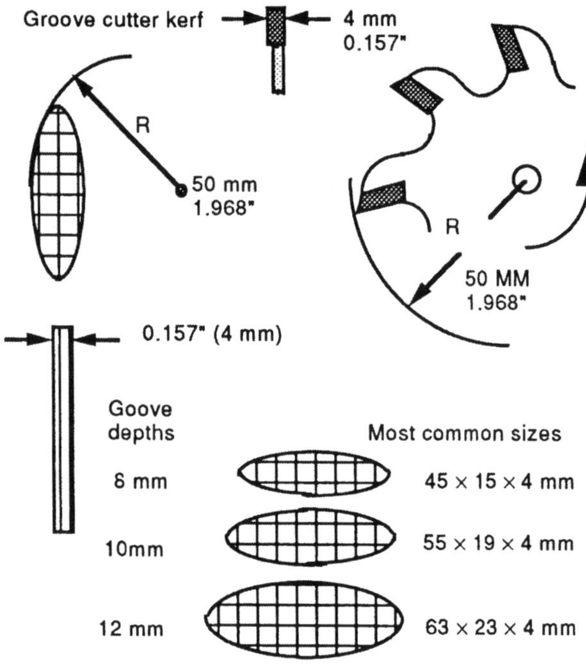

Illus. 9-2—Lamello chip dimensions

The Mini-Spot Repair System

Among the popular hardwoods used in Europe and Asia are several species that often have large pitch and sap deposits or other flaws that would make many cut parts useless in the finished product. A unique cutting tool and repair process, the mini-spot repair system, has been developed to salvage parts that would otherwise have been lost. (See Illus. 9-4.)

The cutting tool and machine process are similar to the lamello system, except that instead of a joinery spline, a small boat-shaped repair section is manufactured to fill the cavity. The repaired section is later sanded and prepared for finish with the main portion of the piece and, without magnification, probably will not be seen at all. There are several similar repair systems of cutting tool and machines available, many of which are not yet seen here, primarily because our cost of raw lumber is dramatically lower than most other areas of the world. The only major wood-repair process that has

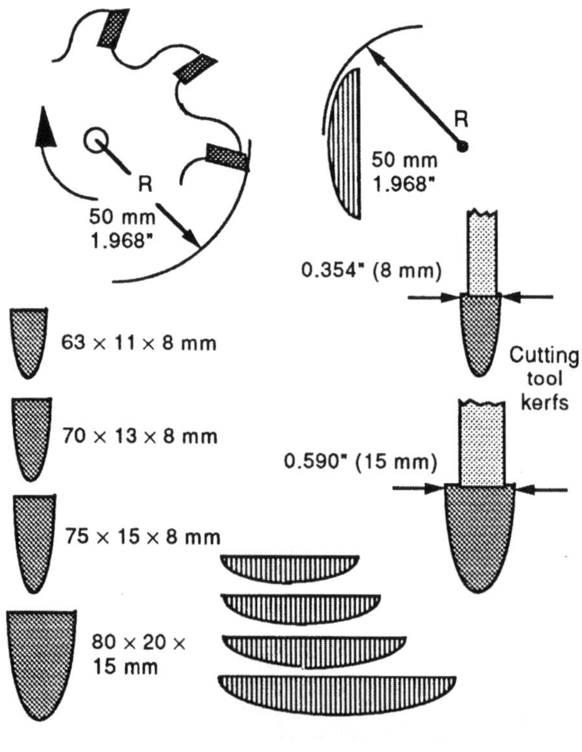

Illus. 9-4—Mini-spot repair system

been ongoing in the U.S. is the plug-cutting repair, used in plywood manufacturing to remove loose knots from veneer sheets.

The lamello spline and mini-spot repair system are but two of the many innovative cutting tool developments seen in recent years that could be expanded upon to satisfy other needs and solve other problems.

Band Saws

Band saws are not special in any sense, but they do have properties that need to be examined and understood. The cutting process, while not dramatic or complex, uses the saw blade in a manner unlike all others. The tool at the same time is flexible enough to be constantly deformed into the radius of the drive and idle wheels and has teeth hard enough to sustain the cutting process. Those would appear to be contradictory properties—flexibility and hardness—but are not. As described in Chapter 3, the characteristics of metals—their plasticity, elasticity, malleability, ductility, etc.—are neither mutually exclusive nor inclusive; each is a unique property. That is, hardness and flexibility are not opposite characteristics nor are they directly related to each other at all. It is a common misunderstanding that hard metals or other materials are also always brittle. Examples of metals having similar metallurgical contents but vastly different properties would be spring steel, which is both very hard and flexible, and a file or wood rasp which is very hard and brittle. Both are made of plain carbon steel.

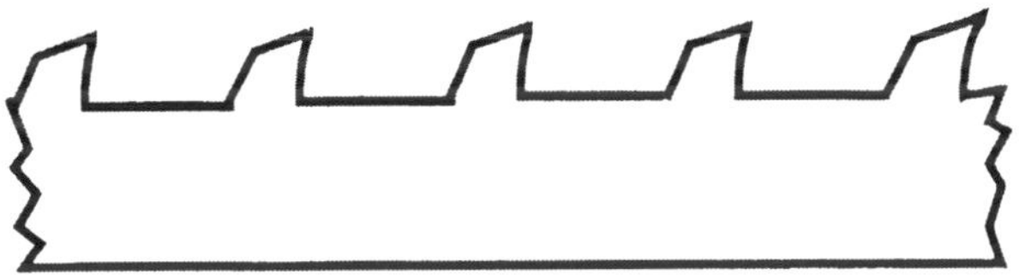

Skip-tooth

Raker-plus set

Wood cutting band saw blades used in the production of furniture are manufactured with the following specifications and features:

Blade width	Teeth/inch	Blade thickness
1/8 to 1/4"	3 to 12	0.020 to 0.030"
3/8 to 1/2"	4 to 10	0.025 to 0.035"
5/8 to 3/4"	4 to 8	0.030 to 0.040"
1 to 1-1/4"	3 to 6	0.035 to 0.045"

Most band saw blades are driven at a rate that varies from 600 to 1,200 feet per minute, depending on the material being cut, feed rate, and characteristics of the cutting teeth on the blade. Wood cutting is accomplished at considerable speed, requiring that the waste chips be gathered and evacuated easily. Specific work loads dictate both the tooth features and number of teeth required on a particular blade.

Illus. 9-5—Band saw blades

The metallic makeup of band-saw blades is as diverse as that of circular saws. There are solid high-speed steel alloy blades, others tipped with hard metals, and yet others made of different metal alloys edge-welded together.The cutting process of a band saw involves a number of chisels joined together one after the other. The teeth are slightly offset to cut a path wider than the base width of the blade, reducing friction or pinching as the blade passes through the material being cut. Another reason for offsetting in the cutting teeth is to reduce contact between the blade and the material being cut. With insufficient lateral clearance, the blade would quickly overheat and be damaged. (See Illus. 9-5 through 9-7.)

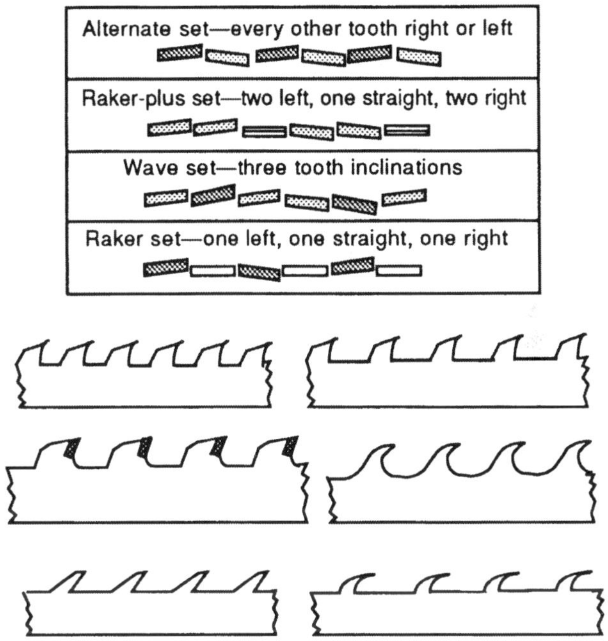

Illus. 9-6—Band saw set and tooth form examples

Successful band-saw cutting requires strict attention to the mechanical features and characteristics of the machine. Of greatest importance is that the blade be

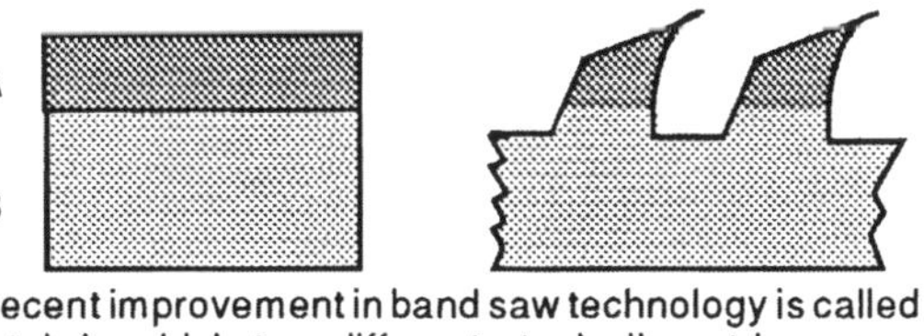

A recent improvement in band saw technology is called a bi-metal, in which two different steel alloy strips are edge-welded together. A is a wear resistant high-speed steel alloy with high percentages of either molybdenum or cobalt into which the features are ground. B is a special low-carbon steel alloy chosen for its strength and flexibility. Band saws are tipped with hard metal alloys to improve resistance to abrasive wear or the effects of elevated temperatures. The most common would be tungsten carbide to resist abrasive wear and the patented alloys called Stellite or Tantung to withstand the high temperatures of cutting wet woods.

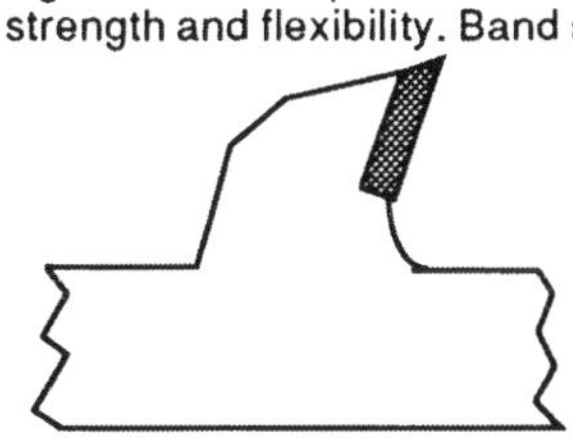

Illus. 9-7—Band saw construction

held with adequate tension and be carefully guided. Of all problems encountered, most will relate to one of these.

A primary difference from one band-saw machine to another is the quality, precision, and strength of the blade guiding mechanisms. There are several variations of band-saw teeth in terms of type of set, number of teeth per inch, tooth shape, and arrangement.

For wood cutting, the chip load per tooth can be relatively high in comparison to that of metal cutting, for instance, because the material does not have great strength. The most popular tooth form is a skip-tooth arrangement that provides a larger gullet space for faster cutting and the easy removal of waste chips. The most effective set, the raker plus, places two teeth to the right followed by a straight tooth, followed by two teeth to the left, and so on. The set and tooth arrangement will be dictated by the strength and density of the material to be cut, blade velocity, and desired feed speed.

There are a considerable number of tooth shapes and spacings for the wide range of materials cut with band saws, from tool steels to foam rubber and nearly everything in between. The performance characteristics for band-saw blades are much like all other cutting tools in terms of chip load, tooth progression, feed speeds, and tool velocity.

Keyhole Slot Cutter

The next tool that we will examine takes advantage of its ability to cut on two planes and with two cutting circles simultaneously. (See Illus. 9-8.) The keyhole slot cutting tool is common to most woodworkers and provides a unique solution that would otherwise require more than a single cutter and process. Initially, the cutter functions as a drill or plunge router bit to enter the workpiece. The process then moves laterally, cutting with the major and minor diameter at the bottom of the tool. At the termination of the process, the tool is returned to exit at the point of entry, leaving a machined cavity having the outward appearance of a keyhole. The design tool provides a simple method of cutting a slot for hanging pictures and similar functions. The cutter bit offers an excellent example of complex tool geometry utilized to perform multiple functions with a single tool.

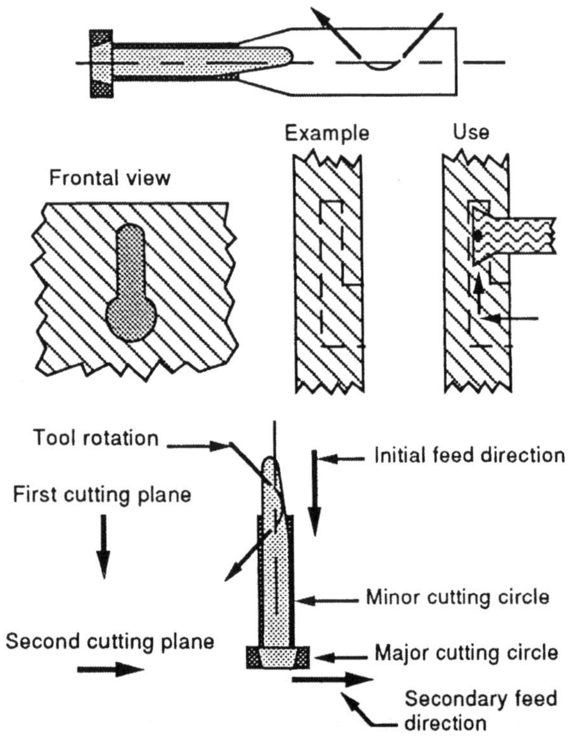

Illus. 9-8—Keyhole slot cutter

There are many different cutting tools available for the most common woodworking machines. Any functional woodworking facility has at least two machines, a table saw and a drilling machine of some kind. The next several examples illustrate some of the less well known and utilized tools for those machines.

Table Saw Radius Cutter

Another cutting tool that is popular in Europe and Asia, though not often seen in the U.S., extends the abilities of a common table saw. (See Illus. 9-9.) The tool very effectively cuts a wide array of radius shapes,

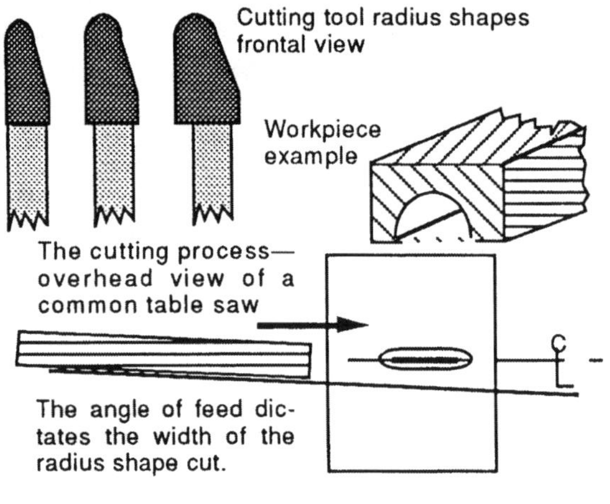

Illus. 9-9—Table saw radius-cutting tool

both in depth and width. It has the appearance of a sturdy, modified saw blade with specially shaped teeth. The cutters are available in several diameters and with different peripheral shapes. The process involves feeding the workpiece past the cutting tool at a slightly oblique angle, as indicated in the illustration. The entire profile depth may have to be cut in several steps since so much material is removed.

A preferred method of using this tool would be with a mechanical stock feeding machine. If the process is to be accomplished by manual stock feeding, considerable caution will have to be exercised because the cutting tool uses a large amount of horsepower that must be controlled. With correct holding devices and fixtures, these cutting tools can produce a wide variety of products, from wooden bowls to exterior trim and wooden rain gutters. Such unique cutting tools can greatly extend the processes of a common table saw.

Dado Groove Cutters

There are several cutting tools that can be utilized on table saw machines to cut grooves and rebates. (See Illus. 9-10 through 9-12.) The first groove/rebate cutter that we will examine is most popular among woodworkers whose primary need is for versatility, reduced cost, and ease of set up.

As shown in Illustration 9-10, the *wobble saw* groove cutter is manufactured in numerous variations and sizes. The basis of the operation of this tool is the tapered collars on the hub of the tool assembly. As the assembly is rotated within its moving parts, the tapered collars are aligned to tilt the saw blade to or from its maximum displacement. The narrowest groove the tool will cut is the width of the saw blade itself; the maximum width can be several times the base width, depending on diameter.

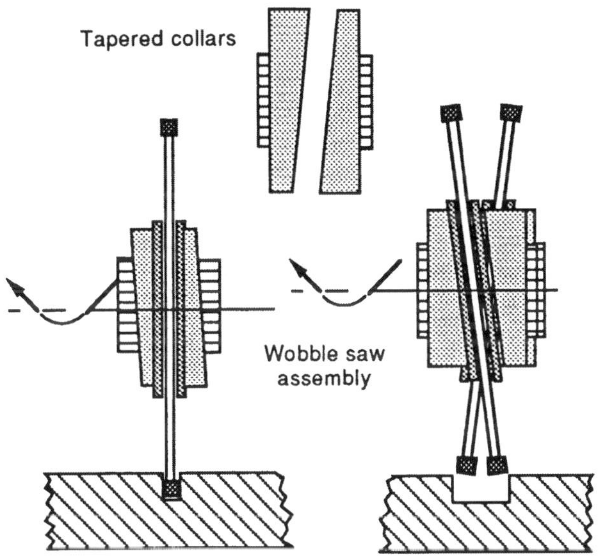

Illus. 9-10—Wobble saw

Generally, a wobble saw groove cutter is not intended for the precise requirements, rigors, and production needs of industry. The tool will function well for the limited needs of occasional use where heavy work loads of long duration are not required.

Care must be taken with all such cutting tools that remove a large amount of material and use substantial horsepower. The feed rate, cutting speed, and all other process criteria have to be clearly understood and controlled.

Illustration 9-11 shows another groove-cutting tool, generally called a *stacked dado cutter set*. It is made up of from six to eight individual cutters that provide the means for cutting different width grooves and rebates. The two outside cutter segments are similar to circular saw blades except that the teeth are slanted inward to ensure that the inside corner(s) will remain square and clean. The remaining cutter segments are called chippers and will have fewer teeth than the outside blades. The chippers are produced in different widths, allowing various grooves or rebates to be cut by simply stacking what is needed between the outside blades. The dado sets are produced in diameters from 6" to 12" and for cutting widths generally from 1/4" to 13/16", in 1/16" increments.

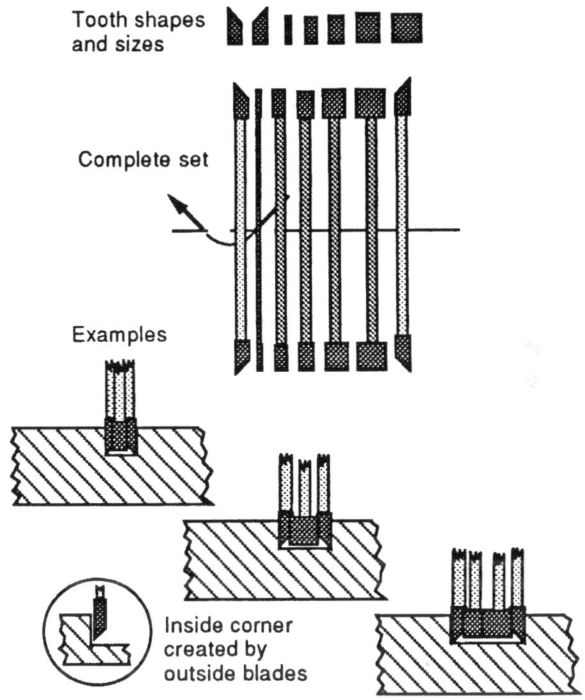

Illus. 9-11—Stacked dado groove cutters

Critical to the safe and successful operation of these cutting-tool sets is their subsequent sharpening and maintenance. The individual cutter segments must be carefully sharpened to maintain balance and concentricity, both individually and as a set. As minor as this may appear, it is not, because the pieces of the set are sharpened individually. A worst possible case would be six or eight cutting circles, all out of balance.

The stacked dado cutter sets are used at all levels of wood product manufacturing because they are versatile, accurate, and strong enough for most tasks. The stacked sets are also used on a wide range of machines beyond table saws, including shapers, moulders, profilers, notchers, tenoners, and others. Their greatest advantage for industry, especially for small manufacturers, is that a single tool set can accomplish a wide range of tasks that would otherwise require a large number of dedicated tools.

Illustration 9-12 contains a list of some other types of groove and rebate cutting tools available in the marketplace. In industry, the production of parts is examined for manufacturing efficiency. Each detail of every part is carefully studied and evaluated. The choice of machining processes and cutting tools is also carefully evaluated for every measure of cost and efficiency.

Among the numerous types of groove/rebate cutting tools there are
1. Fixed knife, fixed width;
2. Fixed knife, adjustable with the use of spacers;
3. Fixed knife, adjustable stack type;
4. Fixed knife, adjustable dial-calibration;
5. Disposable knife, fixed width;
6. Disposable knife, adjustable with the use of spacers;
7. Disposable knives, adjustable dial-calibration;
8. Disposable knife, stackable;
9. Any of the above with or without spurs to protect the inside corners.

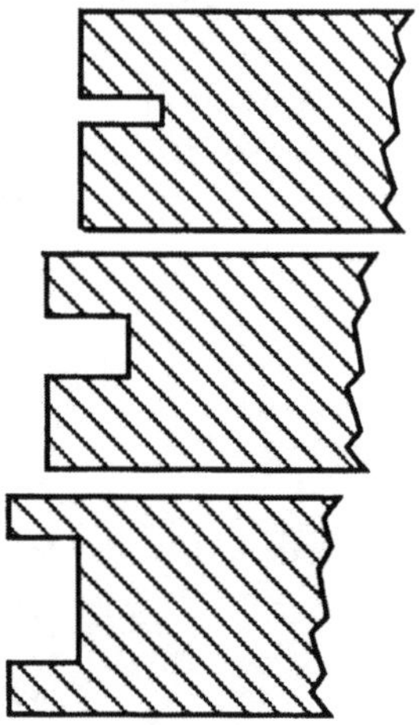

Cutting grooves and rebates on saws, shapers, and other industrial machines most often requires precision, correct finish quality, extended run times, and cost efficiency. The construction joint features produced by these cutting tools demand close tolerance details to provide strength in the products manufactured. Where larger volumes of production are needed, single function, dedicated cutters are used to ensure accuracy. The nature of the material cut and the available machine process dictate the choice of cutting tool.

Illus. 9-12—Dado groove-cutting tools

In industry, the criteria for selecting cutting tools and machine processes begins with identifying the volume of production that will be served by a particular cutting tool or machine to produce a given product or part. The likelihood that adjustable cutting tools would be chosen for high volume, ongoing manufacturing is remote. The choice of cutting tools in industry is seldom based on the initial cost of the cutter itself; rather, the cost is measured in terms of total operating expense over its entire useful life. The wide range of materials and machine processes in industry, along with the other conditions to be met, make the selection of most effective cutting tools an important activity.

Handle/Wood-Shaft Cutters

Within the woodworking industry generally, there are areas of specialization where the products are totally unique. The manufacture of handles for every imaginable tool has developed an entire collection of cutters not often seen in the manufacture of other products. Tool handles for brooms, hammers, axes, shovels, gun stocks, and many other tools are produced in large volumes using machining processes and cutters that have been specially developed for these unique products. (See Illus. 9-13.)

The simple process of shaping the ends of broom handles and similar products has created several cutters to accomplish the work, three of which are shown. See

A—a threading tool for use on a lathe or other similar machine.
B—an adjustable radius cutting tool
C—an adjustable bevel cutting tool

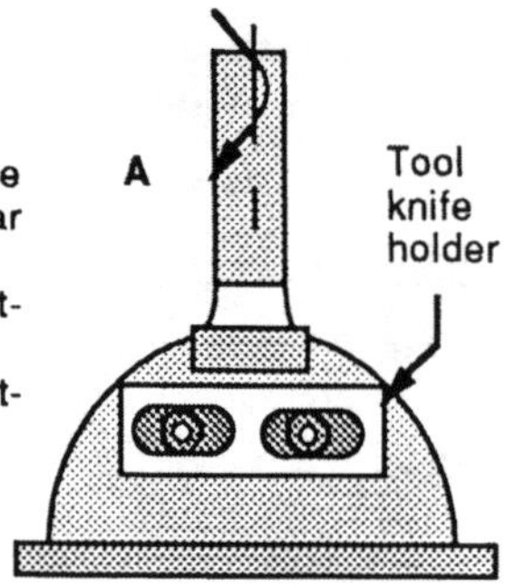

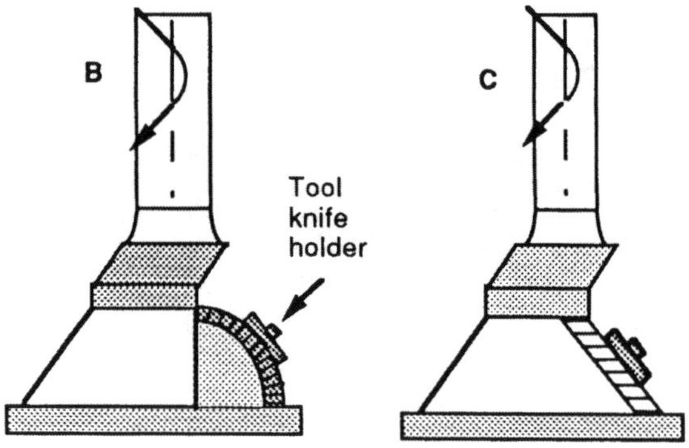

Illus. 9-13—Handle/wood-shaft cutting tools

also Illustration 9-14, which shows a number of common shapes for handle ends. For the more simple shapes, there are numerous cutting tools available for use on different machines, even common lathes and drill presses. The manufacture of more complex wooden objects often requires that the shapes be produced in multiple steps on machines designed to perform specific portions of the needed tasks.

Illus. 9-14—Handle shapes

The cutting tool knives used to shape complex profiles and the cutting process itself are not completely unlike others, except that the workpiece may be manipulated differently. To cut a rounded profile on the end of a broom handle could involve a shaped knife revolving about the stationary workpiece, a workpiece in motion being brought to the stationary shaped knife, or a straight knife moving about the workpiece in a revolving arc to shape the radius detail, with the workpiece in motion. (See Illus. 9-15.)

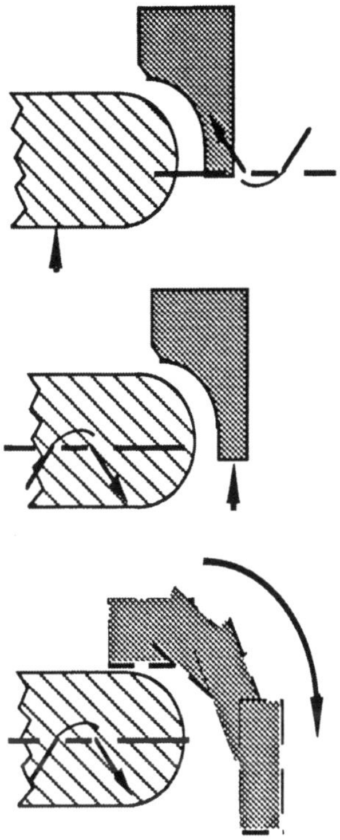

Shaped knife, work-piece stationary

Shaped knife, work-piece in motion.

Straight knife in arc motion, workpiece in revolving motion

Illus. 9-15—Handle cutting: Workpiece and tool manipulation

Carving Tools

Many of the complex shapes that wood is cut into involve the use of machines that manipulate the workpiece as it is brought into contact with the cutting tool. In most other machine processes, the cutting tools revolve in a circular motion, and the workpiece is transported to the cutter only in a linear motion, such as with a circular saw, a router bit, or a feed-through moulder. There are other machining systems that place both the cutting tools and workpieces in rotary motion, as in the mass production of decorative spindles. In others yet, the cutting tools are in rotary motion and the workpiece oscillates only in arc movements of either 90° or 180°, as, for instance, in the manufacturing processes of a mechanized carving/duplicating machine. (See Illus. 9-16.)

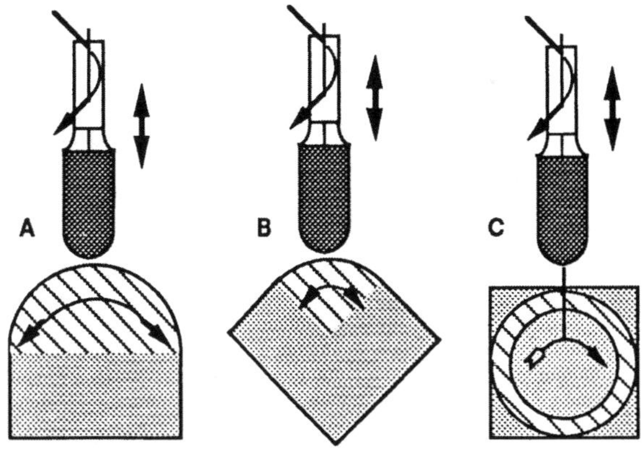

Radiused cutting tools and manipulation of the workpiece make the duplication of nearly any object possible, mechanically.
A. Stock manipulation, rotating oscillation 180°
B. Stock manipulation, rotating oscillation 90°
C. Stock manipulation, rotating 360°
Workpiece manipulation of current technology machines is accomplished with either mechanical or computer control systems.

Illus. 9-16—Mechanized carving processes and cutting tools

The woodworking industry has developed a wide range of automated, intricate machines to shape complex profiles and figures with remarkable speed and accuracy. The time required to complete the machining of a simple but ornate spindle could be less than 10 seconds. To carve a half-figure (180° of relief), three-dimensional statue could take from 10 to 15 minutes, and a 360°, three-dimensional figure could require 20 to 30 minutes. Large, fully automated machines can carve 20 or more such figures simultaneously.

The cutting tools that are used on carving/duplicating machines are nearly as diverse as the machines themselves. (See Illus. 9-17 and 9-18.) The most prevalent carving-machine cutting tool has many of the characteristics of a plunge router bit. The primary difference is that the tool has a more aggressive cutting angle, larger gullet space to allow waste chips to be

Carving tools are most often produced with diameters from 1/8" to 1/2" in several shapes

Shaping burrs are produced in numerous shapes and sizes

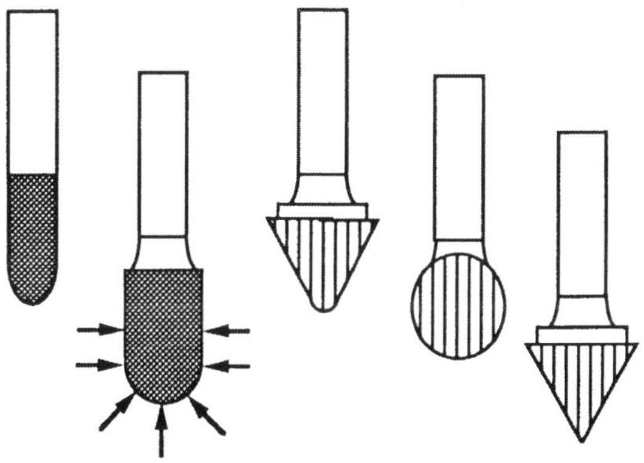

Tools used to cut complex details are manufactured in several standard shapes and diameters, with tool blanks also produced for the end user to shape. The tool knives are similar to router tools. Cutting primarily solid wood, they are most often made of high-speed steel. For cutting very abrasive materials, tools are also available tipped with tungsten carbide.

Illus. 9-17—Carving tools, shaft type

Cup cutters are mounted on numerous bodies, from those that are similar to saw blades to others the size and shape of shaper tools. Cup cutters are produced in various diameters,cut all about their entire periphery, and are simply rotated as they dull. To accomplish the cutting processes, workpieces and cutter bodies are manipulated about each other by complex mechanical systems.

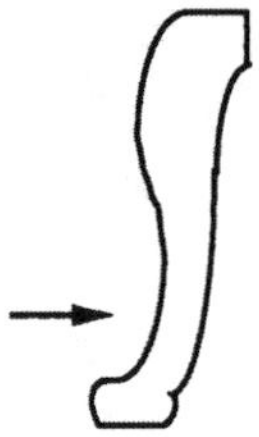

Example of cup-cut workpiece—Queen Anne leg

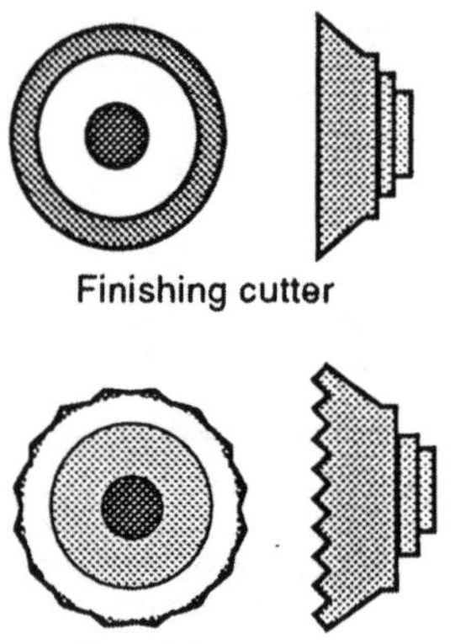

These unique cutting tool knives are available for numerous machine applications in sizes varying from less than 1/4" to more than 2" in diameter. The cutters are made of high-speed steel tipped with tungsten carbide or made of solid tungsten carbide; most can be sharpened.

Illus. 9-18—Cup cutters for carving and duplicating machines

removed easily, and larger, more open relief angles. The cutting tools contact the workpiece on both the end and lateral portions of their bodies. The movements between tool and workpiece are in all three planes of direction, requiring considerable strength, though the chip load most often is not heavy.

The use of cutting tools on carving machines may be progressive; that is, there may be more than one cutter involved in the same process—an initial large, rough, cut-to-size tool and another smaller cutter to complete the details and finish dimension. The logic here is that larger cutting tools can remove more material quickly, providing an overall time saving. The different shapes of cutters provide the ability to complete a wide range of detail features.

The *shaping burrs*, such as in Illustration 9-17, are not widely used, except on unusually dense, hard materials. The cutting edges are small and cannot remove large amounts of material. Their benefit is to reduce the amount of subsequent sanding that may otherwise be necessary. Shaping burrs are more commonly used with manual, hand-held machines to provide intricate detail and clean the mechanically produced carvings prior to finish procedures.

One cutting tool knife that is unlike all others and functions very well as a carving tool is the *cup cutter*. (See Illus. 9-18.) Cup cutter knives are secured to the periphery of tool bodies mechanically and can easily be removed for service or sharpening. The tool body assembly could look like a circular saw blade or shaper tool. As the cutting edge at the periphery of the cutter is dulled, the individual knives are simply rotated to a different position. When the entire edge of the cup surface is dulled, it can then be removed and sharpened.

The cup shape, offers several distinct advantages in geometric features for carving or other similar processes where the primary objective is to remove large amounts of material quickly and provide a tear-free surface. What is first apparent in the description of the geometric features of a cup cutter knife is that the peripheral shape is circular and provides a continuously changing shear cutting surface. (See Illus. 9-19.)

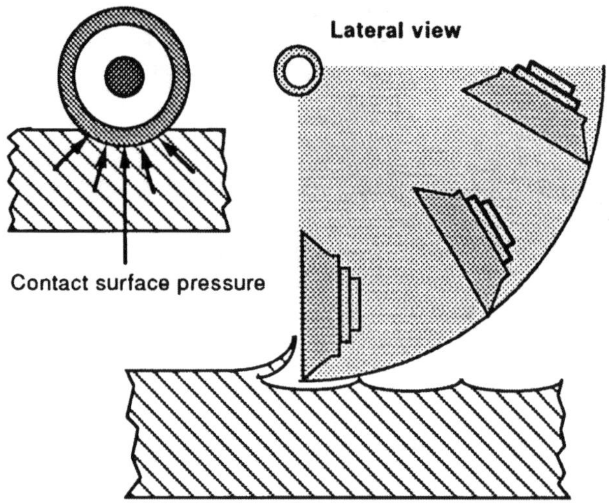

Illus. 9-19—Cup-cutter contact surface pressure

A cutting knife set at a shear angle divides the contact surface pressure along the leading edge of the knife in one direction, but a circular knife divides the pressure in two directions, and the amount of deflection changes continuously along the knife as it progresses in its forward movement. It is the division of contact surface pressure that reduces the energy required to sever the wood fibers and provides a cleaner, tear-free surface.

The function of the serrated cup cutter knife is to alter the shape and size of the waste chips, making them smaller and easier to discharge. Another reason to modify the shape or size of waste chips is to reduce the tendency of the material to tear or split. There are several other cutting tools that use knives with irregular leading-edge shapes with the intent of changing the size or shape of the waste chips; they are generally called chip-breaker knives or teeth.

Dowel Cutters

Among the most interesting wood cutting processes is the manufacture of dowel rods. Wood in the form of round shafts has many uses for woodworkers, as a means of aligning and joining furniture parts, as structural members in numerous constructions, and as trim components. Until the development of refined machines, round-shaft wooden parts were costly and time consuming to produce. In earlier times the only method to produce a rounded object was to lathe turn it and manually shape it to the needed size. (See Illus. 9-20 and 9-21.)

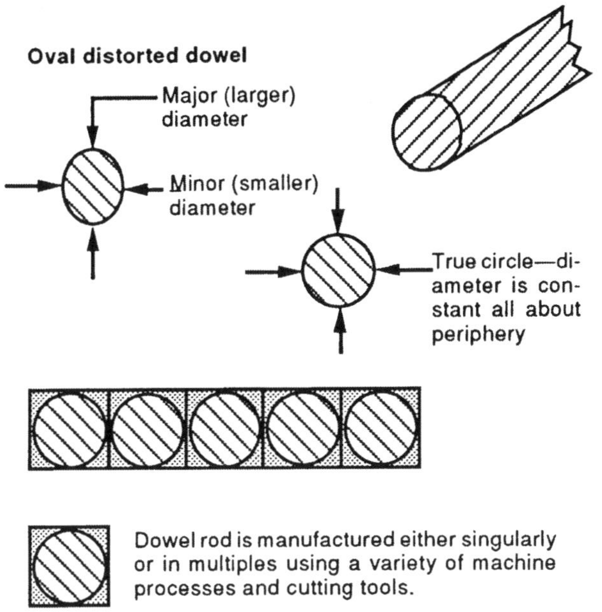

Illus. 9-20—Cutting dowels

Cutting dowel rod is difficult, regardless of the refinement of machinery and cutting tools. To refer to the first chapter of this text, as an engineering and construction material, wood is subject to internal stresses that cannot be seen nor measured. Cutting away a portion of a wood workpiece relieves a portion of the naturally present stresses allowing it to realign and perhaps reshape itself. For this reason, the selection of stock for dowels is as important as the cutting process itself. Further, drying and preparing the stock prior to machining has a more significant impact on the outcome than it does for nearly any other wood product.

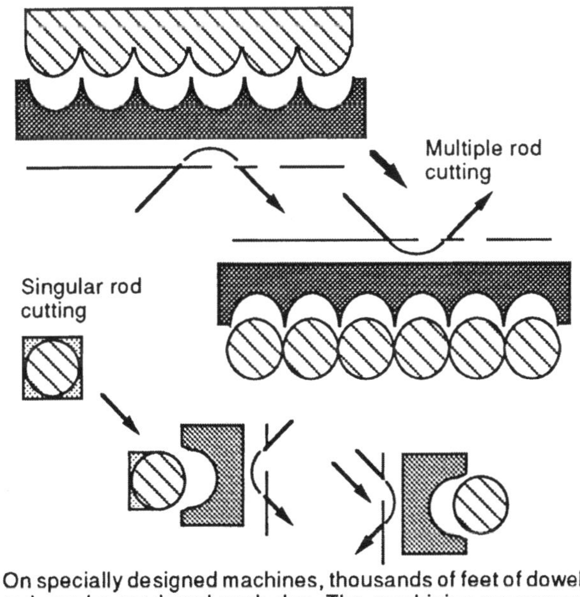

On specially designed machines, thousands of feet of dowel rod can be produced each day. The machining processes and cutting tools have to be carefully controlled and monitored to produce round, straight rods.

Illus. 9-21—Dowel cutting process

The cutting tools and machining systems used to cut dowel have been refined in recent years to the point that both quality and quantity have improved dramatically. Not too many years ago, the loss of production output to scrap was 30 percent and more, caused by a variety of problems inherent in the machining practices and condition of the material being cut.

A last point regarding the cutting and use of dowel stock is that the cutters and machines, if not carefully sharpened and maintained, will produce out-of-round products. The relationship of the geometric features of the cutters are such that if they are not properly maintained, they will cut oval dowels. The use of dowels in the construction and alignment of furniture has an allowable limit of tolerance between the minor and major diameters beyond which the dowels are considered unusable.

Mortise and Tenon Cutters

A rich vernacular has developed in woodworking to describe tools, machines, processes, and products. Over time, the original definitions of some terms have changed or been lost entirely. One example of this is "mortise and tenon," which is the method of assembly in most furniture construction. Originally, "mortise" described a rectangular cavity in a furniture member or a rectangular hole completely through it. "Tenon" originally meant a rectangular protrusion on a member or the end of a structural member, shaped so that it would fit accurately into the mortise. Today, literally hundreds of different furniture assemblies are referred to as mortise and tenon joints. (See Illus. 9-22.)

Numerous machining systems and cutting tools are used to prepare assembly joints. Recent research and development has been directed at examining the strength properties of the materials to be joined together before selecting the joint construction to be employed. This

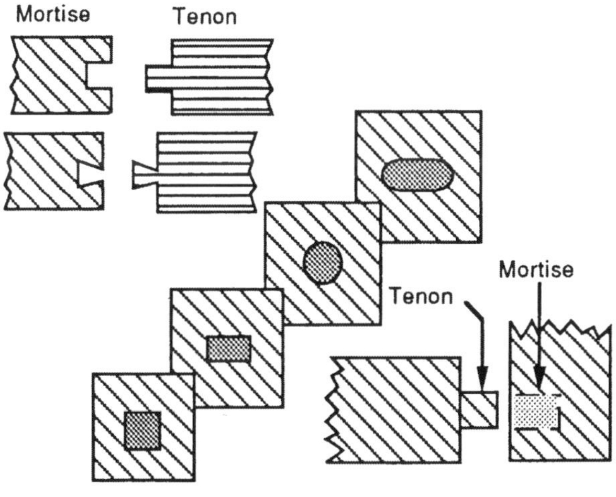

Illus. 9-22—Mortise and tenon joints

represents a historical change; for many years, the same assembly methods were used to produce a particular product—kitchen cabinets were assembled a certain way; traditional furniture was assembled with certain methods; and commercial furniture had its own particular set of normal, accepted assembly systems.

We are having to reevaluate all the assembly methods that we have employed for several reasons: More and different man-made materials are being used that have different strength properties; the use of solid-wood structural parts is diminishing; all machining and assembly methods are becoming more machine intense and less labor intense; the cost of all materials, machines, and tools is increasing; and market competition demands greater efficiency.

For solid-wood products, several joint types will remain effective with regard to strength properties of that material. (See Illus. 9-23.) Several cutting tools and complete machining systems have been developed in recent years to produce round tenon-joint features. They are most widely used in the mass production and assembly of case goods and case-good components such as drawer boxes. These tools are designed for the most efficient use of time and energy. The tools have no extraordinary geometric features except those necessary to control the very close tolerance dimensions called for.

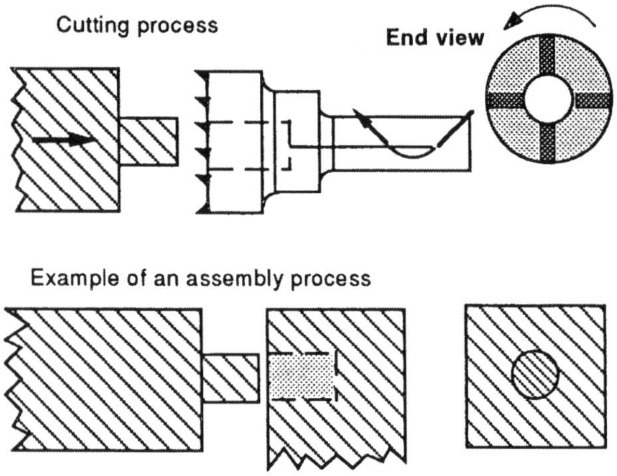

Illus. 9-23—Typical round-tenon dowel cutter

Cutting square and rectangular tenon features is accomplished with common cutting tools that also cut grooves, rebates, and dadoes. These are refined machining systems that completely automate the cutting of the entire detail, regardless of complexity.

Cutting mortise details is more difficult than cutting tenons, most notably because the features are inside the mating workpiece. A rectangular tenon is relatively easy to make with nearly any cutting tool and machine system. To prepare a mating rectangular hole or cavity, though, often requires intensive effort with hand tools.

The ideal, then, would be to design both mortise and tenon features that could be cut in a machine process that preserves the strength properties of the material. There have been many attempts to develop wood-part assemblies that could be completely machine cut, and that effort continues. Among the earliest tools for cutting square tenons was a device that incorporates a drill and a chisel working in unison. (See Illus. 9-24.)

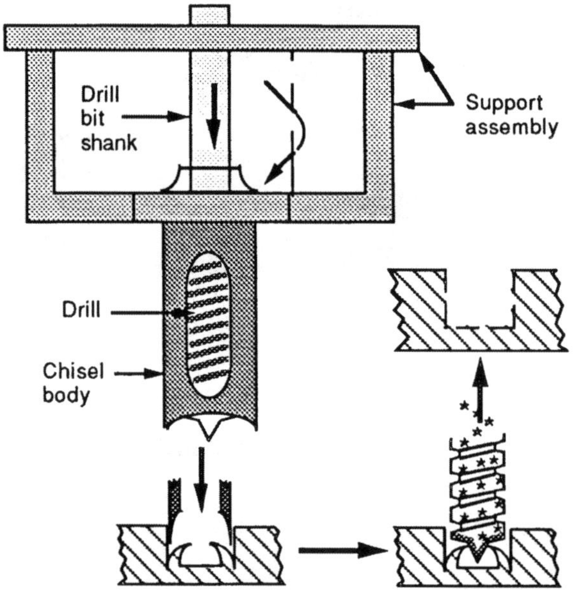

The upper support unit is secured to the drill-press quill assembly, and, as the unit is forced downward into the wood, the outer chisel surfaces curl chips into the path of the drill already operating in the center. Essentially, this is a combination of four angular chisels surrounding a drill capable of producing perfectly square holes.

Illus. 9-24—Drill-press chisel mortiser

The *chisel mortiser* is used on manually operated or power-fed drilling machines. The cutting process is twofold: The downward force drives the leading edges of the chisel that surrounds the drill into the wood, and the drill cuts a circular path, forcing waste from the chisel into the waste stream of the drill to be evacuated. The tool successfully cuts square and rectangular mortise features, though it has limitations. For extended or repeated use, a chisel mortiser of the type illustrated is maintenance intense, requiring frequent and difficult sharpening. Wood species that can be cut are limited to those that are not hard nor dense.

The *rotary mortising cutter*, the most common mortising tool currently in use, is a shaft cutter with the features and appearance of a plunge router bit. (See Illus. 9-25A). The design intention of the chip-breaker configuration is to reduce the size of the waste chips and provide better cut-surface quality. Mortising tools are used in all grain orientations and must be as efficient and clean cutting as possible to give assembly joints the needed strength. Rotary mortising cutters are strong

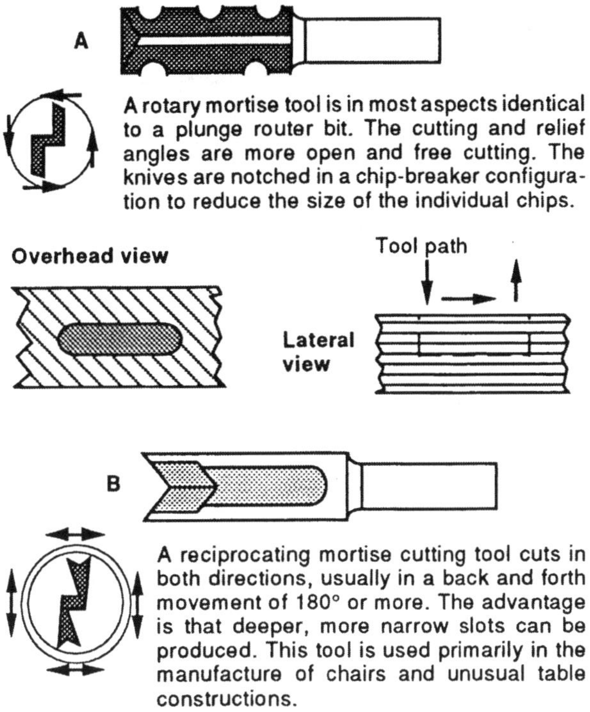

Illus. 9-25—Rotary and reciprocating mortise cutting bits

enough to cut most wood species or man-made materials without problems.

The *reciprocating mortise bit* in Illustration 9-25B, though not as widely used as the rotary type, is more effective for producing narrow, deep cavities. The unusual back-and-forth reciprocating motion of the tool creates short waste chips that are not as likely to plug the gullet spaces and are easily evacuated from the cavity created. The cutting method is slower than the rotary type because of the small chip load and the slower cutting motion.

Another early tool for cutting square and rectangular mortise cavities looks like a miniature chain-saw, which, in fact, it is. The term *chain mortiser* describes both the machine and cutting tool. (See Photo 6.) The cutting tool itself is an assembly that has a closed series of chain links with cutting teeth on their peripheral surface. The tool is manufactured as an assembly, complete with chain, guide bars, idle, and driving sprockets. The tools are produced having any of several different pitches, widths, lengths, and depths. These unusual tools and machines are most used in the manufacture of large furniture pieces—seating, tables, and case goods. Chain mortise cutting tools are designed only for specially made machines.

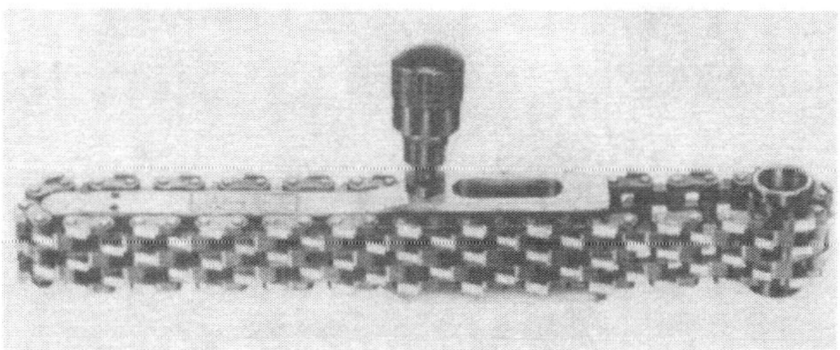

Photo 6—Chain mortiser

For cutting solid woods, preferably without glue lines, chain mortisers are still in use and successful. The machine process of chain mortising, though, is labor intense and costly. Machine manufacturers have not found sufficient demand to refine or futher automate the process in many years.

The *oscillating mortise cutter* is another cutting tool manufactured only for specially designed machines. (See Illus. 9-26.) This tool is simply a series of chisel-shaped teeth that face in alternate directions and are mounted on a common knife segment. The cutting process is identical to that of a chisel: The tool knife segment moves back and forth as it is simultaneously fed into the workpiece. The benefit of this cutting system is that it is very fast, though the finish quality is to be considered structural and not cosmetic. The inside finish quality of a mortise cavity needs to be adequate only to provide a surface that an adhesive can adhere to since it is not seen at all. The irregularities of a surface that will be bonded with an adhesive should not exceed the thickness of the glue film that will be used. A popular misconception is that all adhesives will fill depressions and other irregularities, when, in fact, unless they are specifically designed to both fill and bond, they do not have that ability at all.

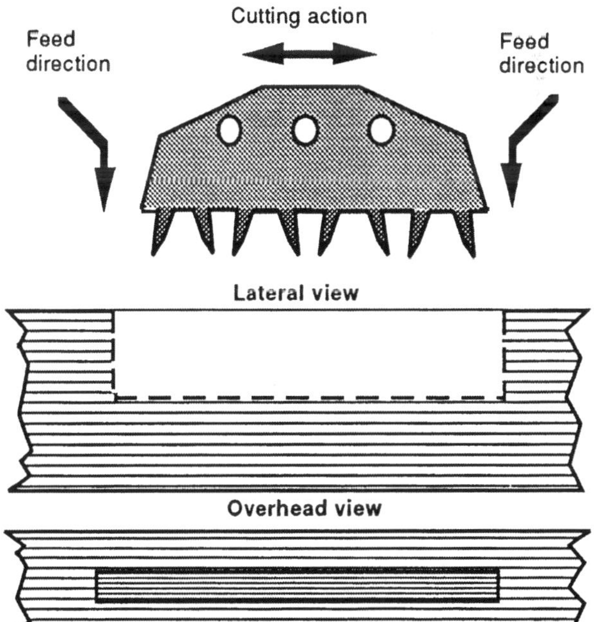

Illus. 9-26—Oscillating mortise cutter

The oscillating mortise chisel machine and cutting tools are widely used in constructing objects such as doors that have long rail-and-stile structural members. There are several lengths and widths of cutting tools available to produce the specific features.

Glue-Joint Cutters

Surfaces of joints that will be secured with an adhesive have two basic forms—structural or cosmetic. In the first instance, the primary concern is directed at the structural integrity and strength of the joint, such as a finger joint or lock-miter joint. In the second instance, the requirement is that the joint have a particular appearance, but is not integral to the objects' structural strength, beyond perhaps extending the contact surface area. Examples are trim on furniture pieces or assemblies of trim components.

Illustration 9-27, shows several glue joints for specific purposes:

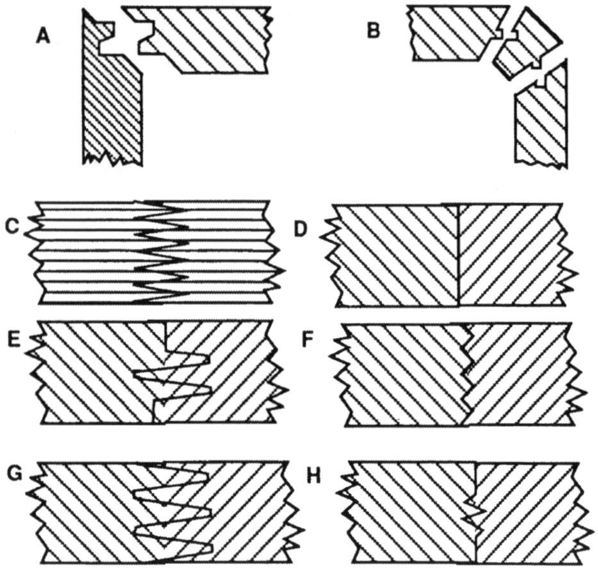

Illus. 9-27—Glue-joint profiles

- A is a *lock-miter* joint that is both cosmetic, in closing the corner to expose no end or side grain, and structural, in that the shape allows the mating surfaces to displace any stress on the joint. A third benefit is that the two profiles are easily aligned in assembly, hence the name.
- B is a decorative three-part joint assembly that, while easily assembled, does not have much more strength than a common butt joint. The length of the mating profiles does not greatly extend the contact surface area to add to its strength.
- C is a *finger joint*, the strongest and most difficult to cut and assemble. As further explained in Illustration 9-29, joining wood parts end grain to end grain with a finger joint will produce an assembled part that has 90 percent or more of the strength of a similar uncut section of the same material.
- D, a simple *butt joint*, is the most commonly used for joining lateral wood sections together, as in a solid-wood tabletop or the vertical wall sections of a solid-wood cabinet. The edge surfaces need only to be smooth and parallel prior to assembly by clamping. The butt joint is the weakest joint relative to others because the contact surface area is only as thick as the two panels joined together.
- E is a glue joint used primarily to join presurfaced solid-wood or man-made panels edgewise, which is the reason for the shoulder at the outer edge. The profile could also be used to join parts together lengthwise if the function was nonstructural, as in using short pieces of trim stock.
- F, the *serrated glue joint*, is used primarily to provide alignment to edge joints and to add strength by extending the contact surface area significantly.
- G, the *square finger joint*, is primarily used to join end grain to end grain for nonstructural parts, similar to the function of profile E.
- H is similar to F, except that it is primarily for stock that has been presurfaced to a particular thickness. The function is to speed the assembly process by providing alignment between the two mating parts and a squared outer-edge surface.

All of these glue-joint profiles, excluding the butt joint, and others add significantly to the integrity of a wood structure by extending the area that makes contact, and they speed the assembly process by providing alignment.

Tool Knives

In Illustration 9-28, cutting tool knives A and B are the same lock-miter cutting tool in different orienta-

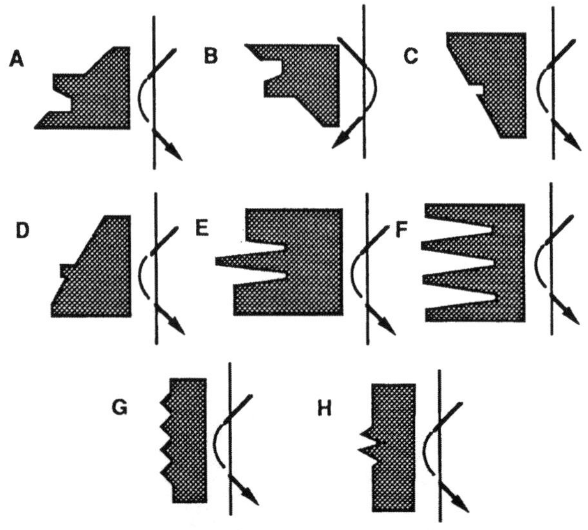

Illus. 9-28—Glue-joint cutting knives

tions as they would function to cut mating surfaces. Critical to their success is careful alignment between the cutting tools and the work pieces, even though the same tool cuts both profiles with the work pieces in different orientations. C and D are the tool knives needed to accomplish the three-part mitered corner profile. Note that the cutters or work pieces will have to be in different orientations to complete all four profiles needed. In E, F, G, and H, the cutting tools cut in the same orientation and direction to complete both mating portions of the profiles, differing only in height location from the profile to counter-profile.

In Illustration 9-29, the finger-joint details show many of the critical features of the process. Used primarily to salvage short rail parts for structural purposes, the cutting and assembly process returns 90 percent and more of the wood's original strength. Finger jointing is the only successful means, without any mechanical fastener, to join end grain to end grain to withstand structural use.

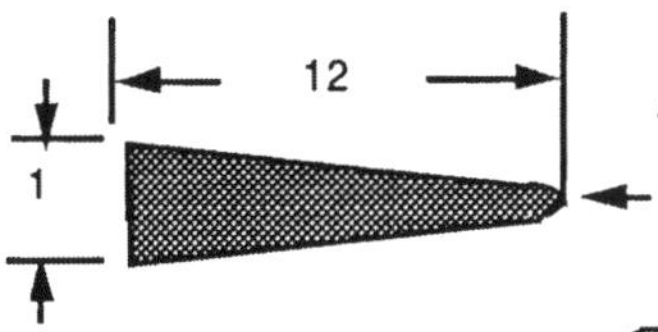

A The ideal finger joint cutting knife detail

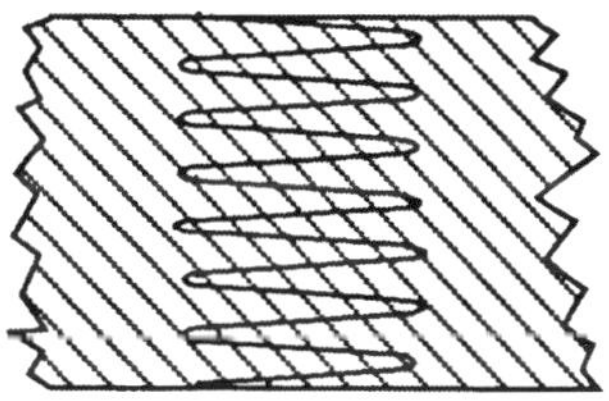

B Lateral view of finger-jointed stock as seen in many products

Of all wood-joint cutting, none requires more precise attention to detail than finger jointing. The same cutting tool performs both the profile and counter profile of the process. The purpose is to extend the use of short pieces of wood that would otherwise be wasted. In A, the cutting tooth shown has a length to height relationship of 12 to 1, whch will accomplish the greatest mechanical advantage of the joint. At a pitch less than 12 to 1, the joint will be weaker; at a greater pitch there is no further gain. The small radius at the tip provides two benefits: to extend the life of the cutting tool and reduce the likelihood of stress being transmitted to the adjacent part. To assemble a sound structural finger joint most often requires from 800 to 1,200 pounds of force, depending on wood specie.

Illus. 9-29—Finger jointing end grain

Critical to this success are the following factors:

1. The profiles must be assembled as soon as possible after cutting to avoid effects of oxidation or dimensional changes.
2. The joint must be assembled with sufficent force to close completely; the force necessary will vary with specie and strength differences.
3. The machining process must be precise and free of vibration.
4. The cutting tools have to be kept sharp, clean, and well serviced.

The machining process of cutting finger joints can be accomplished even on a small shaper fitted with correct holding devices, but it is most often performed on large, specially designed machine systems. Unless a wood product manufacturer generates a large volume of short offal pieces that can be salvaged, the cost of producing finger-jointed stock is high relative to the worth of the final product.

Advancement of Technology

Throughout the woodworking industry, there are hundreds of cutting tools and machines designed to perform tasks that are unique to specific products. Further, there are cutting tools that serve to solve problems not often seen outside large industrial manufacturing facilities. The differences between large and small manufacturers are not limited to the physical size of the facility or the volume of product. Many of the machining processes, material-handling systems, and manufacturing methods are radically different also.

Just the past 15 years have seen dramatic changes in technology and manufacturing methods in many areas of woodworking. These advances are moving the industry very quickly from labor intense to machine intense; that is, more of the tasks to produce a finished product are being done in machine processes that require less manual labor.

Machine-intense technology advances much more easily in a large manufacturing facility because the individual tasks are easier to segregate into work units that can be done with machine processes. For smaller facilities, many tasks have to be grouped together and, thus, are beyond the ability of most machine processes. The rapid advancement of multiple-function, computer-controlled machining centers, what are commonly called CNC (computer numerical control) routers, which can also drill holes, cut to size, and perform many other machining processes, is making that distinction less prevalent.

An example of machine-intense manufacturing can be seen in the differences between how large and small manufacturing facilities address the cutting of lumber stock or panel materials they need in their product to net size for subsequent machining and other operations:

- The small manufacturer will probably feed material manually into a circular saw simply to cut away the unwanted stock. Afterward the offal will have to be retrieved for disposal, and the floor swept.
- The large manufacturer will most likely conveyor feed the material into an automated saw or tenoner machine to cut it to

dimension. During the cutting process, the waste is pulverized with special cutting tools into chips and sawdust to be evacuated by a vacuum dust-collection system.

The example here is only a bit oversimplified. The small facility absorbs the cost of manually handling the workpieces and waste, for instance, that larger manufacturers do not. The larger manufacturer can produce disproportionally more product with fewer employees simply because more machine processes can be dedicated to individual tasks.

Numerous tools have been developed to reduce the labor of machining processes. A prevalent problem is the disposal of waste of all types, in particular small pieces of any material that has been cut, such as the long narrow strips left from cutting of dimension lumber.

Edging Saws and Hoggers

A solution to having to handle and later dispose of narrow waste pieces is to pulverize them into chips and sawdust that can easily be gathered into a dust-collection system. The terms edging saw and hogger are used interchangeably, but the tools are significantly different in construction, even though they perform identical cutting processes. (See Illus. 9-30 through 9-32.)

An edging saw pulverizes the narrow waste of rip-

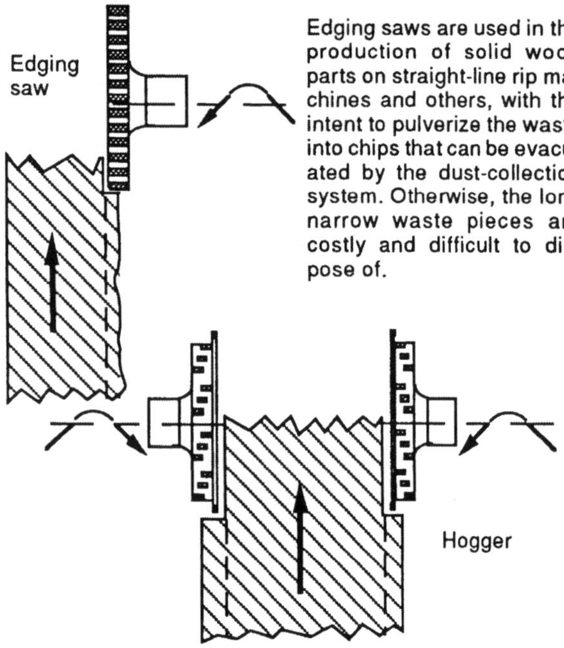

Edging saws are used in the production of solid wood parts on straight-line rip machines and others, with the intent to pulverize the waste into chips that can be evacuated by the dust-collection system. Otherwise, the long narrow waste pieces are costly and difficult to dispose of.

Hoggers are used primarily on tenoner machines. There are several designs currently in use; the most prevalent uses either disposable knives of solid tunsten carbide or other alloy that incorporates saw blade segments to pulverize the waste into chips.

Illus. 9-30—Edging saw and hogger

cutting raw lumber to rough-size dimension. The cutter has to have strength to withstand the effects of cutting

Hoggers and edging saws differ significantly in construction. Edging saws require much greater strength to withstand cutting solid-wood materials. Hoggers cut panel materials that do not have great strength, though the processes are often much faster and require different features. The result of using either tool is identical—pulverizing waste into sawdust that can easily be evacuated.

Neither edging saws nor hoggers run exactly parallel to the direction of workpiece travel, as there would be no lateral clearance. The offset though, is minimal, approximately 1.2°.

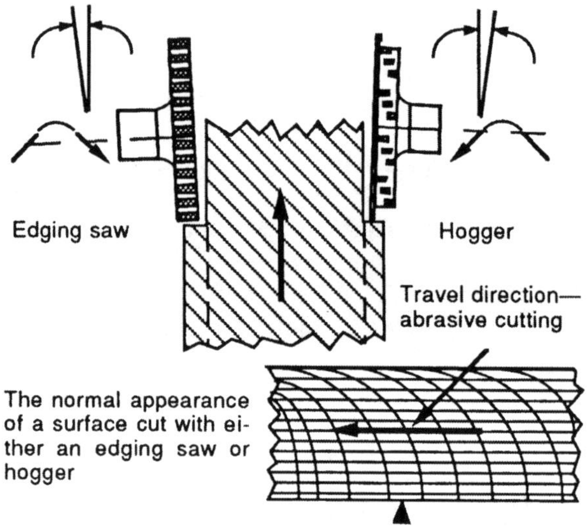

Illus. 9-31—Edging saw and hogger functions

all solid-wood species, both soft and hardwoods. The cutting widths of these tools will vary from 1/2" to 2" or more.

A hogger is a later development of cutting tool technology, designed to size to dimension man-made panel materials. The cutting processes are most commonly much faster and the materials have less strength than those encountered by edging saws. The construction of hoggers, then, does not require the great strength

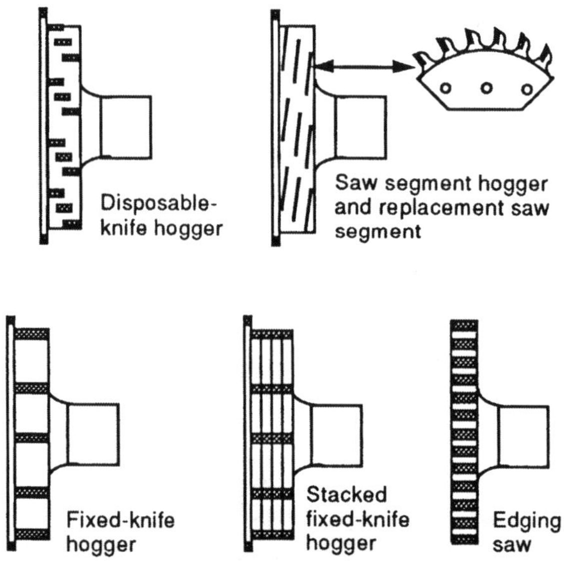

Illus. 9-32—Types of edging saws and hoggers

of edging saws, even though the cutting processes are identical. The cutting widths of hoggers vary from 1/2" to more than 3".

The finish quality of the cut surface created by hoggers and edging saws requires subsequent machining to be even reasonably smooth. The tools are effective in part because there is a slight angle of inclination in the cutting process, as shown in Illustration 9-32. The design intent of the entire machining process is to pulverize waste into chips at considerable speed. The finish quality created by both is not intended to be even glue-joint smooth.

Throughout the woodworking industry, there are many other examples of cutting tools that have been developed to reduce or alter the process times and labor content of machining procedures. As design features and manufacturing practices continue to change, there will be more advances, with cutting tools having even greater feed speeds and other improvements to their performance.

Other Special-Purpose Tools

A contributor to this text suggested a listing of unusual and interesting cutting tools that we did not examine and discuss; to wit, the following in no particular order:

- Plug cutter;
- Log joinery cutters;
- Dove-tail cutter;
- Unusual drilling tools;
- Unusual router tools;
- Unusual groove- and box-joint cutting tools;
- Unusual miter cutting tools for different machines;
- Unusual cutting tools designed for specific machine types, such as buck shapers, profilers, and carving machines;
- Unusal door and window cutting tools;
- Tenoner machine tools;
- Powered lathe tools, both manual and automatic;
- Cutting tools for unusual materials such as veneers, soft metals, hard plastics, and many others;
- Unusual tools for moulders, shapers, and similar machines,
- Unusual tools for hand-held drills, routers, and circular saws.

By now the point is made that there is a significant number of cutting tools beyond drills, router bits, and circular saw blades.

Summary

Throughout, this text has attempted to show the current level of technology available in all areas of woodworking. The unusual cutting tools and machining processes examined here are but a small cross-section of the many developed in recent years. Today, within our industry, a remarkable transformation is taking place, and woodworking technology has in the span of a few years made tremendous advances. Cutting tools and machining systems readily available today did not exist only ten years ago.

Advances in metallurgy, mechanics, and tool and machine design have all contributed to how we perceive and accomplish nearly every manufacturing process. The technologies have touched every imaginable area within woodworking industries, from log harvesting to packaging a finished product for transport to the customer.

With these advances come new tasks and responsibilities for woodworkers—all of us. The first is to completely understand the technology well enough to use it. What we have to understand is that, as the technical content of our work changes and expands, we have the responsiblity to seek information and comprehend it. A current buzz word in industry is "world-class manufacturing," an idea which directs that every employee in a manufacturing facility has the responsibility continually to seek improvements to every facet of the manufacturing processes. This includes investigation of all technologies that touch a particular enterprise—machining systems, cutting tools, materials, material-handling systems, and all others.

There is no way to conclude before the fact that "world-class manufacturing" will solve every problem that we could encounter, but the idea is clear enough that we all have more to learn. The competitive nature of all manufacturing and marketing is such that each participant must continually seek success. If the definition of success is improved quality and precision, or greater production, or greater profits, that will be a stated goal.

For the foreseeable future, a definable task for the entire woodworking industry will be to catch up with the technology that has been developed and delivered to it.

Chapter 10

Safety in Woodworking

Photos courtesy of Leitz Tooling Systems, Inc.

Safety in Woodworking

Safety as it relates to wood cutting tools is often more an examination of processes and machine functions than a concerted effort to evaluate all elements involved. There are essentially three elements to consider: the cutting tools themselves, the woodworking machines that hold and drive the cutters, and the many manufacturing processes required to complete a finished product. Since this book lacks sufficient space to pursue all three in great depth, we will concentrate primarily on cutting tools and address the other concerns as they directly affect the performance of the cutters.

Wood cutting tools are the "end point" of all the processes we use. They make contact with and cut materials into the necessary size, shape, and profile. They are responsible for the quality and precision of all details of every cut surface. However, until cutting tools are secured to the machine that will hold and turn them, we can only describe them in terms of their design features. (The point here is that while tools have undergone many innovative changes in recent years to make them more safe, it is not until we put them into motion that safety concerns actively arise.)

The features that support the intended functions of cutting tools are different measures of strength and size, correct geometric forms, and appropriate metal alloys, among other things. The performance characteristics that permit cutting certain materials at given feed speeds requiring a predictable amount of horsepower are calculated into the design features. However, all of the design features, well engineered and thoughtfully pursued, will be extended dramatically when placed on the machine that will actually perform the needed work. The machines and the processes we use in the course of our work are more important as concerns safety than the cutting tools themselves, even though the cutters are at the end of the "chain of events."

Safety factors:

1. Relate exclusively to features of cutting tools,
2. Can be identified as concerned with appropriate use of the cutting tools,
3. Are an integral part of the machine process involved,
4. Pertain to the integrity of the machine used in the process,
5. Relate to the physical characteristics of the wood material being cut (whether hard, dense, strong, or not so),
6. Relate directly to elements of human error, oversight, and carelessness.

The first five of these can always be improved upon through careful study and attention to detail. The last, though, often defies complete solution.

It may be of interest to note that each year in the U.S., of the numbers of injuries recorded in all areas of industry, most relate to "falls" or attempts to "lift a heavy object." Injuries that involve cutting tools (in woodworking and other areas) make up a very small portion of the list of causes. This indicates that most cutting tools are well made and that there is a healthy respect for them and care given to their use.

In all areas of woodworking, there are many thousands of machines that vary by brand name, features, sizes, and functions. The cutting tools that are manufactured for use on these machines are equally diverse, which makes the range of variables very complex.

Our first task is to define safe woodworking tool use. Safe tool use involves a process that:

1. Does not pose any physical danger to the machine operator, within normal constraints;
2. Can be sustained without harm to the cutting tool beyond normal wear;
3. Will not harm any part of the machine in the normal operation of that process, beyond normal wear;
4. Successfully completes the machining of the wood material with respect to accuracy and finish quality.

Any departure from these concerns should be examined carefully.

If a problem is found in any part of a cutting tool process (i.e., in the tool, machine, or procedure), no complementing circumstance or action will correct it. For example, a saw blade that is out of balance or concentricity will not be corrected or have its performance improved if it is placed on a machine spindle that is perfect in every regard. The same will be true if a perfect saw blade is placed on a machine arbor that is "out of round" or wobbles.

Elements of the Cutting Process

All elements present in a given cutting process must be correct and appropriate. They include:

1. The features and properties of the cutting tool,
2. The application of the cutting tool,
3. The condition of the cutting tool,
4. The condition and properties of the machine to which the cutting tool is attached,
5. The parameters of the process (feed speed and so on),
6. The condition and properties of the material to be cut relative to the cutting tool and machine process,
7. All means used to hold the material or bring it into the cutting process,
8. All safeguards for the machine operator and others in the immediate area (good lighting, lack of clutter, etc.),
9. Rehearsal of the process before it actually takes place, to further examine any possible problems,
10. Adequate training of the machine operator in all aspects of the process to be performed.

We will not attempt to address responsibility for safety of a machine or a feature of a cutting tool that is faulty or not within a standard intended or described by its manufacturer. Nor can we identify specifically what a manufacturer hopes to convey to tool and machine users regarding the manufacturer's view of appropriate product use. *Responsibility* is a large and complex topic regarding the use of any product from automobiles to cutting tools—and nearly any other that can be imagined. A woodworker at any level of experience is responsible for knowing as much as possible about all facets of every cutting process and all tools, machines, and materials to be cut.

Seeing all of the characteristics of any process is difficult in that most often several elements of the process take place simultaneously. Consider a process as simple as cutting lengthwise, or ripping, a piece of wood on a common table saw. The process includes:

- Set-up procedures,
- Selection of the correct tool (saw blade) for the process,
- Correctly securing the tool to the machine,
- Inspecting the machine,
- Setting the fence to guide the workpiece,
- Turning on the motor,
- Examining the revolving tool for any discrepancy,
- Examining the process elements,
- Carefully pushing the workpiece toward the saw blade,
- Completing the cutting process while keeping hands away from the revolving tool with a push-stick or other means,
- Turning off the motor,
- Inspecting and measuring the product(s) of the process.

Most woodworkers perform these steps by habit and reflex, without having to consciously go through them.

Even in the simple process just described, many variables may or may not have been present during the actual cutting process. For instance:

- Did the workpiece push easily into the cutting process?
- Was there any vibration in the machine, cutting tool, or workpiece that could be seen or sensed?
- Did the motor speed decrease dramatically during the process?
- Was the stock pushed with the same force during the entire length of the cut? Less or more at the beginning or end?
- Was the feed speed appropriate for the process? Was there any unusual sound?
- What is the appearance of the chips produced by the cutting process?

- Is the finish quality of the cut surface appropriate and correct?
- Is there any discoloration or disfigurement of the cut surface?
- Are the products the correct size? Are the edges parallel?

The point here is that there is no simple cutting process. All are complex, and each has many variables. Moreover, all woodworking cutting processes take place at speeds much greater than those involved in other processes in our everyday environment.

Vibration

Most aspects of safety that relate directly or indirectly to cutting tools are often difficult to diagnose for cause and effect. Consider as an example a table saw that exhibits considerable vibration. The causes could be (but are not limited to):

1. Improper fit between the bore of the saw blade and the machine arbor (bore too large, arbor too small),
2. Arbor hole in the saw blade placed eccentrically,
3. Machine arbor worn eccentrically,
4. Bearings that support the machine arbor worn or broken,
5. Belt transmitting power from the electric motor to the tool spindle torn or worn,
6. Misalignment of the machine arbor,
7. Saw teeth or body out of balance and concentricity,
8. Saw blade body distorted or bent,
9. Saw blade missing teeth,
10. Part of the entire assembly not secured (i.e., loose).

Vibration in all of its forms poses the greatest danger to all cutting processes. When put in motion, a cutting tool takes on the added characteristics of the machine that holds and drives it. The effects of the work load placed on the cutter, the physical properties of the material being cut, and even the varying rate at which the cutting process takes place also create different vibration mechanics to deal with.

Another characteristic of vibration mechanics that has significant consequences is the elevated temperatures at which most cutting processes take place. (The range of temperatures that cutting tools encounter on different parts of their bodies can vary from room temperature to over 1,000 ° F.) The study of vibration mechanics is beyond doubt the most complex topic pursued by toolmakers and metallurgists over the past 20 or so years. The effort has been aided with computer-controlled test equipment that has been developed in just the last several years. Vibration is of such critical importance, not only to the cutting tools themselves, but also to all moving parts of every machine, that everything possible must be tried to reduce it.

For cutting tools and the machines that hold and drive them, there are essentially three forms that vibrations take: (See Illus. 10-1, 10-2, and 10-3.)

1. *Torsional*—The twisting action caused by the impact of the cutting tools against the workpiece;
2. *Transverse*—Any flex to the machine spindle or tool body, caused also by the impact of the cutting tools, balance, work load, or wear in any machine part;
3. *Longitudinal*—Any movement along the axis, 90° from the rotation of the tool spindle, caused by wear factors, elongation and distortion of any part of the machine assembly, and other factors.

Torsional vibration takes place in every cutting-tool machine process, though the actual movement can not generally be seen or measured without very refined instruments. The primary cause of torsional vibration is the impact of the cutting tool striking the material being cut, whether the tool has a shaft or is mounted on a machine spindle. The frequency and amplitude of the vibration is a relationship of the physical characteristics of the spindle or shaft, the tool's mass, rpm, balance, and concentricity of the entire assembly.

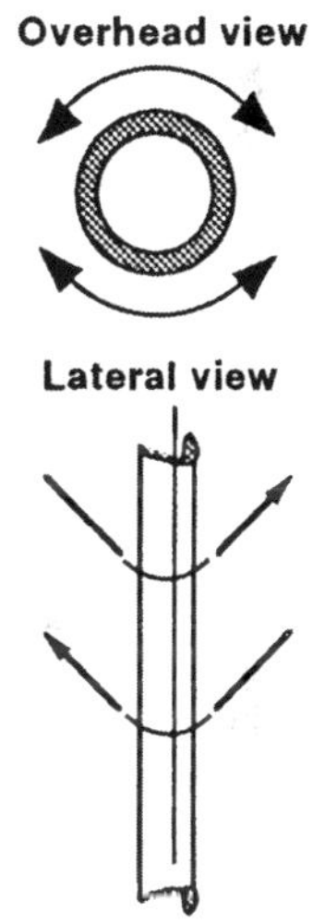

Illus. 10-1—Torsional vibration

The first element of machine and cutting tool vibration to examine is the frequency of movements, or how many times per second the moving parts oscillate. The frequency of vibration movements most often is simply the rpm of the tool spindle multiplied by the number of cutting knives/teeth on a given tool. For example, a shaper cutting tool that has four knives being turned at 6,000 rpm will have a basic frequency of 400 cycles/second (4 X 6,000 = 24,000 + 60 = 400).

It must be clearly understood that there are other frequencies present, which are created by friction, strength of the parts involved, and other features of the machine and the machining process. Other cutting tools with more or fewer cutting teeth/knives operating at different rpm will have other basic vibration frequencies.

Frequency-of-vibration mechanics (in terms of cycles per seconds) is more important to toolmakers

This kind of vibration is the most noticeable and most common of the three. The possible axis movements, flexing (transverse), twisting (torsional), and elongating (longitudinal), all have detrimental and safety effects on any cutting process, and all are present. Transverse vibration is the result of work-related stress, impact shock, and the fit of all moving parts in a machine process. Saw blades can flex in reaction to temperature changes, work stress, or machine vibrations. Router tools can flex due to work loads, eccentricity, balance, feed rates, and especially dullness. Transverse vibration is not only most common but causes unusual effects that are difficult to diagnose and prevent. A flexing saw blade, for instance, may be the result of dramatic changes in the moisture of the material being cut or in the alignment of a part of the machine.

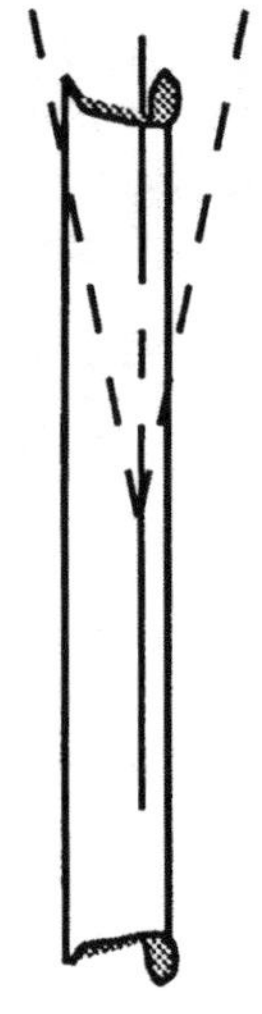

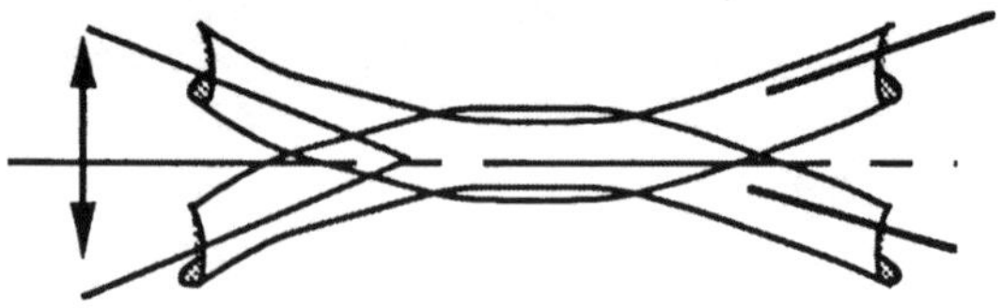

Illus. 10-2—Transverse vibration

and metallurgists than to woodworkers. For toolmakers, vibration characteristics affect design criteria and the choice of alloys used to make the cutters. Metallurgists consider vibration mechanics in all their decisions regarding both the choice of alloys used in machine parts and the heat treatment processes used to achieve needed characteristics.

The identification and measurement of vibration in moving (revolving) objects must be accomplished on two planes that are 90° from each other. Of greatest importance is the amount of vibration movement. *It is, after all, the amount of vibration movement that accounts for most machine wear, metal fatigue, and breakage of cutting tools and machine parts.* The amount of vibration movement is generally called *amplitude*. (This terminology parallels the way we describe "loudness" of sound and certain measurements of electricity.)

Vibration amplitude and its effect on tools and/or machine parts result from how far out of balance the moving parts are, or eccentricity. This involves

- The strength of the supporting parts to withstand the movement,
- The tolerance fit of the parts involved in the assembly,
- The frequency/rpm of the machine process,
- The amount of horsepower present,
- The character of the metal alloys present.

The instruments used to measure vibration become more refined and accurate as computer systems are implemented to support and direct them. Characteristics of vibration are measured in the same manner as sound and electricity. To see the makeup of vibration, its features must be identified in terms how much movement takes place and how frequently it takes place. (Though "how frequently" is not the primary concern, in that it is a controlled variable.)

Illustration 10-4 indicates how we can relate vibration movements to either the pendulum used in a clock or the metronome used to keep time while playing music. The characteristic oscillations of the pendulum, metronome, and vibration can be illustrated by the wave-like movements they produce—movement back and forth, amount of movement, and number of move-

Longitudinal vibration takes place more between parts of an assembly than within any single part. Elongation occurs due to temperature changes and tension put on the individual parts of cutting tool and machine assemblies, but for safety the greatest concern is the vibration between parts. Examples would include any movement 90° from the rotation of a cutting tool or assembly. Most of it is the result of normal or abnormal wear of bearings, machine spindles, tool holding/clamping systems, poor tolerance standards, and other wear and fit characteristics.

Tool systems to be particularly cautious of, for all forms of vibration, are those that are subject to high work loads and great temperature changes.

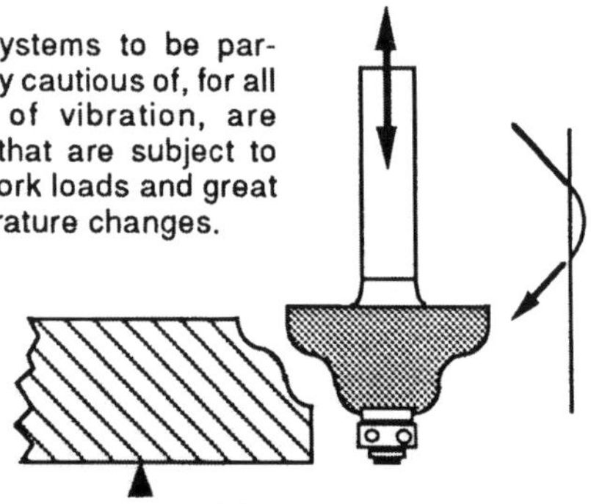

Illus. 10-3—Longitudinal vibration

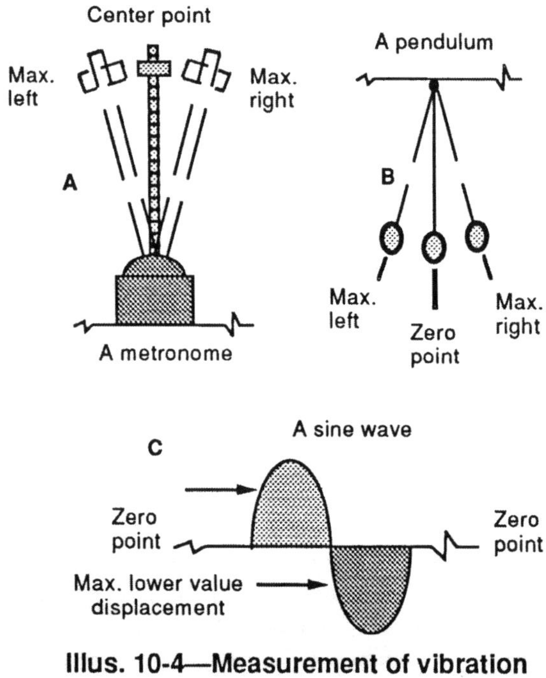

Illus. 10-4—Measurement of vibration

ments per given time period. The wave created by these movements is called a *sine wave.* It can accurately describe all of the oscillation's features, no matter how complex or involved.

To measure vibration, the instruments used create a visual image of the movements with an accurate sensing device. As the vibration present in any cutting tool or moving part of a machine is made visible, the instruments used can do much of the necessary diagnostic work. For example, as the vibration present is made visible, many of its features are shown: the amount or strength, the frequency, and whether more than one vibration is present.

An individual sine wave represents one complete cycle of vibration movement. The total height of the wave, above and below a center (zero line), describes the strength or maximum value of the movement. How many such complete waves occur in a given period is expressed as *frequency* , which is most often recorded as cycles per second. (For sound and other measurements the term *hertz* represents one cycle per second.)

Illustration 10-5 further shows the features of sine waves. The instrumentation now used to identify, measure, and correct the dangers and problems of vibration relies heavily on its refined ability to create visual images. The more advanced vibration-testing instruments use both visual and sound images to speed and simplify the procedures. As computer technology continues to advance, test instruments will further improve.

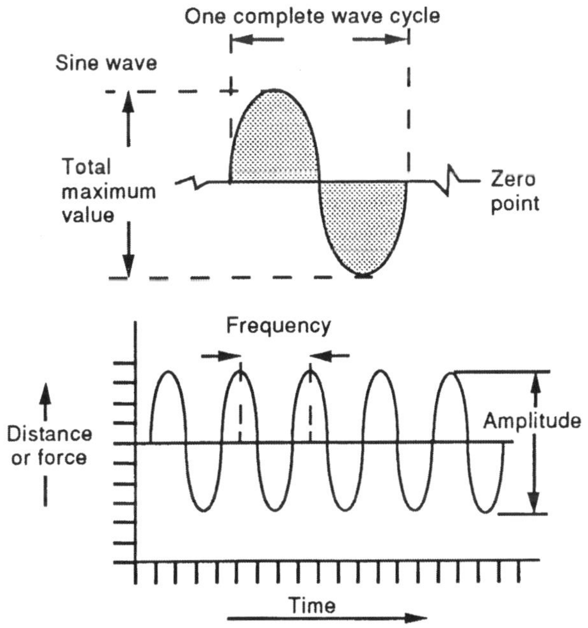

Amplitude of a given wave is the maximum value of the displacement.
Frequency of a number of waves is the number of waves occurring per each second.

Illus. 10-5—Sine-wave measurement of vibration

In further examining vibration and its consequences, it must be clear that vibration is an effect, not a cause. While it is not a normal nor desired effect, vibration is present in virtually every machine movement

Reducing Vibration

Despite the fact that there is a certain amount of vibration present (often very difficult to detect and measure) in every machine movement, the engineering challenge has always been to remove as much vibration as possible. (Though admittedly there are other engineering challenges such as developing faster and more precise wood- and metalworking machines and manufacturing products that are price competitive and functional.) As technology in all manufacturing areas has progressed, many improvements have been made to reduce dramatically both the amount and effects of vibration present in all machines.

Regarding cutting tools, the search for a way to reduce the number of causes of vibration and its effects has taken several paths in recent years. One of the scientific approaches taken has involved close examination of how cutting tools function—determining what each cutter tooth or knife is doing during the time it is in contact with the material being cut. It is possible now, with the use of computer-controlled high-speed photography and strobe-light technology, to study what each cutting knife/tooth does in the course of its work while *in motion.* Different materials and all operating parameters are studied to determine exactly what the cutting process is. As each minute detail is examined, the geometric features can be and are further refined. For example, the size and shape of the spaces provided on a cutter body to collect the waste (the gullets) have been carefully made smaller and more precise, to allow adequate space for the waste chips while being small enough for the tool to be made more circular in shape and easier to balance.

Another way sought to reduce the effects of vibration has been through use of new tougher alloys of steel for making cutting tool bodies and woodworking machines. In the past, for instance, the machines relied on large, heavy cast-iron parts to absorb and contain any movement or vibration. Recently, manufactured woodworking machines are lighter in weight, better designed, and made of several alloy steels. Nearly all cutting tool bodies are made of highly refined steel alloys specifically designed to withstand the work loads placed on them. The research on newer knife materials continues to add alloys that have features that better fit the needs of cutting the materials brought into woodworking industries.

Perhaps the most significant development regarding machinery is the means employed to support moving parts to make them more precise and free of vibration. Advances in metallurgy and manufacturing standards in the production of all types of bearings has in only the past 30 years brought forth systems and methods that are much more accurate, long lived, and able to endure all types of environments. An example of this is the now common "sealed" bearings that are made for nearly any application and require no maintenance.

The evolution of woodworking cutters, with exact processes identified, has seen the development of tools dedicated to more narrow, specific task ranges. For example, we know after careful study of a cutting process that is hand fed, that the fastest an average worker can move the workpiece is about 18 feet per minute. Consequently, all cutting processes that are hand fed need only have as many knives and other features as are required to sustain a feed rate of 18 feet per minute.

This kind of refinement permits establishment of many manufacturing standards, with more precise details. Today there are, for example, circular saw blades designed to cut only solid woods and others to cut man-made panel and composite materials. There are shaper cutters intended for hand-fed operations and others that can perform much faster power-fed cutting processes. The overall point here is that cutting tools are becoming more refined, precise, and "task" oriented.

As cutting tools have become more precise, well balanced, and free of vibration, they have become more circular in shape. In many instances, they appear to be a completely closed circle. (See Illus. 10-6.)

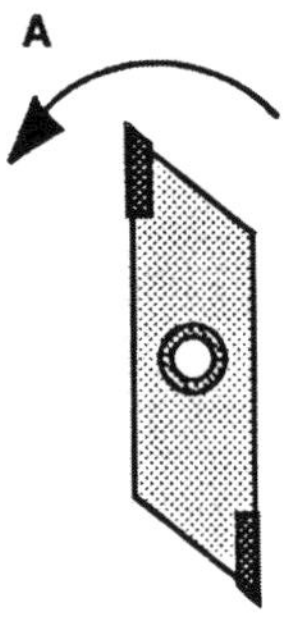

Cutting tool design continues to progress toward better balanced cutters. A and B illustrate cutting tools that were prevalent 35 years ago. They were crude and difficult to maintain in concentricity and balance. Tools manufactured today are more circular, balanced, and have details that are scientifically designed to perform the needed work (C). Where the knives are exchangeable, all ancillary parts are carefully weighed and shaped to maintain the balance and precision. That is, even wedges, screws, and other clamping devices are produced to very close tolerances to sustain balance. Fixed-knife cutting tools are produced to equal and even better dimensional standards that will support high rpm and feed speeds.

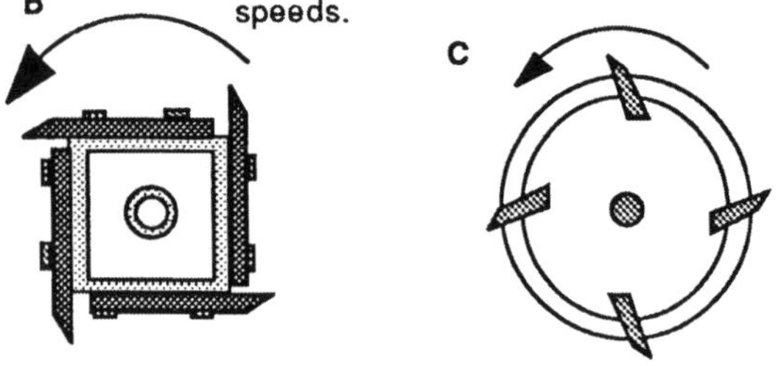

Illus. 10-6—Design progression

As more refined metals and metalworking technology have been developed, cutting tools have improved dramatically. The first shaft-mounted cutters were little more than flat bars or square steel blocks that took on the function of cutting tools. In some instances, the knives were fashioned of the same material as the body; in others, the knives were attached mechanically.

Balance

Balance has always been difficult to achieve and even more difficult to maintain as tools were subsequently sharpened. In the earliest attempts at exact concentricity with cutting tools, several mechanical means were used to measure the tools for both balance and concentricity. However, without modern instruments and measuring systems, the earliest cutting tool manufacturers could not possibly produce the cutters that are common today.

Balance has properties that must be made more clear in terms of what it is and how it relates to safety and cutting performance characteristics. In physics, balance is described under the laws that describe the effects of force, force vectors, motion, velocity, and gravity. Balance for cutting tools can be examined, measured, and described in two ways: while the tool body is stationary and while the tool is in motion. (See Illus. 10-7, 10-8, and 10-9.)

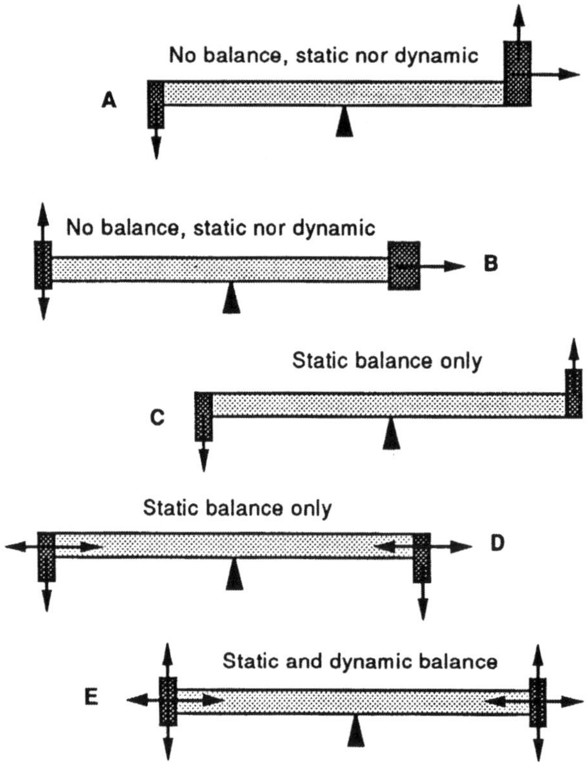

Illus. 10-7—Elements of balance on two axes

Balance, when measured on one plane, while the object is stationary, is called *static balance*, as shown in Illustration 10-7. C and D are each in balance, but they will not be in balance if turned 90°. An example of this would be a saw blade with one or several cutting teeth that protrude farther from one side of the body than from the other. If we laid the saw in a horizontal position, it would be in balance. However if placed on the machine arbor it would wobble and vibrate, and consequently be a very dangerous tool. While motionless, a cutting tool or any object can display balance, or be in balance, on one axis and not the other. However, when in motion, the effects of force, gravity, velocity, and the other axis of balance will effect the blade's ability to run accurately and safely.

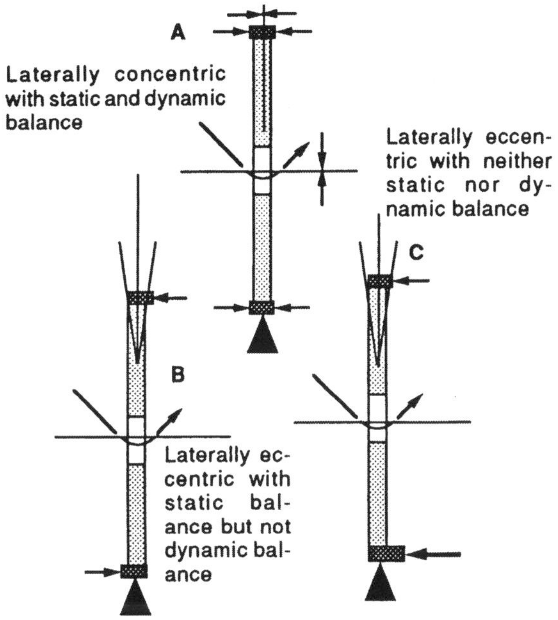

Illus. 10-8—Static and dynamic balance: vertical edge view

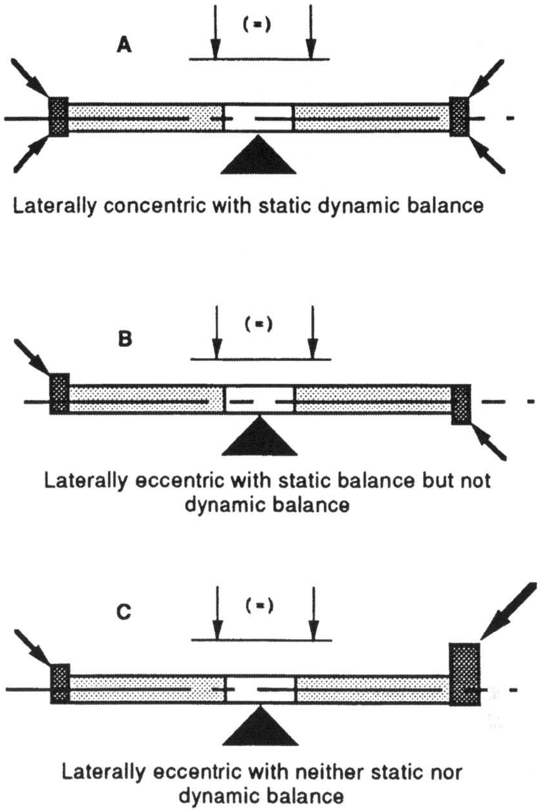

Illus. 10-9—Static and dynamic balance: horizontal edge view

In Illustration 10-7, A, B, C, D, and E are as follows:

A—An edge view of a saw blade or similar tool having teeth at opposite sides of the periphery of unlike size (weight) and not centered on the tool body;
B—A similar tool with unlike size (weight) teeth, but centered on the tool body;
C—A similar tool with identical size (weight) teeth, but placed off center (opposing) in relation to the tool body;
D—A similar tool with identical size (weight) teeth that are both placed parallel to each other and off center in relation to the tool body;
E—A similar tool with identical size (weight) teeth, both of which are centered on the tool body.

Illustration 10-7 could also be used to explain how to balance the tires on an automobile: first they are *static* balanced, then put into motion to complete *dynamic* balancing. (The description of balance on one axis is called "static" and on both axes is called "dynamic".) To put any cutting tool or object into rotary motion, it must be in dynamic balance. To measure both axes for balance, especially with objects of unusual size and shape (such as saw blades and other cutting tools), requires refined instruments that have been in use for many years. Recently, use of computer-aided logic systems has made the measurement of balance relatively fast and very accurate. For cutting tools and similar objects that have small cross-section dimensions, it is more practical to measure the second axis while they are moving. Instruments in current use have been developed to perform such testing with great precision.

Progress through Tool Design

As cutting tools are further refined to improve all performance and safety characteristics, the element of vibration and the means to reduce it are continuously studied. Exactly how each cutting knife/tooth is brought to the process is being changed and improved. The extent to which tool design can be altered to reduce vibration makes it difficult to determine the *circle geometry* involved and the actual cost of producing a given tool.

For all cutting tool functions, as the teeth/knives are brought into the material, the entire body of the cutter—and all of the machine parts directly involved—

have to absorb the contact shock of the individual impacts. For instance, if a shaper cutter tool has four knives and turns at 6,000 rpm, each minute there are 24,000 individual impacts between tool and material. The severity of these impacts relates to the nature of the material being cut, the sharpness of the knives, the amount of material being cut by each tooth, the feed rate of the process, and so on. The relationship of rpm and the number of knives present in a given process dictate the frequency of the vibration created in this particular cutting procedure. If, in our example, the individual knives were not all cutting exactly the same amount of material (that is, were in different cutting circles), it would be possible to have four different amplitudes of frequency operating at the same time. Further, other tools having more or fewer knives and operating at different rpm will obviously have different characteristics of vibration mechanics.

Recently, those studying cutting tool technology have examined more closely the effects that the knives have on the process while actually in the act of cutting, as well as using the duration of contact with the material as a means to reduce vibration. We have long known that placing the knives at other than 90° to the direction of circular travel, which is generally called "shear," will improve the characteristics of finish and the amount of horsepower used.

Placing cutting knives at a shear angle also prolongs the time that each individual knife is in contact while cutting. Directing the contact surface pressure by knife placement also will improve both the surface and edge quality of the cut surfaces. (See Illus. 10-10, 10-11, and 10-12.)

Having the cutting knives placed at 90° in relation to the circular movement makes the individual contact of each knife/tooth segment as abrupt as it can be. Turning the knife only a few degrees dramatically affects all performance characteristics by prolonging the time of the contact period and displacing the pressure created along the knife by impact/contact. Knowing this, we can conclude that all cutting tool knives should be placed at a shear angle—except that circle geometry is not quite that easy to solve, and the cost to the toolmaker is most often very high.

Illustration 10-10 shows four knife arrangements that each offer different characteristics of finish, horsepower use, and vibration. In all these drawings, the cutting/machine process is a simple jointing operation. The differences from one to the other lies in how the knives are placed on the cutter body in relation to the direction of travel. Knives placed squarely, that is at 90° from the travel direction, will have the harshest impact and amount of vibration due to the abrupt contact/impact. Other knife arrangements that include shear-set knives or a mixture of knives will have significantly different performance features. The arrangement in the last drawing, indicating the cutting knife

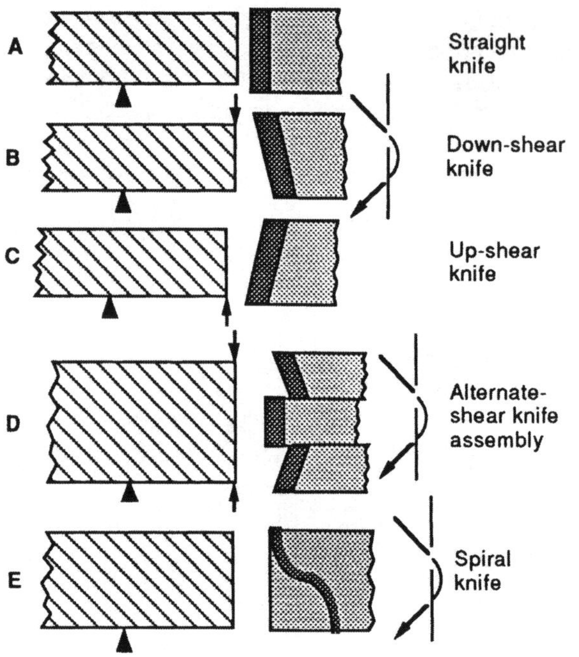

Illus. 10-10—Cutting tooth/knife arrangement to reduce vibration

arranged in a spiral about the cutter body, will produce the best results.

Illustrations 10-11 and 10-12 show the effects of having the cutting knives in the cutting process without interruption. *Since it is interruption of the cutting process that is responsible for much of the vibration and hazard present, completely closing the cutting circle*

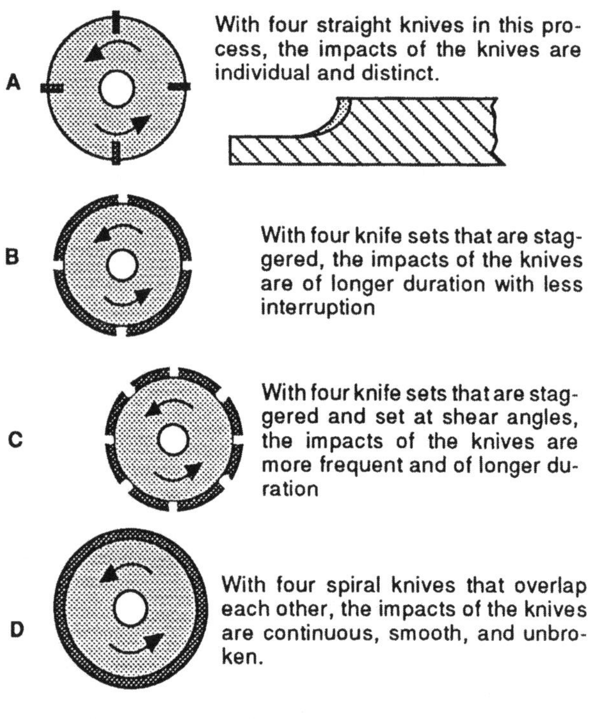

Illus. 10-11—Vibration mechanics of knife arrangements

provides a means for improvement. The arrangement of the cutting knives as shown has the effect of dramatically altering all performance characteristics.

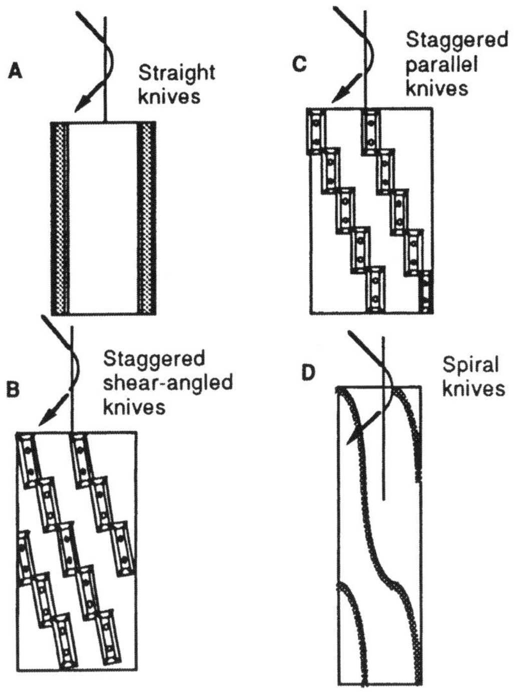

Illus. 10-12—Knife arrangements for reduced vibration and improved machining efficiency

Where the cutting tool uses small individual knives, they can be arranged in a spiral around the periphery of the tool body,which prevents interruption of the process. Technology related to closing the cutting circle first appeared in metal working industries and is just now being applied to woodworking. Cutting tools that utilize spiral-shear and spiral-incremental knife arrangements are very expensive to manufacture. However, they can reduce both energy required and vibration. They will also cut at greatly increased speeds.

In summary, regarding vibration and its effects, it is important to understand that each tool, machine, process, and material to be cut behaves differently. In considering all woodworking procedures, there is no situation where vibration is an accepted characteristic. Occurrences of vibration are dangerous and must be corrected. As mentioned earlier, the particular frequency of a cutting tool is related to the number of knives/teeth present, the rpm of the machine spindle, the material being cut, and other factors. A saw blade having many teeth will have a natural operating frequency of vibration much higher than a shaper cutter with only four knives.

Router tools are compact and do enormous amounts of work, travel at considerable speeds, and are perhaps the most mistreated cutting tools of all. Because router tools have such a small cutting circle and compact features, the importance of their concerns with balance, concentricity, and vibration mechanics is often overlooked—though they should not be.

Cutting tools that are themselves an assembly require particular care regarding vibration characteristics. Most new cutting tools seldom present balance, concentricity, and vibration problems. However, too often, methods and quality of work involved in repair and sharpening can create problems.

Machine vibrations are often more difficult to diagnose and correct than problems related to the cutting tools themselves because they involve more moving parts, which wear—indeed, wear unpredictably. The normal wear of any moving machine part will create vibration in a cutting tool process that can present a safety hazard. As with cutting tools, any vibration in any machine parts presents a danger that is not always immediate, but a danger nonetheless. As technology improves in all areas of woodworking, vibration and the dangers it presents will continue to diminish.

Other Concerns

Within different cutting processes, some mechanical features of the procedures must be examined for safety. (See Illus. 10-13.)

Among the most dangerous errors made by anyone who uses a power machine is underestimating the effect, or force, of horsepower. (One horsepower is defined as 550 foot/pounds per second.) Most stationary and many portable woodworking machines have more than one-horsepower motors. One horsepower is simply a unit of measurement of work energy and admittedly does not seem large—except by comparison. A well-conditioned athlete riding a bicycle to the best of his ability generates only about 0.65 horsepower. So the likelihood of any individual being able to overcome the power of even a small woodworking machine is remote. Most machines have considerably more than one horsepower, with even portable hand-held routers having three or more horsepower. Some large industrial machines have motors of 100 horsepower and even larger. Since there is no possibility of containing the energy present in any given cutting process, we must understand and control all of the energy. From the moment that an electric motor is started on a woodworking machine, we have to understand exactly what we are unleashing and be able to control it safely.

Dull cutting tools are dangerous and more costly to use for several reasons:

- They require more horsepower to cut a given material;
- They require more energy to be fed into the cutting process;
- They generate much more heat and vibration;
- They require more work from all of the

machine processes (motor stresses—bearings, support mechanisms, etc.; hand- or machine-powered feed mechanisms; stock holding and clamping systems);

- Increased danger to the machine operator, the material being cut, the machine, and all its parts;
- Deterioration of process in terms of cutting finish quality, dramatically increased cost and time requirements, and diminished tool life and machine operating life.

To examine the forces present in a process, a common table saw is easy to understand and visualize. What forces are present to safely complete the process of cutting, or ripping, a wood sample lengthwise? (See Illus. 10-13.)

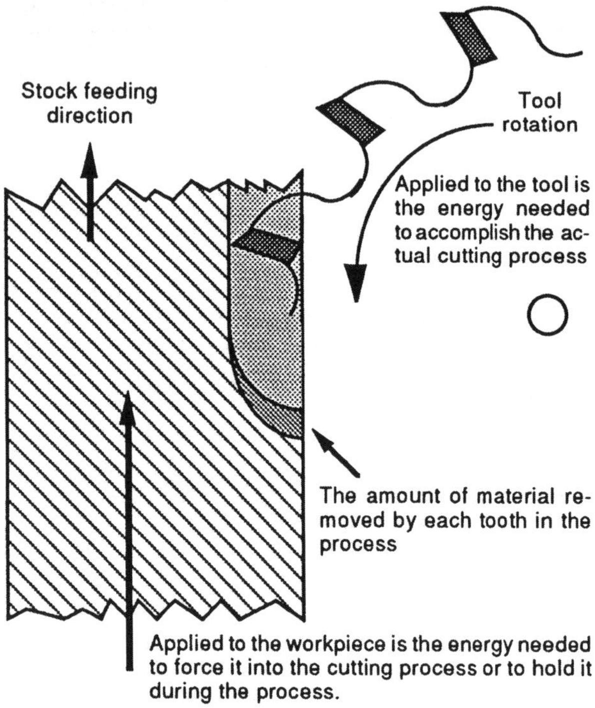

Within a given cutting process, the total energy needed is divided between the amount of horsepower taken by actually cutting the material and energy required to force or hold the material in the process. The condition of the material and the sharpness and geometric features of the cutting tool in some relationship to each other account for all of the forces present.

Illus. 10-13—Stock feeding

1. How much energy/horsepower does the motor that is turning the cutting tool have?
2. How much energy is required to force the wood into the cutting process?
3. How much available horsepower does the actual cutting process consume?
4. Does forcing the wood into the process use all available horsepower and stop (stall) the motor on the machine?
5. What is the relationship between the energy needed to push the wood and the energy needed to cut it?

Only the first question can be easily answered by reading the information on the motor tag. The other four have too many variables to be easily answered, and there are two forces present that effectively work against each other to some degree.

The two opposing forces are the energy needed to cut the material (that is, turn the saw blade) and the force required to push the wood into the cutting process as shown in Illustration 10-13. These two forces are present in every cutting process and are secondary only to vibration mechanics in posing the greatest danger to woodworkers. They present a danger by changing constantly in relation to each other and creating other effects that pose unpredictable consequences. Most of the forces needed to cut the material and bring the material into the process relate to the geometric features and condition of the cutting tools themselves, with sharpness being the single most important concern. (See Illus.10-14 and 10-15.)

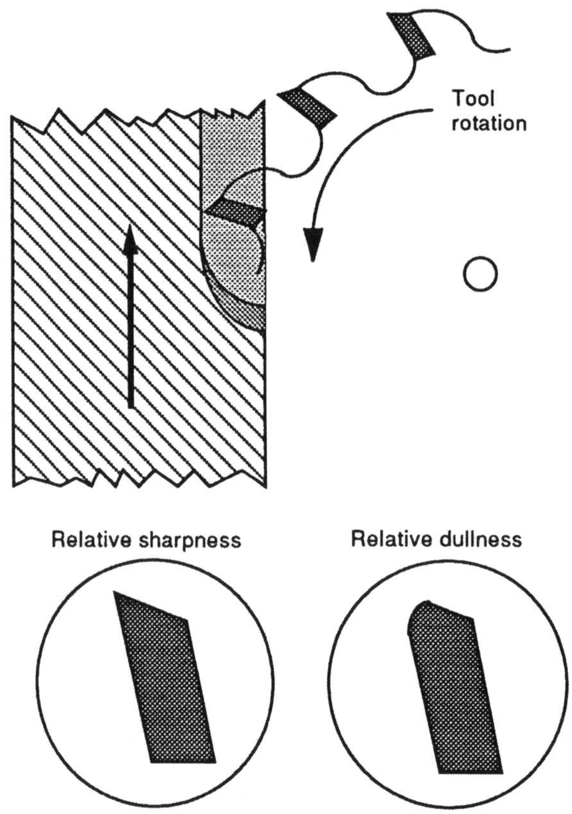

Illus. 10-14—Tooth/knife sharpness

Cutting tool knives/teeth dull during the cutting process, most often at predictable rates for material of known properties (at different rates for natural woods and man-made materials). As conditions such as moisture content change, we have learned through experi-

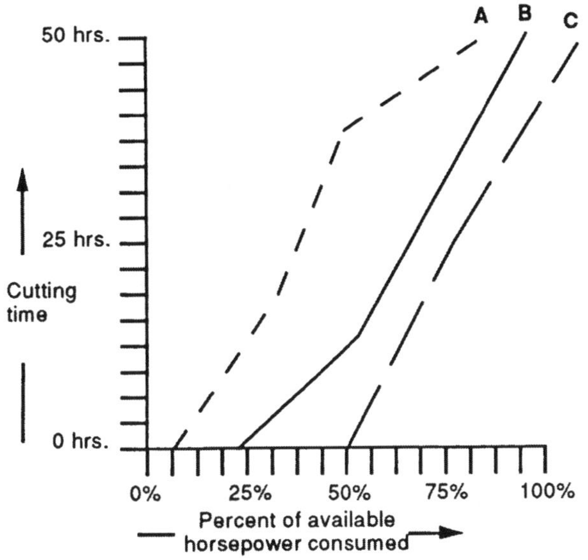

Expressed in this graph is the idea that cutting tools have a more or less definable sharpening life cycle. The variables in each process are the physical characteristics of material being cut, the metallurgical and geometric features of the cutter, and all of the mechanical properties inherent in the process itself. As cutting tools dull, they consume more horsepower, generate more heat and vibration, and, consequently, are more dangerous. A, B, and C express the rate at which cutting different materials may dull the tools in different processes.

Illus. 10-15—Cutter sharpening life

ence or formal training that we will have to deal with different performance characteristics.

Referring back to the initial reference to controlling horsepower, it must be understood that most often several actions take place simultaneously even in the most simple process. In our rip-cut procedure on the table saw, for instance, we must:

1. Assure correct tooling and setup procedures,
2. Inspect the workpiece for squareness and the like,
3. Rehearse the procedure and check for correctness,
4. Carefully push and guide the workpiece along the guide fence to the cutting tool,
5. Listen to the cutting process for any unusual sounds,
6. Continuously monitor the force required to push the workpiece,
7. Monitor the relationship of the workpiece to the saw machine table, guide fence, and cutting tool,
8. Monitor the placement of hands or devices used to push the workpiece through the cutting process,
9. Monitor and gauge the placement of hands or devices used when stopping to adjust placement to complete the process,
10. Monitor the placement of upper and lower body to make the process as safe and stable as possible.

By now it must be clear that old sayings may have some merit, such as, "Nothing is as simple as it seems." The complete description of everything that is taking place during any cutting procedure, such as the seemingly simple procedure of rip-cutting a small piece of wood, includes:

1. The exact mechanics of how the wood fibers are severed, curled into chips, and carried away to be discharged;
2. The chemical reactions that go on during the cutting process, which involve water and chemicals found in the wood;
3. The physical properties and effects of the water, water vapor, and steam generated in the cutting process;
4. The thermal mechanics and thermal properties of the cutting process;
5. The effects of the atmospheric gases and pressures present during the cutting process, and so on.

The five mentioned individual sets of properties within the cutting process are real enough, though a woodworker would not have to pursue them in depth to complete the needed work.

The point is that there are elements of chemistry, physics, and mechanics present in the cutting process that cannot be seen, yet must be considered. Safety relates to understanding the visible and physical properties of the cutting processes as well as those that cannot be seen.

Machine Processes

Within the machine processes used in woodworking, there are unique characteristics that need to be examined concerning safety for the machine operator, cutting tools, or the machine itself. Among the first features of a cutting process to be examined is the contact arc length of the tool during its contact with the material being cut. (See Illus.10-16.)

Tools of different diameters that cut on the periphery of their bodies have different contact arc lengths that will affect many performance and safety characteristics. Where this has greatest importance is with the cutters that process high work loads relative to their size. Safety factors that relate to contact arc length are more important at the extremes—that is, with very short and very long arcs. A short contact arc for a router tool has the danger of deflection and increased vibration, while a full 180° contact arc presents problems of high temperatures and work loads, as well as reduced operating life.

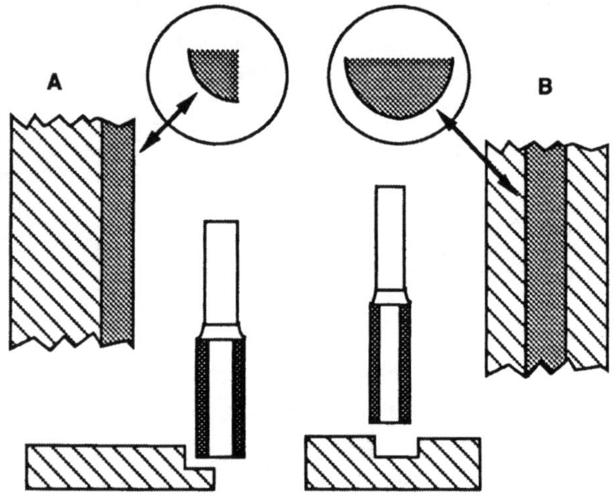

For router tools and others that cut on their periphery, a part of their performance is dictated by the length of cutting arc used in a particular process. The two tools above show cutting arc lengths of, relatively, 90° and 180°.

A—Cutting with a 90° arc, the tool will be more subject to deflection and transverse vibration.
B—Cutting with a 180° arc, the work load will be higher, changes due to elevated temperature will be greater, and the tool wear life will be shorter.

Illus. 10-16—Cutting-tool arc lengths

There is no ideal contact arc length for all cutting tools or cutting procedures because a complex relationship exists between cutter diameter, workpiece thickness or profile depth, and the geometric features of the tool.

There are many safety factors related to normal wear of common machine parts. An increasing problem is the means used to hold and drive router tools: *compression collets.* (See Illus.10-17.) A router-tool compression collet works by closing around the shaft of the cutter as it is forced into the mating surface of the motor spindle. The high work load placed on the tool creates unusually high wear and fatigue on the metal from which the collets are made. The more or less normal wear of such a device elongates the opening, allowing the tool to wobble and vibrate. Even when made of the best materials, a compression collet has a normal operating life of about 1,000 hours. In industry, that may only be two to three months; for a hobby worker, operating life will no doubt be much longer.

A common source of vibration is the normal wear and metal fatigue of the devices used to hold and drive cutting tools. Of the numerous sizes, shapes, and styles of compression collets produced, all have predictable working lives. All moving parts and many assembly parts of every machine wear, affecting the vibration mechanics and safety of the movements.

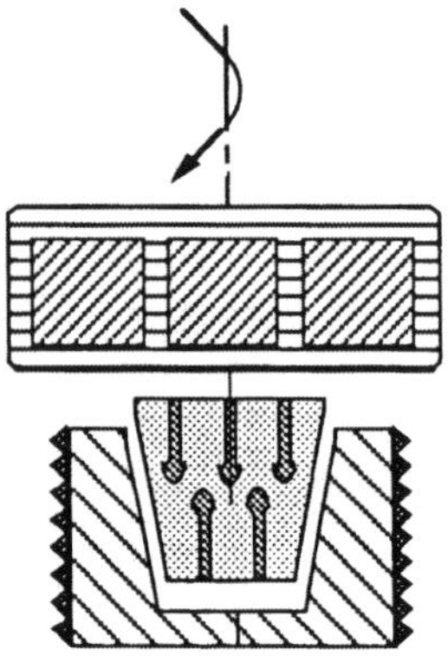

A router tool compression collet when new is completely concentric.

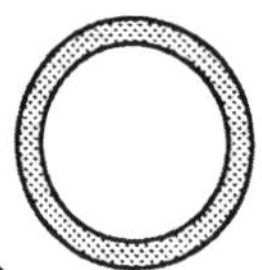

The internal shape of a router collet becomes eccentric from normal wear and metal fatigue.

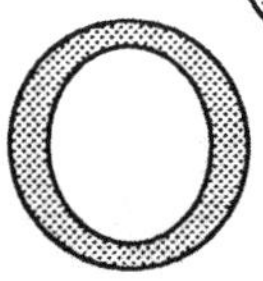

Illus. 10-17—Router tool compression collet

For all machine parts and cutting tools, especially those that operate at high speeds and have high work loads, normal wear is expected. Wear, in this context, includes any change of physical structure that affects ability to perform within design parameters—fatigue; loss of hardness, temper, or any strength characteristic; or actual change in physical dimension due to being worn away.

The router machine/tool compression collet is being driven at 15 to 20,000 rpm and has a 1,000-hour work life. How much work does it do? To comprehend the magnitude of what causes metal fatigue and work effects, we have to examine what takes place during each hour of operation. The tool makes approximately 1,200,000 revolutions each hour. During each revolution, the tool withstands the stresses of workpiece impact as many times as the number of teeth/knives it has; assuming two knives, the total per hour would be 2,400,000.

Another common machine part that is overlooked as a source of danger to safe tool operation is the *saw collar* used on all circular saw machines. (See Illus.10-18.) The function and purpose of a mated pair of collars is to provide additional strength and rigidity to cutting with a circular saw blade. A circular saw blade has such a thin cross-sectional dimension that it actually gets most of its strength from being in a state of motion, from centrifugal force. The mechanical forces encountered by a circular saw blade are very complex, created in part by the fact that such a small percent of the blade's periphery is in contact at any time with the material being cut. Great forward momentum is impeded for a brief period of contact, then allowed to continue. This is easier to visualize and measure with a cutter that has fewer knives/teeth, but it exists in every cutting process.

Saw-stiffening collars provide benefit by absorbing any lateral movement of the tool body, and by keeping it moving as straight as possible. The recurring problem with saw collars is that after being put on and removed from the machine tool spindle repeatedly, the torque/tension used to secure them also can distort and warp their shape. If not completely flat and parallel, saw collars can create the effect they are intended to protect against. Frequent inspection is required with replacement as necessary.

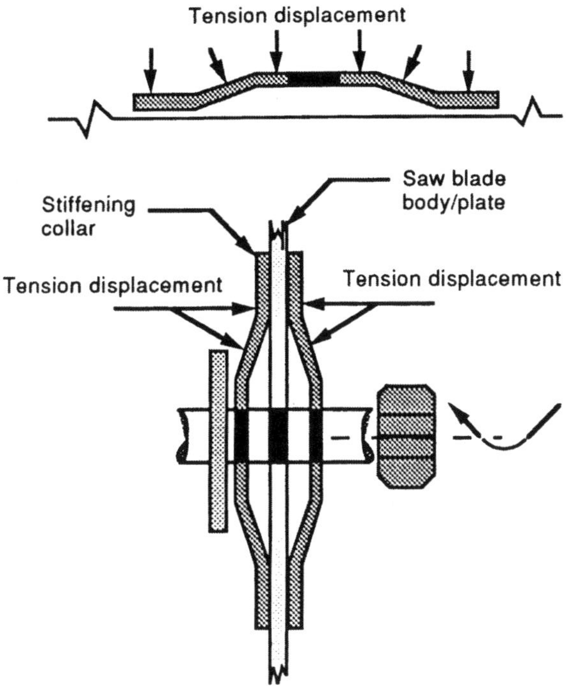

Saw collars provide support and rigidity to circular saw blades. To perform well, the collars must be precise and completely parallel. If warped or deformed at all, the collars will create the very effect they are intended to prevent. For best use of these very important devices, inspect them on a regular basis.

Illus. 10-18—Saw blade stiffening collars

Illustration 10-19 shows further the mechanical effects of a circular saw. Of all cutting tools, the circular saw is no doubt the most common, and it may also be the most frail and complex. A saw blade is most dangerous when conditions of the tool, the machine that holds and drives it, or a feature of the material being cut causes it to become unstable and wobble. During such an extraordinary condition as wobbling, the normal progression of one tooth after another entering the cutting zone is altered. This allows whole groups of teeth to make contact at once (by the lateral protrusions created), transferring enormous energy to the workpiece. Such force will do something like throw the workpiece backward, tearing sections of the tool body or teeth away. A point here is that most saw-cutting procedures use only a small percentage of available horsepower, but the power is always there. In the event of a "crash" (such as a saw blade becoming unstable), remarkable force will be seen.

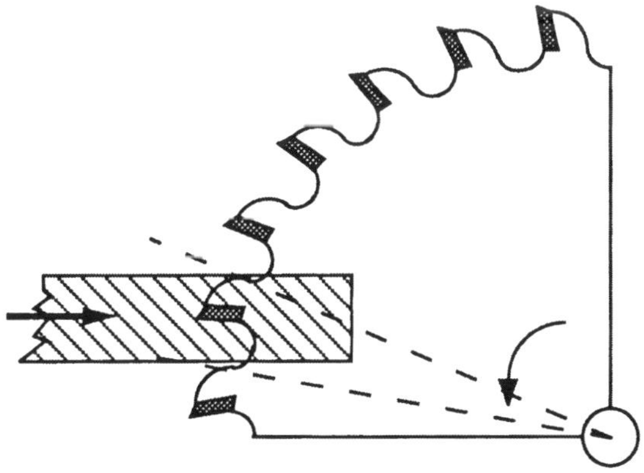

All circular saw blades derive most of their strength from the force of their momentum, that is, from the fact that they are in motion. The rpm, peripheral speed of the revolving body, gives the tool remarkable strength on one plane that requires accurate details of balance and precision to sustain. That is, saw blades have considerable strength if measured on their vertical plane but are relatively frail if measured horizontally. For safety, the use of saw blades requires a clear understanding of their strength in a single plane of operation and that they can become unstable and very dangerous with any lateral movement or vibration of either the tool or the material being cut.

Illus. 10-19—Circular saw blade mechanics: Workpiece contact

Illustration 10-20 gives a brief description of the basic safe-use guidelines for circular saws.

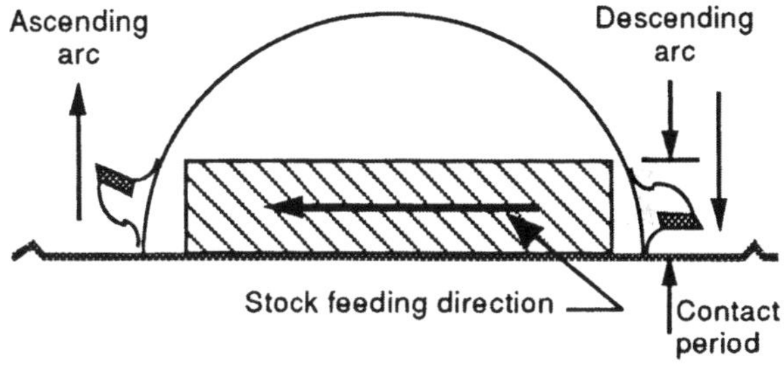

In all circular saw blade cutting processes, there are basic safety principals:

1. The saw blade should be correct for the work process, well made, and in good condition.
2. All machine features should be correct, accurate, and free of any vibration.
3. Clamping, guiding, or feeding devices should be correctly aligned and secure.
4. Material to be cut is clearly identified by specie or make-up and condition such as moisture content.
5. Stock to be cut is not distorted, warped, or twisted.
6. Cutting distorted stock requires unique, positive holding devices.
7. Special holding devices are necessary to cut small or odd-shaped stock pieces; never free-hand cut.
8. Make use of any device that guides and brings the stock safely to the cutting process while keeping the operator's hands away from the tool.
9. The work area must be well lighted and uncluttered.
10. Rehearse any process before attempting it, if only for a moment. You may solve an unforeseen problem. Do not wait to solve a problem until you are in the midst of it. Cutting tools and processes are not that forgiving—ever.

Illus. 10-20—Circular saw blade cutting

Design for Safety

For the past 20 or more years, several European countries have been the site of study, design, and testing of woodworking tools and machine processes and enforcement of rigid laws, all intended to make them more safe. Two basic principles have been pursued:

1. Cutting tools are more refined and task oriented.

2. There is not a wide range of workmanship or materials used to make cutting tools. All cutting tools are subject to industrial safety and performance standards.

Since 1967, cutting tools have been subject to strict design criteria that identified their intended use. Of these, cutting tools that produce shaped profiles and certain circular saw blades have the most restricted-use design features. The primary concern has been keeping the machine operator as safe as possible while still allowing completion of the necessary work.

Through studying and measuring the work tasks in woodworking—determining how well and how fast a person can safely feed the machining processes—a "real-time" rate was established. The fastest that anyone can possibly push wood workpieces into a table saw, shaper, or similar machine is approximately 18 feet per minute. This applies to any machine process that is manually fed. (There are entirely different standards for machining processes that are enclosed, away from the operator, and fed by a mechanical means.)

A cutting tool that will allow only a certain maximum feed rate uses an additional peripheral feature called a *chip limiter*. (See Illus. 10-21.) The cutting knife protrudes only a small distance above the chip limiter, which regulates how much and how fast the tool can cut. If too deep a cut is attempted, the material makes contact with the limiter and it simply rubs. As an added benefit, the possibility of having the workpiece thrown back at the machine operator is reduced because each knife can only have a very small chip load. The most common use of this cutter design is for shaper tools that have either fixed or movable knives. A circular saw blade that has chip limiters would be limited to rip-cutting processes. The primary intent is to avoid

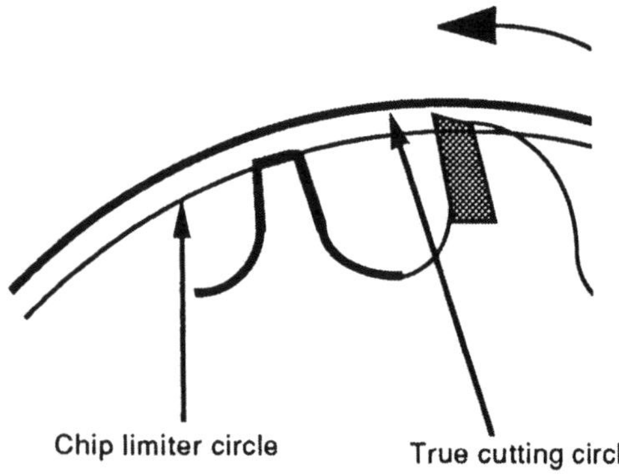

The basic principle of this tool design is that the blank tool post ahead of the cutting tooth will not allow force feeding of the process. The cutting tooth will cut a given amount of material only; then, if the stock is forced into the process faster, the top of the blank post will rub, not allowing a deeper cut. This is based on a nominal manual feed rate not to exceed 18 feet per minute for shaper/profile cutting tools and certain circular saw blades. Other, more open designs will allow much faster, power-fed feed rates.

Illus. 10-21—Chip limiter

feeding the tool too quickly, which could elevate its temperature, plug it, or in some way make it unstable. There may also be benefit to having more mass at the saw blade's periphery.

In recent years, technological progress has moved at a remarkable rate in all areas of industry. In the near future, we will see increased refinement of cutting tool design as more is learned about the precise details of the cutting process. In just the past several years, we have determined how to measure many of the temperatures present during operations, and learned more about the forces that are created and how they are dissipated, as well as much more about chemical reactions that take place. The area of cutting tool manufacturing that will surely change is the refinement of geometric features. For example, we already know the benefits of locating the knives in an unbroken spiral. We also know that the most effective relief angle is not an angle at all, but a log periodic spiral. Manufacturing methods will have to be found to not only produce improvements, but to keep their costs down.

Summary

When a person is injured using a cutting tool, the results are horrific. Human flesh does not offer any resistance to the enormous forces present in even the smallest woodworking machine. Cutting tools are for the most part superbly designed to effectively reduce *anything* in their path to small chips.

To date there has been no way to completely remove the human elements from any mechanical task, process, or procedure—that is, to keep humans from being human. At most, we can make every effort to provide the best possible design criteria, metallurgy, and mechanical features for cutting tools and machinery. Of equal or greater importance is the amount of information that we can provide that will describe the functions, processes, and procedures needed to complete each task involved.

As regards the specific use of cutting tools and functions, a list of concerns includes:

1. Is the tool selected for the task appropriate and in good working condition?
2. Is the machine that will hold and drive the tool in perfect condition in every feature that is important to the process?
3. Is the material to be cut a known species and are all of its properties known?
4. Does the condition of the workpiece pose a danger? (For example, is it twisted or deformed in some way?)
5. Are all elements of the work environment safe, uncluttered, well lighted, and so forth?
6. Has the procedure involved been studied and rehearsed?

No element of a machining process is more important than another, because all relate to each other in some way. A small vibration is a very large problem if it makes a cutting tool unstable. A small deformity in the material to be cut could cause it to pinch, jam, and be thrown back toward the operator. A slight misalignment of a machine part will not correct itself, and so on. In every instance, there is no substitution for "attention to detail"—everything is detail.

Regarding details, during the 1988 summer Olympics, the American 10-meter platform diver, Greg Louganis, when asked to explain his incredible abilities, replied: "Each dive that I perform, I really do not have to consciously think about while I am in motion. I have performed, studied, and rehearsed each move thousands of times, from the platform and in the gymnasium. From the time I leave the platform, in competition, I am simply doing all the intricate moves as a conditioned response. A particular dive is just tying a series of rehearsed movements together."

As woodworkers, our problem is just the opposite. We must perform thousands of operations differently every time. The number and magnitude of variables present in our work tasks change constantly, even for similar procedures. The best safety advice is to learn as much as possible about tools, machinery, materials, and procedures.

We have all heard the "classic excuses" (see below) for overlooking safety concerns—and we've probably uttered a few distinctly our own. An excuse can be either an explanation or a refusal to accept responsibility—concerning safety, neither is acceptable.

Miscellaneous Concerns

Design of cutting tools is most often directed at the most commonly available machines on which the cutting tool is intended to be used. For instance, a small circular saw blade made for use on a hand-held portable saw will perform any task within the rpm and horsepower range (1-1/2 horsepower, for instance) of that machine. The use of such a small saw blade on a much larger table saw machine (one having 5 horsepower, for instance) would not be allowed by the tool manufacturer because the available horsepower is so much greater than the tool was designed to withstand.

Number of teeth involved in the cutting process can cause problems:

- Too few will overload the individual teeth;
- Too many will elevate tip temperatures;
- Too fast a feed rate in relation to the number of teeth will require more horsepower.

A direct relationship exists between the amount of *energy needed to cut a given material* and the *energy needed to hold or force the tool (or workpiece)* into the process. A portion of the horsepower is transmitted to the workpiece—and not into the actual cutting away of the material.

The total energy needed to complete a given cutting process is divided between the horsepower the cutting knives require to cut away the material and the energy/force needed to hold or push the material into the cutting process. This is a complex relationship that includes all of the cutting tool geometric features and the physical characteristics of the material being cut. In some instances, parts of the machine involved in the cutting process are also involved.

The importance of *correct and efficient evacuation of saw dust and chips* cannot be overstated. Without careful removal of the waste/offal from the cutting process:

1. The chips can remain in the cutting circle near the teeth/knives, where they are repeatedly cut and rubbed against adjacent surfaces;
2. The sharpening life of tools involved can be reduced by as much as 50 percent;
3. Horsepower consumption can increase by as much as 30 percent;
4. The temperature of the process will dramatically increase;
5. Feed rates will be reduced;
6. Finish quality can be greatly altered or diminished; surfaces can take on the appearance of having been burnished, glazed, or scorched;
7. Vibration will increase significantly;
8. There will be increased wear to all moving machine parts involved in the process.

Classic Excuses:

- I didn't have time to...
- The correct tool was being sharpened, or was in use elsewhere...
- I know it was the wrong tool, but I only had one to process...
- I know it was really dull, but I had to get this one done...
- I know the feed speed was too fast, but I had so much to do...
- The guard gets in the way...
- With the guard in place, I can't see ...
- I thought it would work...
- It works better this way...
- I thought it was strong enough...
- I've done it like this before...
- Someone told me to do it this way...
- I didn't think that would happen...
- I thought someone else checked that...
- I didn't know that...
- It was all right yesterday...
- We had it checked just last week...

- I've only cut X number of pieces...
- Do I have to check all the time?...
- We had it sharpened a month ago...
- It's always done that...
- We've always done it this way...
- Of course I checked that. I think I did...
- That noise is common for that machine...
- I know I should, but...
- I know I shouldn't, but...
- Besides, vibration doesn't hurt anything...
- It always pushes really hard on that machine...
- I've never seen that happen before...
- I didn't think that it would break the...
- I wasn't pushing that hard...
- It was all right when I started...
- Of course it was tight ... straight ...whatever...
- It's the wood's fault—it was crooked ... it was wet ... it was too thick ... strong ... hard ... soft...
- I heard it would work this way...
- No one ever told me that...

Chapter 11
Ready Reference

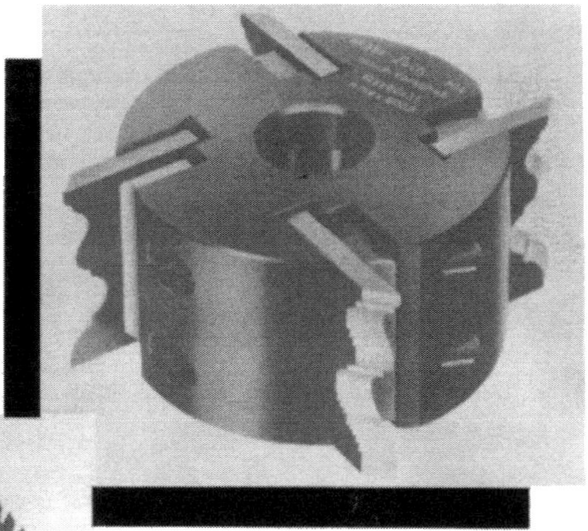

Photos courtesy of Leitz Tooling Systems, Inc.

Ready Reference

This chapter will provide a review of performance characteristics, detail features, and use of the most common wood cutting tools and processes. The recommendations for feed speeds, tooth/knife progression, etc., are conservative, and many of them are exceeded in areas of industry that have either unusual machine applications or cutting tools specifically designed for certain tasks.

There is no all-inclusive way to identify before the fact exactly how every cutting tool or machine process will react to all materials. The first task in every cutting process is the careful description of the properties of the material to be cut. Density, strength, moisture content, grain orientation, if present, and all others will dictate most of the features of the cutting tools and machine processes that will follow. We cannot have too much information regarding the use of cutting tools.

The five formulas that we generally call the cutting tool performance calculations will tell us most of what we need to know regarding the work being done or to be done. Of the five, the first three are the most important and commonly used, and the last two relate more to tool design features:

1. Tooth/knife progression tells us how far each tooth progresses in relation to the movement of the workpiece. This progression dictates the appearance of the cut surface and is often expressed in terms of knife marks per inch. We use this figure to calculate feed speeds, work load, and finish quality.
2. Feed speed tells what the progression or the workpiece is or should be based on a desired finish quality or work load. Feed speed and tooth/knife progression can be balanced with each other to achieve a desired outcome.
3. Rim speed is the speed of the cutting tool on the periphery and identifies what all of the performance characteristics can be. It is the relationship of diameter to pi (Π) to rpm, expressed in terms of circular distance and time.

Cutting Tool Performance Calculations

Tooth/ Knife Progression (Sz)
Feed Speed (U)
Rim Speed (V)
Cutter Mark Depth (chord) (t)
Mean Chip Thickness (hm)

$$Sz = \frac{U \times 12}{N \times Z} = \text{progression per 1/1000"}$$

$$U = \frac{Sz \times Z \times N}{12} = \text{ft./min.} \qquad V = \frac{D \times \Pi \times N}{12 \times 60} = \text{ft./sec.}$$

Other symbols:
D = diameter in inches
Π = 3.14159
N = rpm
Z = number of teeth/knives
a = depth of cut in inches

$$t = \frac{Sz^2}{4 \times D} = \text{depth of cut as 1/1000"}$$

$$hm = Sz\sqrt{\frac{a}{D}} = \text{average thickness as 1/1000"}$$

The formulas are precise and assume a perfect cutting circle, with all teeth/knives in complete alignment. There is no compensation for deviations in machine parts, vibrations, or any stock feeding or clamping system. Further, there is no allowance in the formulas for normal tolerance between the cutting tools and the machine spindles and methods employed to hold and drive them. Normal possible deviations will not affect the working accuracy of the calculations. (See also Illus. 11-1.)

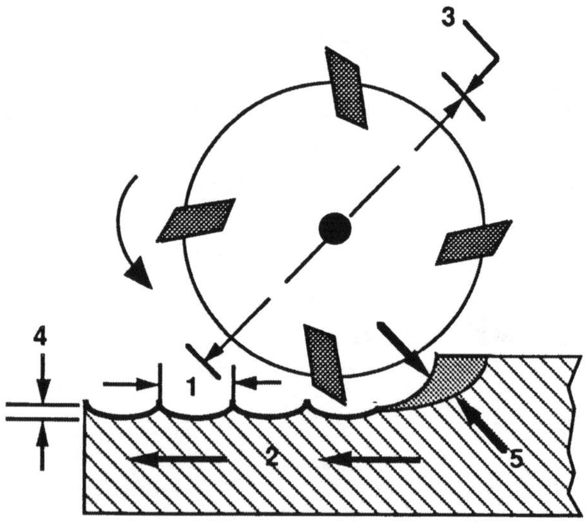

1. Tooth Progression (Sz)—the progress of each tooth/knife
2. Feed Rate (U)—the movement of the workpiece
3. Rim Speed (V)—the peripheral speed of the cutting tool
4. Cutter Mark Depth (t)—the path-depth of the tooth/knife
5. Mean Chip Thickness (hm)—average thickness of each chip

Illus. 11-1—Tool performance calculations

Illustration 11-2 shows the most basic geometry problem in woodworking. We describe all surface areas in terms of a square inches, square feet, etc. However, we describe a circle only in terms of its diameter, and too often that can create dangerous conditions when we are examining the features of cutting tools. Without carefully describing all physical dimensions of circles, we often miscalculate the size, strength, and other properties of cutting tools.

For instance, the shaft of a 3/8" router tool is not half the size of a 3/4" shaft but a fourth the size in cross-sectional area. It is of great importance that circles and the features of complex circles that make up the details of cutting tools be carefully described. A given cutting tool may have several circles that must be described and maintained carefully. Another example would be the mistaken description of the diameter of a cutting tool body as the cutting circle, which is, in fact, larger.

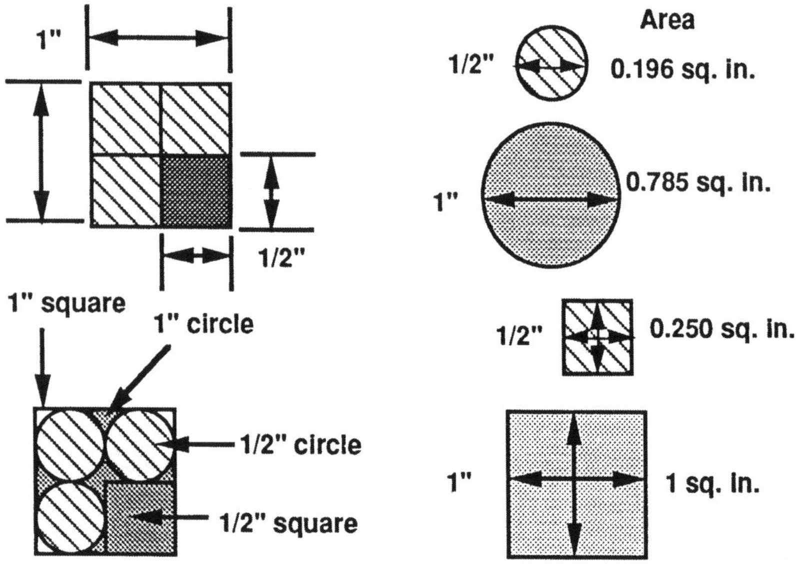

All cutting tools for powered equipment are defined in terms of circle geometry. Estimating their strength and size, however, is hampered somewhat by a simple misconception. The relationship in area between a circle and square is a seeming paradox. We express all areas in terms of a squared dimension. But we describe a circle in terms of its diameter and a square in terms of the length of a side, which causes the confusion. A 1" square is 4 times larger in area than a 1/2" square. This is equally true of a circle if we correctly express it; that is, a 1" diameter circle is 4 times larger than a 1/2" diameter circle.

To find areas and circumferences:
1. The area of a square or rectangle—multiply the width by the height.
2. The circumference of a square or rectangle—add all of the lengths.
3. The area of a circle—multiply the square of the radius by Π.
4. The circumference of a circle—multiply the diameter by Π.

Illus. 11-2—Description of circular and square areas

Performance Characteristics

Abrasive Cutting vs. Climb Cutting

Abrasive cutting

- Operates against the direction of stock feeding;
- Builds chips as the tooth advances, beginning thin, ending at the thickest dimension;
- Does not dull the knife/tooth as quickly as climb cutting;
- Utilizes horsepower smoothly as the chip size increases gradually;
- Creates less vibration than climb cutting because the individual tooth/knife impacts are not abrupt;
- Provides poorer finish quality than climb cutting;
- Can be accomplished by either mechanical or manual stock feeding.

Climb cutting

- Operates with the direction of stock feeding;
- Builds chips that are largest at the beginning and continue to diminish in thickness;
- Works with abrupt impact of the knife/tooth against the workpiece which can create significant vibration;
- Provides superior finish quality because the knife/tooth rubs only the surface at the end of the cutting contact arc;
- Dulls the tools more quickly because of the rubbing effect;
- Allows faster feed rates due to the better finish quality;
- Must be accomplished with only mechanical stock feeding mechanisms.

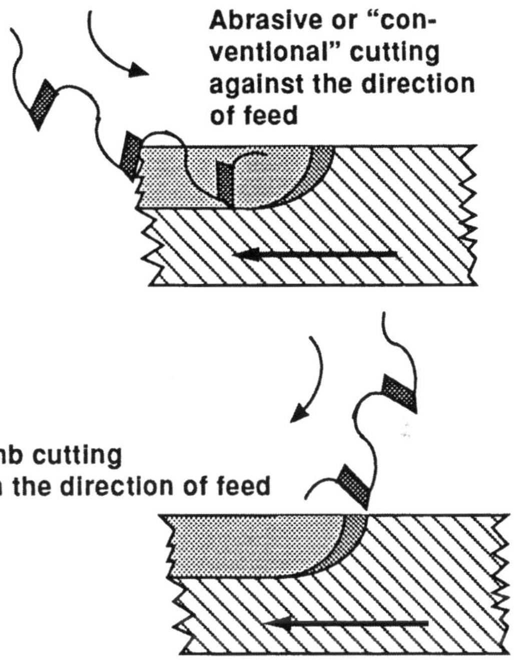

Illus. 11-3—Abrasive vs. climb cutting

Grain Orientation

1. Cutting with grain requires the least horsepower and gives the best finish quality.

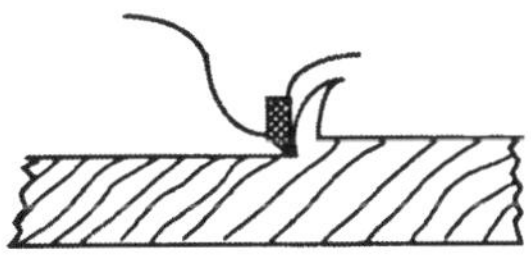

2. Cutting along the grain gives a smooth finish and requires very little horsepower.

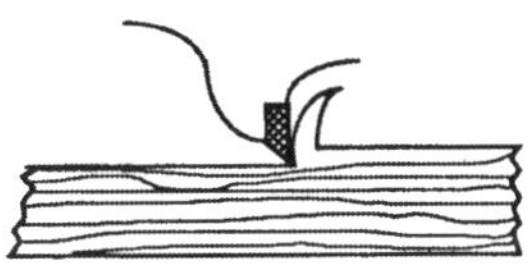

3. Cutting across grain is easy, but the cut surface most often rough.

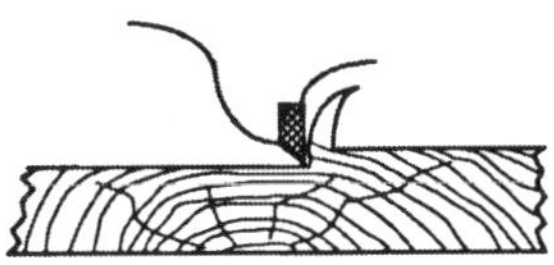

4. Cutting against grain requires more horsepower and is difficult; fibers tend to lift and tear, giving a rough surface finish.

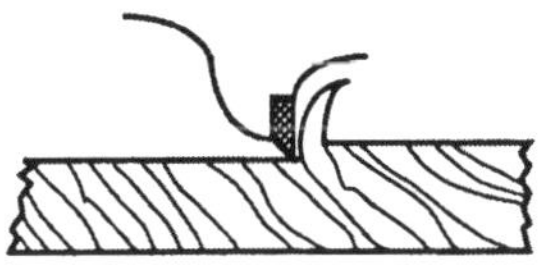

5. Cutting end grain is difficult, yields a rough surface finish, and requires the most horsepower of any solid wood cutting.

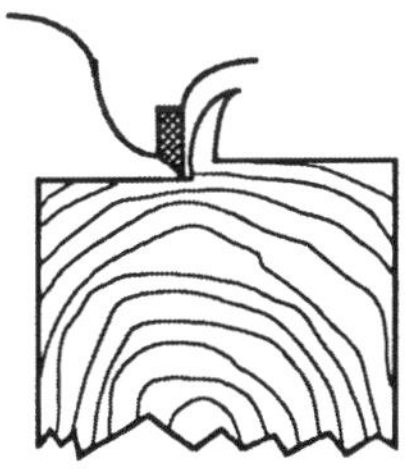

Illus. 11-4—Cutting solid wood: Grain orientation

Performance Recommendations: Circular Saw Blades

Tooth/Knife Progression (Sz): Saw Blades–Different Materials

- Softwood along grain .... 0.007 - 0.035"
- Softwood across grain .... 0.004 - 0.010"
- Hardwood .... 0.002 - 0.006"
- Chip/particle board .... 0.004 - 0.012"
- Hardboard (dense fiberboard) .... 0.004 - 0.008"
- Plywoods .... 0.004 - 0.012"
- Veneered panels .... 0.002 - 0.010"
- Light metal alloys .... 0.001 - 0.003"
- Plastic laminated panels .... 0.001 - 0.006"

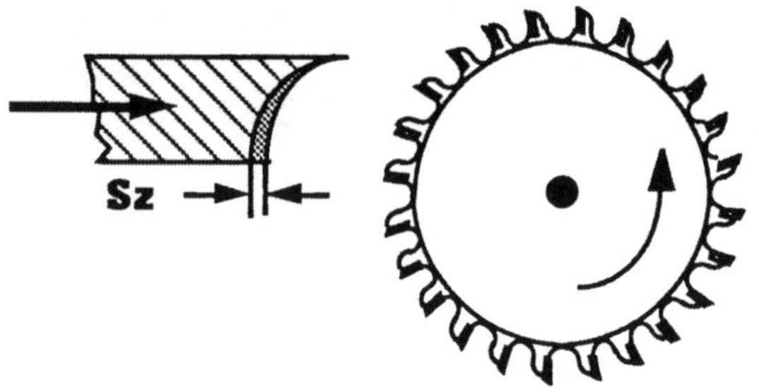

Note: To convert any of the given figures to metric dimensions, simply multiply the number by 25.4, e.g.:

0.007 × 25.4 = 0.177 mm
0.035 × 25.4 = 0.889 mm

Saw Blade Elevation

Distinct and dramatic geometric changes take place when the height relationship changes between the saw blade and the material being cut. In this illustration, the two workpieces are identically dimensioned, though they are placed at different heights.

At Elevation 1, there are four teeth in the cut; the cutting angle is acute; and the contact arc is long.

At Elevation 2, there is one tooth in the cut; the cutting angle is shallow; and the contact arc is short.

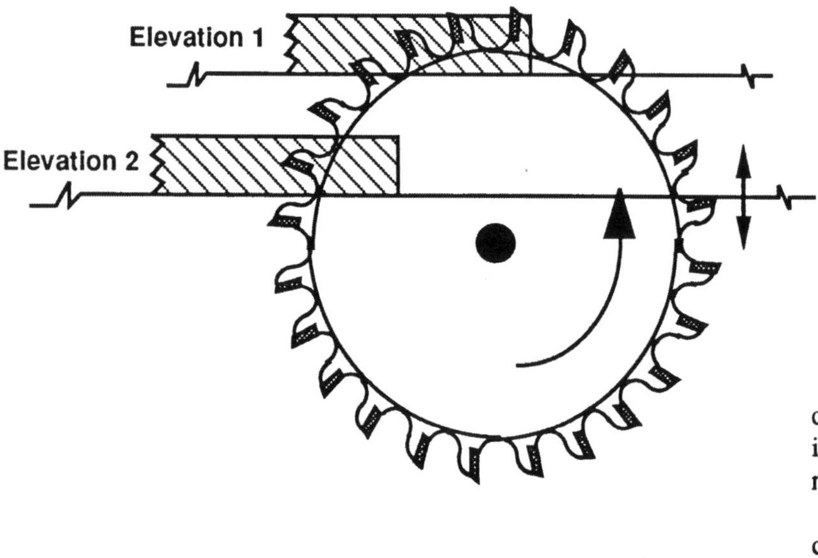

Illus. 11-5—Circular saw blade elevation

Adjusting the height of the saw blade will have the following effects:

1. Altering the number of teeth involved in the cut,
2. Altering the entrance and exit angles of the teeth by as much as 30°,
3. Altering the length of contact arc for each tooth,
4. Affecting the feed rate, finish quality, and energy required.

There is no "ideal" elevation because there are too many variables to interpolate: number of teeth, nature of material, tool diameter, etc.

Basic rules: never cover the teeth completely, and always have at least one tooth in the material being cut.

Saw Blade Tooth Spacing

The choice of tooth spacing relates to finish quality, feed rate, horsepower consumption, and the nature of the material cut:

1. Wide tooth spacing, 1" to 2-1/2 ", is normally for rip cutting solid woods and other material of large physical dimensions and relative low strength. Finish quality is diminished, and feed rates can be high.
2. Medium tooth spacing, 5/8" to 1", is for cutting most materials in the woodworking industry, from natural woods, hard and soft, in any grain orientation, to man-made panels. Tooth shapes and groupings dictate many specific applications.
3. Narrow tooth spacing, less than 5/8", is for materials with small cross-sectional dimensions, high strength, or other properties that require a very small chip load. Dense, thin fiberboards, thermoplastics and duroplastics, and fine wood veneers examples.

To find tooth spacing (x):
Diameter × Π ÷ No. of teeth

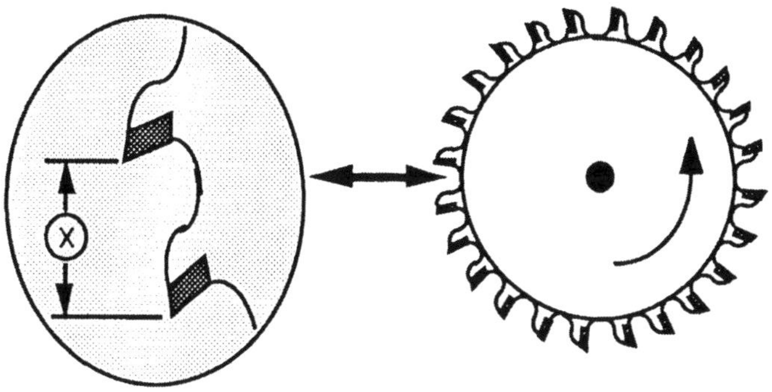

Illus. 11-6—Circular saw blade tooth/knife spacing

Basic rules: for thin, dense, and strong materials, more teeth; for thick, less dense, and weak materials, fewer teeth. Always one tooth in the material as being cut.

Saw Blade Activity Zones

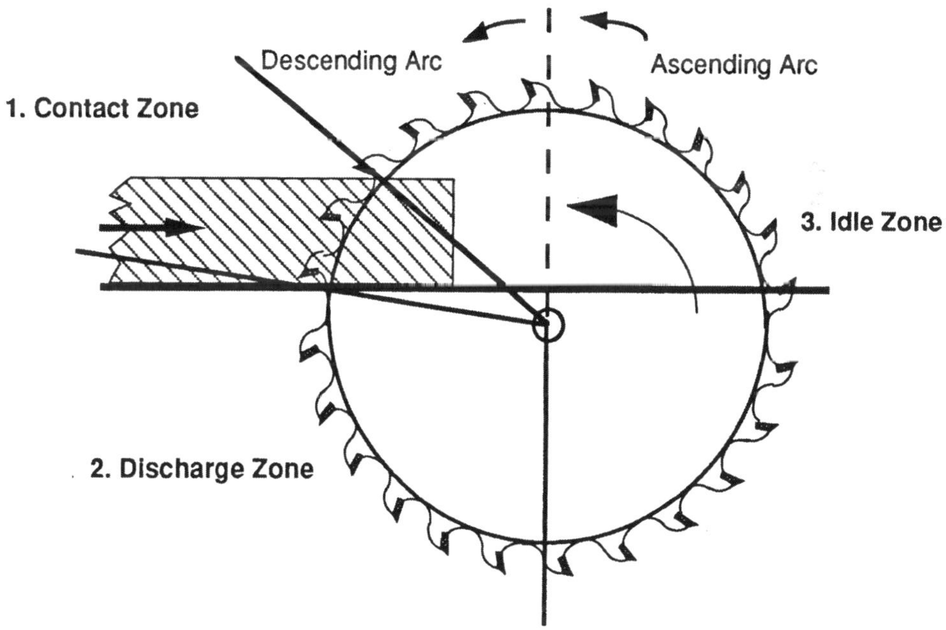

Illus. 11-7—Saw blade activity zones

1. Contact zone, where the cutting actually takes place;
2. Discharge zone, where the waste is evacuated;
3. Idle zone, where the tool cools and becomes stable, prior to entering the contact zone again.

In estimating the physical effects of an ongoing cutting process, it is necessary to understand that about the entire periphery of a saw blade, or other cutter, there are different zones of activity. If there were but a single activity and zone, the effects would be easy to measure and manage, but we have to comprehend that activity in each zone has to be complete or all activities will eventually cease. If the waste is not completely discharged, the cutting process will not continue; if the tool does not stabilize, irregular movement will increase, destablilizing the process.

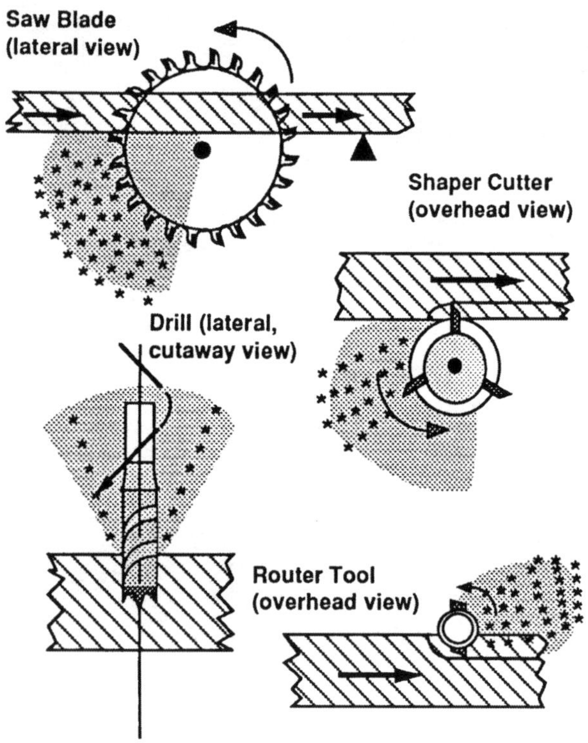

Illus. 11-8—Waste discharge

Controlling Waste Discharge

The importance of correct and efficient evacuation of the sawdust and chips cannot be overstated. (See Illus. 11-8.) Not carefully removing the waste from the cutting process can have a number of negative effects:

- The chips can remain in the cutting circle near the knives/teeth and be repeatedly cut and rubbed against all adjacent surfaces.
- The sharpening life of the tools can be reduced by as much as 50 percent.
- Horsepower consumption can be increased by as much as 30 percent.
- Temperature of the process will dramatically increase.
- Feed rates will be reduced.
- Finish quality can be greatly diminished. The surfaces can take on a burnished or glazed appearance.
- There will be a significant increase in vibration.
- There will be increased wear to all moving machine parts.

Saw Maintenance

Most industrial circular saw blades are manufactured with sufficent knife material to provide from 20 to 40 sharpenings, depending on use and maintenance. The correct sharpening sequence does not include material being removed from both the top and face of the saw blade tooth in the same cycle, except in the unusual instance where the tool has been run for a prolonged period or has cut very abrasive material. The size and shape of current circular saw blade teeth requires the sequence to be 2 to 4 top sharpenings for each face sharpening. (See Illus. 11-9.)

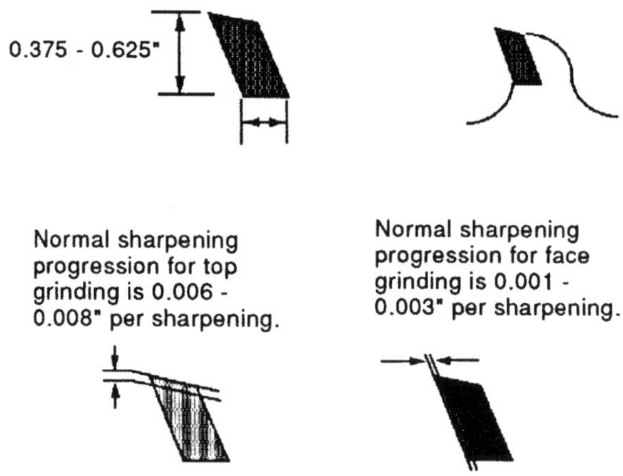

Illus. 11-9—Service recommendation: Tipped circular saw blades

Buying Guidelines: Circular Saw Blades

Information necessary for selecting saw blade features and characteristics include descriptions of the following:

A. The machine and machine process involved:

1. Machine characteristics—type, size, rpm, horsepower, arbor size, rotation, etc.;
2. Machine process—ripping, cross-cutting, miter-cutting, etc.;
3. Holding, clamping, manual or power feeding, and guiding systems.

B. The material to be cut:

1. Specie, strength, condition, and moisture content; if man-made, describe properties;
2. Physical dimensions and grain orientation.

C. Needed tool features/abilities:

1. Tool type and physical dimensions such as diameter arbor, plate thickness, and kerf;
2. Number of teeth needed relative to spacing, feed rate, stock thickness, and strength;
3. Tooth shape relative to the nature of the material—solid wood, panel material, other.

D. General information.

1. The process—an ongoing, high-volume production or limited, infrequent use for a unique material;
2. True cost indices of the cutting tool—initial cost, service costs, and total operating life.

In sum, is the tool the best choice for the machine process, for the material to be cut, and for cost justification?

Performance Recommendations: Other Cutting Tools

Rim Speed Recommendations

The table below provides recommended rim speeds for tools of various types and different materials. All speeds are expressed in feet per second. To convert to metric dimensions, divide the figures by 3.28, e.g.:

235'/sec + 3.28 = 71.6 m/sec.

To find rim speed (V) for a specific process:

$$\frac{D \times \Pi \times N}{12 \times 60} = \text{ft./sec.}$$

	Cutters—All Types		Circular Saws
Material	Solid High-Speed Steel	Carbide-Tipped	Carbide-Tipped
Natural Softwoods	160 - 220	180 - 280	220 - 300
Natural Hardwoods	120 - 180	150 - 220	210 - 280
Exotic Woods	100 - 160	140 - 200	200 - 260
Veneers			210 - 280
Treated Woods		120 - 180	160 - 220
Glued Wood Solids		180 - 240	180 - 240
Plywoods		150 - 220	160 - 240
Chip/Particleboard		180 - 240	180 - 240
Fiberboard		120 - 200	180 - 240
Dense Fiberboard		150 - 230	150 - 230
Thermoplastics		90 - 220	100 - 140
Duroplastics		50 - 120	50 - 120
High-Pressure Plastics		150 - 230	150 - 230
Nonferrous Metals (Aluminum, Copper, etc.)		45 - 120	45 - 220
Materials with Mineral Additives			
Plasterboard			120 - 200
Rock-Wool Board			40 - 120
Concrete-Bonded Particle/Chipboards			120 - 200

Recommendations: Drills

Tooth/knife progression. For most wood and wood-based materials the normal range of progression, which for drills is also feed rate, will vary from 0.0015 to 0.010" per revolution. The criteria are the diameter of the hole drilled, the strength and density of the material, and the rpm used to perform the work.

General Rules:

- Hard, strong materials—lower feed rates;
- Soft, weak materials—higher feed rates;
- Large drills—lower rpm;
- Small drills—higher rpm.

Range of rpm for most applications:

- 500 to 1500 for large diameter drills,
- 1500 to 2500 for medium diameter drills,
- 2500 to 4000 for small diameter drills.

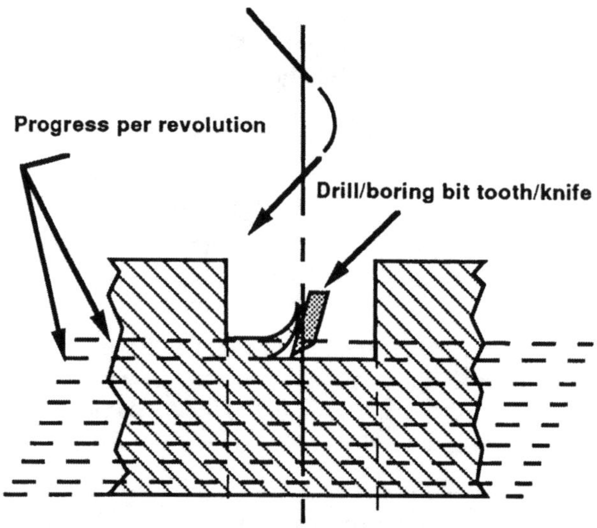

Illus. 11-10—Tooth/knife progression (Sz): Drills and boring bits

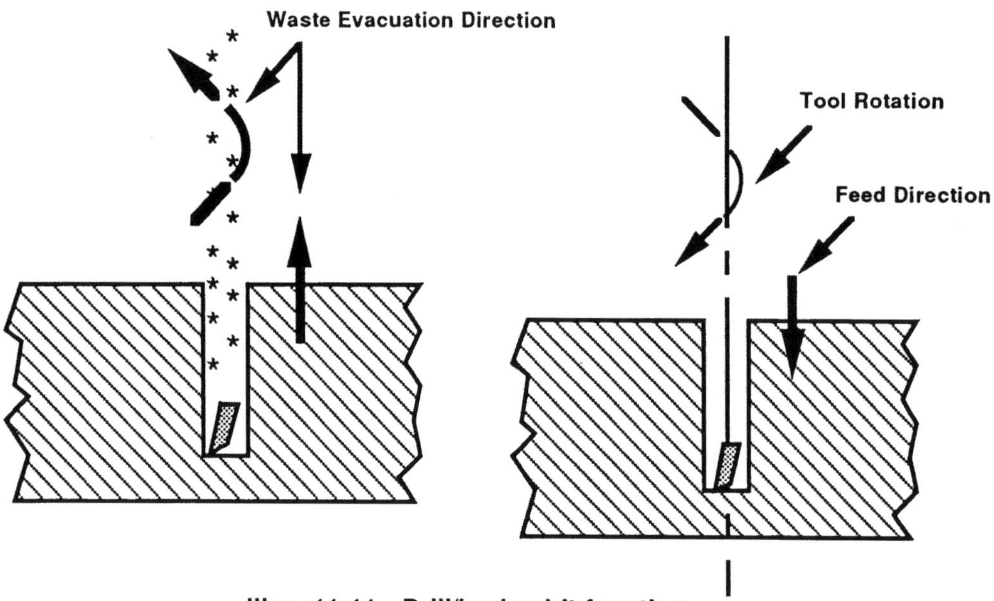

Illus. 11-11—Drill/boring bit function

Drill Function

Performance characteristics of drill- and boring-bit cutting processes are unique among cutting tools:

- The heat generated in the process is dispersed by the evacuation of the waste, which must not be impeded.
- Flute design and effectiveness is critical to the tools' function, especially for narrow deep holes.
- All geometric features are compact and must be well balanced and concentric, and knives kept sharp.
- To perform well, the relationship of rpm to progression must be carefully observed. Revolution that is either too fast or too slow can present serious dangers to the tool.
- Unlike other cutting tools, drills and boring bits cut only on the end of a turning shaft, creating unique geometric problems of defining and controlling the properties of the cutting circle. At the very center of the cutting circle, there is no rim speed, while at the periphery, the speed can be high, hence the purpose of a center point.

Recommendations: Routers

Router Knife Progression. There is no completely accurate or inclusive way to make router tool usage recommendations, given the enormous number of variables. There would have to be a set for each cutting diameter, shank dimension, and type of material cut. For an example, the table at right shows the variance in knife progression for a router tool having a 1/2" shank and two straight knives, turning at 20,000 rpm, conventionally cutting the materials listed with the diameters indicated.

Precision routers of solid tungsten carbide with advanced design features and spiral knives can easily double these figures with most materials. The condition of the machines that hold and drive routers is paramount. Because of routers' small features and high speeds, the single most important mechanical detail is that they be as free of vibration as possible.

The formulas for router knife progression and feed rate:

$$V = \frac{\Pi D}{12} \times \frac{rpm}{60} = \text{rim speed (in ft./sec.)}$$

$$Sz = \frac{U \times 12}{N \times Z} = \text{progression per 1/1000"}$$

Sz = knife progresson in 1/1000"
U = feed rate in ft./min.
N = rpm
Z = No. of knifes
12 = inches per foot

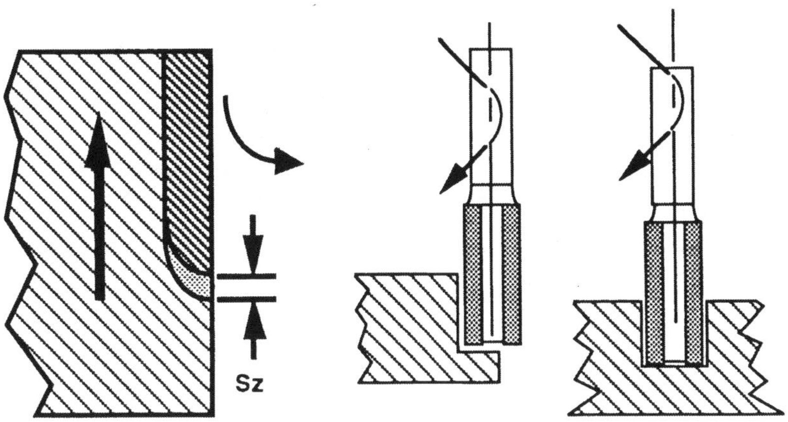

Illus. 11-12—Knife progression: Routers

Tool Diameter →	1/8 - 3/8"	3/8 - 5/8"	5/8 - 1"
Material ↓	Knife Progression		
Natural Woods	0.001 - 0.003"	0.003 - 0.006"	0.006 - 0.010"
Man-Made Panels	0.001 - 0.004"	0.004 - 0.008"	0.008 - 0.12"
Plastics	0.001 - 0.003"	0.003 - 0.006"	0.006 - 0.008"
Soft Metals	0.001 - 0.002"	0.002 - 0.004"	0.004 - 0.006"

Router Cutting Circle

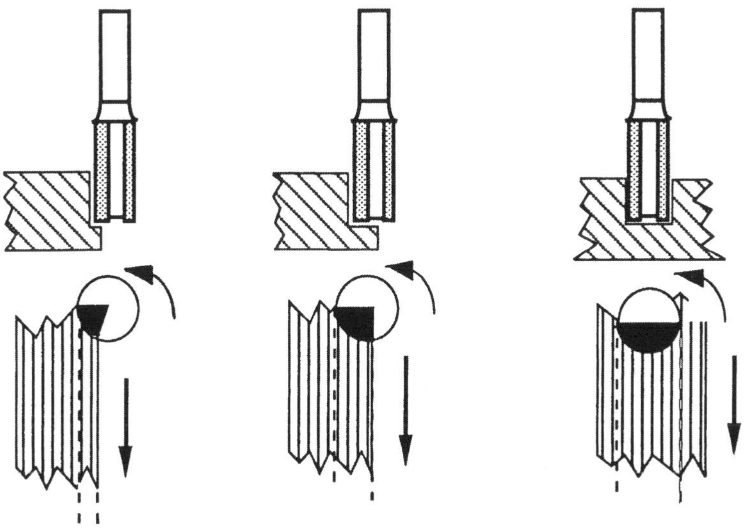

Illustration 11-13—Router Tool Cutting Circle

With router tools, it is important to observe the arc length of the process. These tools, having small, compact features, are not capable of a wide range of work loads. Estimates of sharpening life, horsepower consumption, and similar considerations can be made by examining the contact arc length of the cutting circle of each tool and process.

When cutting the same material, the wear factors and energy requirements are in direct proportion to the portion of the cutting circle involved in the work process. Within the mechanical constraints of the machine and process, the best tool life and performance will be gained by using the tool with the largest possible diameter for the particular process.

Recommendations: Shaper/Profile Cutter, Planer, and Jointer

Shaper/Profile Cutter Tooth Progression. Recommended length of cutter marks (Sz) for shaper and profile cutting tools relates to a very wide range of materials, feature details, and machine processes:

- Fine chip—0.012 - 0.03"
- Medium chip—0.031 - 0.078"
- Coarse chip—0.078 - 0.125"

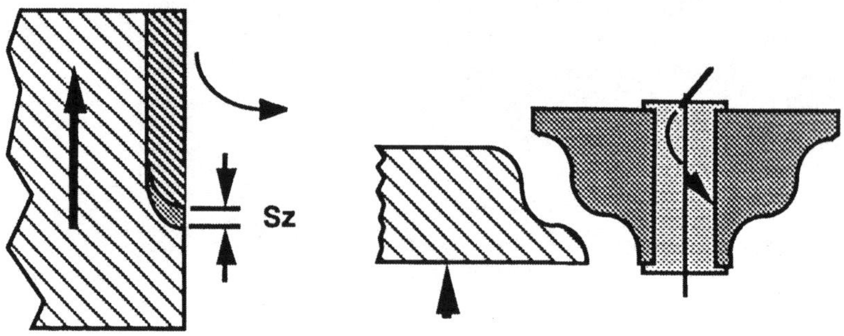

Illus. 11-14—Shaper/profile cutter tooth progression (Sz)

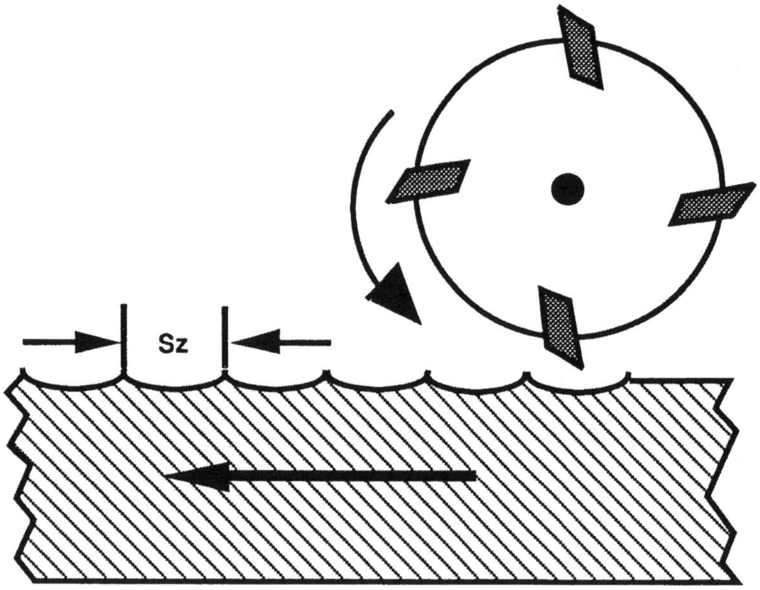

Illus. 11-15—planer and jointer tooth progression (Sz)

Planer and Jointer Tooth Progression. Recommended length of cutter marks (Sz) for planers and jointing tools relate to a very wide range of softwood and hardwood characteristics. Both machine processes involve sizing and surfacing operations:

- Fine chip—0.008 to 0.024"
- Medium chip—0.024 to 0.066"
- Coarse chip—0.066 to 0.096"

Cutter Marks and Finish Quality

In the U.S., a common expression that relates finish quality to cutter mark length is knife marks per inch,with more marks per inch equating to a smoother finish. The table at right provides a summary of the relationship of knife marks per inch and length of cutter mark (Sz).

Note: to convert any of the figures to metric dimensions, simply multiply the number by 25.4, e.g.:

0.1666 times 25.4 = 4.23 mm
0.0500 times 25.4 = 1.27 mm

Knife Marks/inch	Length of Cutter Mark (Sz)
100	0.010"
80	0.125"
60	0.166"
40	0.250"
20	0.500"
10	0.100"
8	0.1250"
6	0.1666"

Tool Mechanics and Machine Processes

Horsepower Consumption of Woodworking Processes

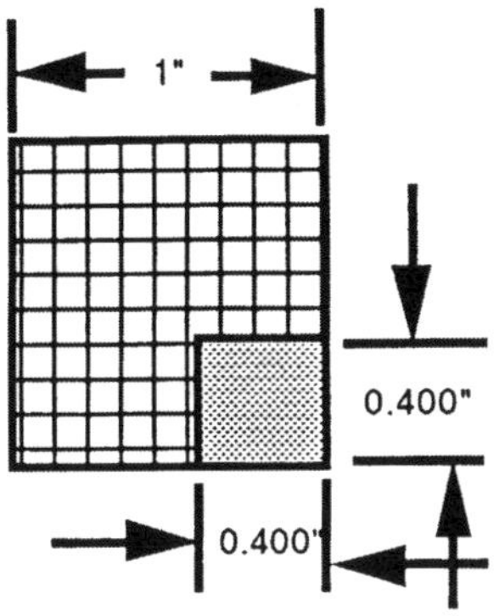

There is no all inclusive way to tell exactly how much horsepower a wood-cutting process will require. For most processes, a safe estimate would be that the removal of a square measuring 0.400" by 0.400" will require one horsepower (746 watts of electrical power). Note: The area of the square is 0.400" × 0.400" = 0.160" sq. in.

Elements that affect horsepower consumption include the nature of the material cut, sharpness of the tool, vibration mechanics of the process, feed speed, cutting speed, and others. The figure of 0.160 sq. in./1 hp for most cutting processes is a conservative-high average.

The machining processes in Illustration11-16 may be done with a shaper, moulder, or tenoner. The figures in the list below indicate the amount of material removed in square inches and the horsepower needed.

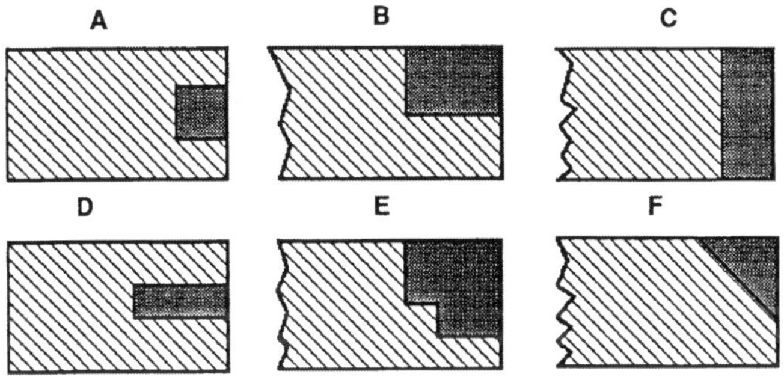

Illus. 11-16—Horsepower consumption

A. Groove—area 0.160 sq. in.—requires 1.0 hp.;
B. Rebate—area 0.373 sq. in.—requires 2.3 hp.;
C. Sizing/jointing cut—area 0.500 sq. in. —requires 3.1 hp.;
D. Deep groove— area 0.187 sq. in.— requires 1.1 hp.;
E Large complex rebate—area 0.797 sq. in.—requires 4.9 hp.;
F. Bevel profile—area 0.265 sq. in.— requires 1.6 hp.

Concentricity and Parallelism

In the construction, use, and maintenance of cutting tools, it is eminently important that all features remain balanced, concentric, and parallel. These design features are vital for the safety and efficiency of every cutting process. There are most often a large number of details to be aware of, and each contributes to the viability of the cutter as a whole. Design characteristics that are subject to any change during service or maintenance must be carefully observed. For example, the amount of knife material removed for sharpening must be exactly the same from one knife to another.

The five parts of Illustration 11-17, each having six segments, illustrate the basic relationships of parallelism and concentricity. (See also Illus. 11-18 on the following page.)

A is neither parallel nor concentric.
B is neither parallel nor concentric.
C is parallel but not concentric.
D is concentric but not parallel.
E is both concentric and parallel.

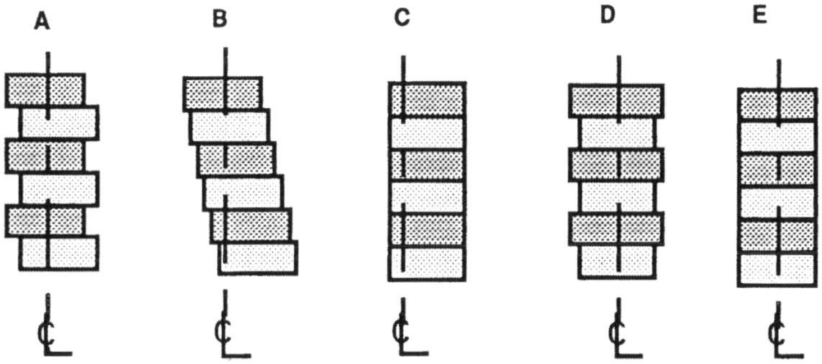

Illus. 11-17—Concentricity and parallelism

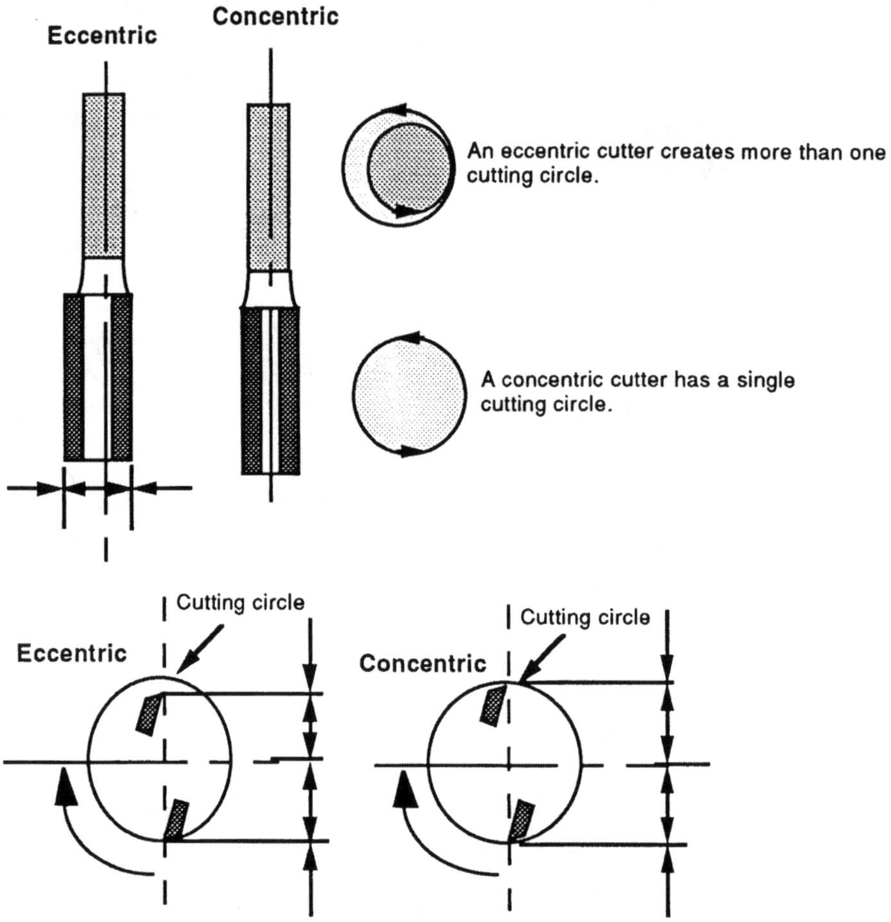

Illus. 11-18—Concentricity

Compound Shear Angles

Among recent tool advances is the arrangement of cutting knives to improve corner edge surface quality by controlling the contact surface pressure.

In Illustration 11-19, the two knives cut in the same cutting circle and are parallel to the tool shaft, though they are set at a shear angle. The advantages of the compound shear angles of the knives is the superior edge finish quality and the requirement of less horsepower to accomplish the work.

At the upper and lower edges of the workpiece, the contact surface pressure is driven against the material, leaving the corners smooth and free of tears and roughness. The same effects will be seen in all similar shear angle and alternate shear-angle cutting tools. There are numerous tools available with this knife arrangement, including planer/jointer heads, router tools, and others.

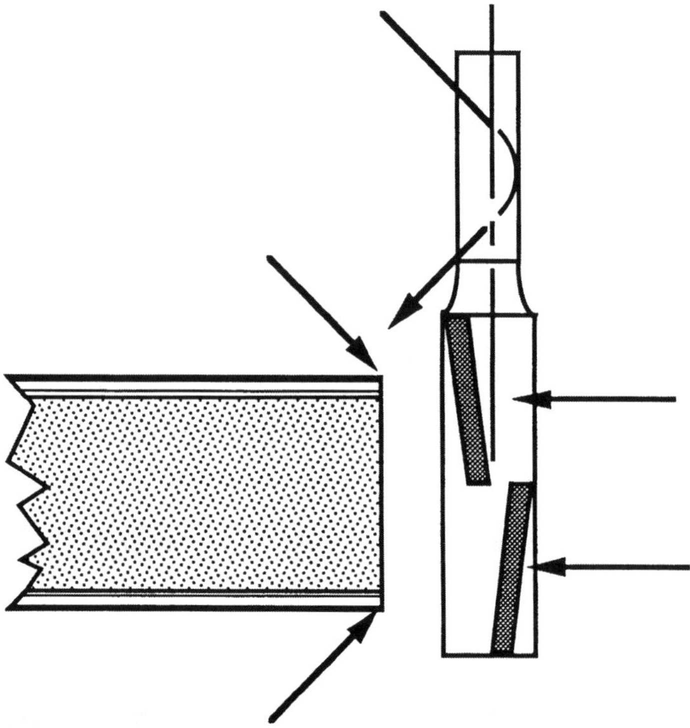

Illus. 11-19—Compound shear angles

Progressive Shear or Spiral Knives

Illustration 11-20 A shows a cutter with a progressive shear or spiral configuration of knives. The advantage of arranging knives in this manner is that the knives do not individually impact the material being cut, thus reducing the vibration usually present. Arranging the knives in either a progressive shear or spiral makes the cutting process continuous and divides the contact surface pressure along the plane of rotation among the individual knives.

In B, the knives are arranged 90° apart on the tool body, are parallel to the axle, and impact the workpiece individually, possibly creating mechanical vibration, especially when cutting hard, dense materials.

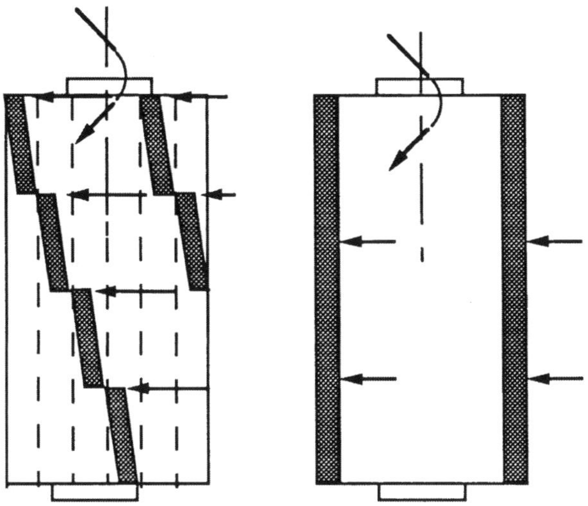

Illus. 11-20—Progressive shear or spiral knives

Tool Rotation

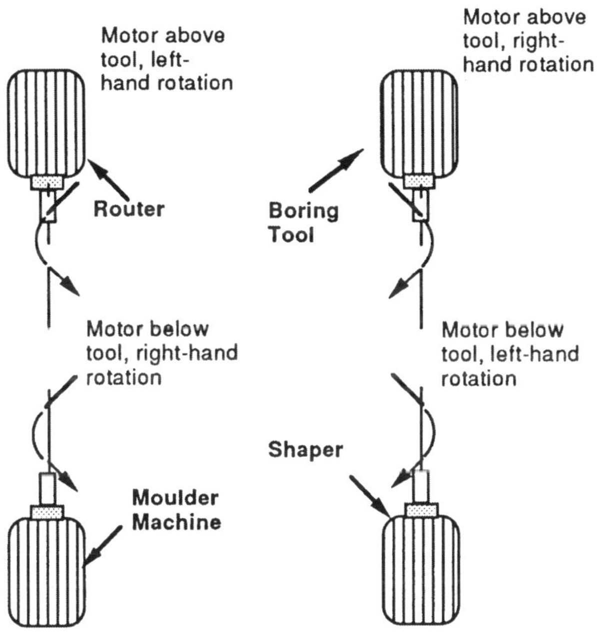

Illus. 11-21—Tool rotation 1

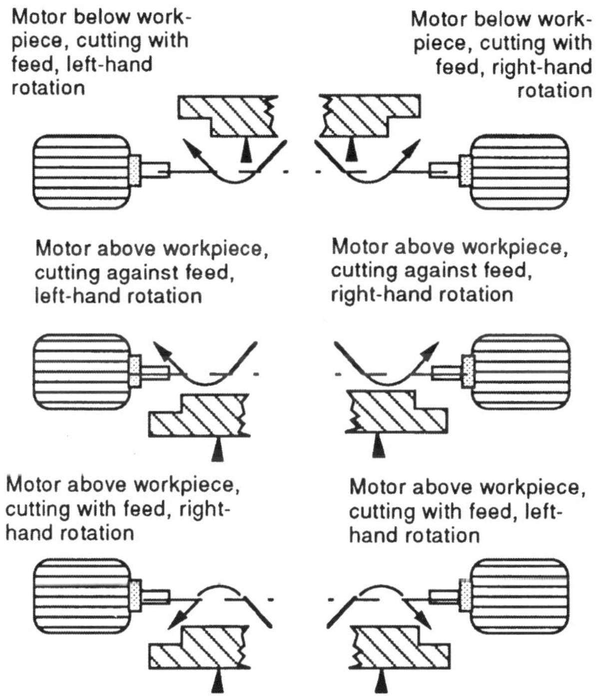

Illus. 11-22—Tool rotation 2

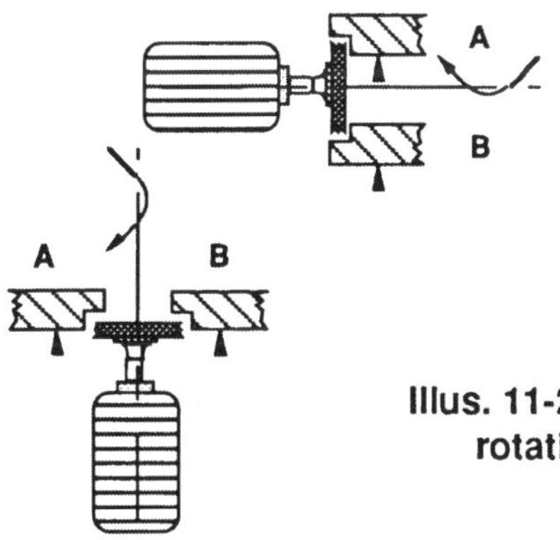

Illus. 11-23—Tool rotation 3

The examples in Illustration 11-21 can have rotation opposite to those indicated. Most commonly, we identify tool rotation as clockwise or counter-clockwise while viewing the tool "through the motor."

Illustration 11-22 shows what could be any of several industrial machines including tenoners, feed-through moulders, or edge bander/tenoners.

In Illustration 11-23, the motor spindles are turning in left-hand rotation, and the workpieces are moving away from us; therefore, the examples labeled A are climb cutting with the direction of feed, while those labeled B are abrasive cutting against the feed.

Cutting Tool Materials

Tungsten Carbide Alloys

Several systemsfor classifying metals have evolved in this country and elsewhere. Trade asssociations, professional and governmental organizations, and private industries, among others, establish and use these numerical codes to identify the general contents and technical or commercial requirements that metal products must meet. One such system commonly used in Europe to identify materials for industrial applications is the DIN (*Deutsche Industrie Norma*) or German Industrial Standard.

The table below shows characteristics of the tungsten carbide alloys that most commonly make up wood cutting tools, according the DIN standards. The materials classified here generally conform to U.S. standards established by the American Society for Testing and Materials (ASTM)—specifically, ASTM B-406-70 for contents and ASTM B-294-76 for hardness.

Hardness is measured here on Rockwell scales A and C. The A scale results from testing metal with a diamond-tipped penetrator and a 150 kg load, the test used for the hardest materials; the C scale uses the diamond-tipped penetrator with only a 60 kg load.

In the table, alloy composition refers to percentages of wolfram (tungsten) carbide (WC) and cobalt (Co). The tools made from these alloys are either attached mechanically or brazed to steel carrier bodies for woodworking. Alloys K-01 through K-05 must be attached mechanically because the chemical process of brazing requires at least 5 percent cobalt in the alloy to join the metals of the tool and the carrier body together.

DIN No.	Alloy Composition (%)		Nominal Grain Size Range	Rockwell Hardness		Density	Transverse Rupture Strength
	WC	Co		C	A	(gms./cc)	(psi)
K-01	97	3	Fine	81	92.9	15.5	245,000
K-05	95.5	4.5	Fine	80.5	92.2	15.05	270,000
K-10	94	6	Ex. Fine	80	92.5	14.95	270,000
K-20	94	6	Medium	79	91.7	14.95	300,000
K-20	94	6	Coarse	78	90	14.90	320,000
K-30	93.5	6.5	Medium	76	88	14.80	320,000
K-30	91	9	Medium	76	90	14.65	360,000
K-40	89.5	10.5	Medium	72	88	14.5	380,000
K-40	88	12	Medium	71.5	87	14.3	400,000

Tungsten Carbide Alloys: Wear Characteristics

At right is a table comparing DIN standards with those of the American metal producer Multi-Metals with reference to wear characteristics of tools made of tungsten carbide. Because Multi-Metals products are trade protected, the exact content and manufacturing methods used to produce them are not readily available. Such products need only comply to performance standards specified by the patent holder or manufacturer and demonstrated during the patent examination and granting process.

The table, however, does give a general comparison from which we could extrapolate alloy contents and characteristics as specified by other manufacturers of similar materials for cutting tools.

Impact Resistance: lower (top) → higher (bottom)
Wear Resistance: higher (top) → lower (bottom)

Multi-Metals	DIN
OM-4	K-01
OM-3	K-05
XM-2	K-10
OM-2	K-20
1M-2	K-20
2M-2	
3M-2	
OM-1	K-30
1M-12	K-40
2M-12	
2M-12	
1M-13	K-40
2M-13	
1M-14	
2M-15	
1M-20	

Cutting Tool Purchasing Guidelines

The decisions involved in purchasing any cutting tools extend beyond the features and characteristics of the cutters themselves. Information on four elements of the cutting process is necessary for careful selection of tools:

A. The machine that will hold and drive the cutting tools:

1. Machine characteristics: type, size relative to maximum workpiece size or maximum tool diameter, horsepower, tool-holding method, arbor or tool-holding capacity, rotation, etc.;
2. Workpiece clamping, holding, or guiding system(s), manual or power assisted, etc.;
3. Machine process: cross cutting, rip cutting, shaping along grain, shaping across grain, or exactly what the machine cutting process is to accomplish;
4. Unusual machine characteristics: ability to tilt, to reverse, to accommodate exchangeable tool-holding devices, or other features and abilities;
5. Spindle speed—rpm, fixed or variable, etc.

B. The material that will be cut:

1. Solid wood specie, man-made material, panel material, or other;
2. The condition and physical properties of the material including strength, moisture content, or other pertinent information;
3. Physical dimensions—length, width, and thickness; if multiple, list all dimensions involved in the machine process;
4. Workpiece make-up such as surface coverings (veneered or plastic lamination), glue-block construction, or other details.

C. The cutting tool features:

1. The tool type and physical dimensions pertinent to the process or range of processes the tool is expected to accomplish, such as diameter, cutting width, shaft or arbor dimension, etc.;
2. The number of teeth, tooth spacing, desired feed rate, rim speed, relative to stock dimensions and physical properties of the workpiece material;
3. Tooth shape for parting tools such as circular or band saw blades;
4. Cutting tool rotation and direction of the cutting process—against or with the feed direction;
5. Tool knife alloy content—high speed steel, tungsten carbide, or other material as dictated by the abrasive or thermal cutting properties of the workpiece material to be cut.

D. The machine process:

1. Volume of production—ongoing, high production or infrequent use for a unique material or process;
2. True cost indices of the cutting tool —initial cost, service costs, and total operating life expectancy;
3. Demands on the tool and or tool system for the available dedicated machine and machine process, material to be cut or shaped, and finish quality, feed rates, and operating and total costs;
4. Careful evaluation of all machine processes and cutting tools capable of the required task(s) to be done.

Appendix

References

Hoadley, R. Bruce. (1989). *Understanding wood.* Newtown, CT: Taunton Press. *

Horton, Holbrook L., Jones, Franklin D., & Oberg, Erik. (1988). *Machinery's handbook.* New York: Industrial Press. *

Johnson, Carl G., & Weeks, William R. (1977). *Metallurgy.* Homewood, IL: American Technical Publishers. *

Koch, Peter. (1964). *Wood machining processes.* New York: Ronald Press. (Out of print, but available in microform from University Microfilms, Ann Arbor, MI)

Suchsland, Otto, & Woodson, Geo. E. (1986). *Fiberboard manufacturing practices in the United States* (Ag. Dept. Handbook #640). Washington, DC: U.S. Dept. of Agriculture-Forest Service. *

U.S. Forest Products Laboratory. (1974). *Wood handbook: Wood as an engineering material* (Ag. Dept. Handbook #72). Washington, DC: U.S. Govt. Printing Office. *

The wood book '83. (1983). San Francisco: Avery-Phares.

(Items marked with an asterisk (*) proved particularly helpful, and I recommend them for further study.)

Additional Reading

Allen, D. K. (1969). *Metallurgy theory and practice.* Homewood, IL: American Technical Publishers.

Barger, Vernon D., & Olsson, Martin G. (1973). *Classical mechanics: A modern perspective.* New York: McGraw-Hill.

Boothroyd & Knight. (1984). *Fundamentals of metal machining and machine tools.* New Yŏrk: Dekker.

Breneman, John W. (1960). *Mechanics.*New York: McGraw-Hill.

Bucksch, H. (1986). *Dictionary of wood and woodworking practice.* Holz-Woerterbuch.

Clason, A. (1988). *Elseviers' dictionary of metallurgy and metal working.* New York: Elsevier Press.

Griffiths, D. L., & Spence, W. P. (1981). *Woodworking tools, materials and processes.* Homewood, IL: American Technical Publishers.

Groneman, Chris H., & Glazner, J. (1981). *Technical woodworking.* New York: McGraw-Hill.

Hansen, Peter. (1986). *Physical metallurgy.* New York: Cambridge University Press.

Hibbler, Russell. (1985). *Mechanics for engineers.* New York: Macmillan.

Kane, G. (Ed.). (1982). *Modern trends in cutting tools.* Dearborn, MI: Society of Manufacturing Engineers.

Petrovich, J. (1990). *Custom tools for woodworkers.* Stockpole.

Sainhart, J. (Ed.). (1988). *Cutting tool material selection.* American Society of Tool Manufacturers and Engineers.

Sanderson H. (Ed.). (1985). "Progress in powder metallurgy." *Metal powder.*

Scott, Ernest. (1980). *Working in wood: An illustrated encyclopedia.* New York: Putnam Publishing Group.

Spence, Wm., & Griffiths, D. L. (1989). *The woodworkers benchtop reference.* Blue Ridge Summit, PA: Tab Books.

Summit, R., & Sliker A. (1980). *Handbook of materials science: Wood.* CRC Press.

Zimmerman, Fred W. (1981). *Exploring woodworking.* South Holland, IL: Goodheart-Willcox.

A Partial List of Cutting Tool Innovators

Amana Tool, USA
Byrd Tool
Cascade Tool
Cherokee Tool
DML/Lineberry Div.
Delta Corp.
Demps Saw
Dimter Tool—USA
Drake Corp.
Ekstrom, Carlson & Co.
Eagle America Corp.
Everlast Saw
Excel Saw
F-S Tool Corp.
Fantacci Spa—USA.
Freeborn Tool
Freud Tool—USA
Gladu Tool—USA
Graycon Tool Div.—Deihl
Great Lakes Tool
Greenlee/Forest City Tool
GUHDO—USA Tool
Henderson Saw
Hickory Saw
Howema Tool—USA
Industrial Carbide & Saw
G. Kohnle Tool—USA
Lach Diamond Tool—USA
L.R.H. Tool
Leiser-USA Tool
Leitz Tool—USA
Lemmon & Snoap Tool
Leuco Tool—USA
Markay Tool
Newman-Whitney Corp.
North American Products
Northfield Products
Oertli Corp.—USA
Omega Saw & Tool
Onsrud Tool
Pacific/Hoe Co.
Pacific/Western Machinery
Parsons Diamond Tool
Powermatic
Precision Tool
Precision Diamond Tool
Ridge Carbide Tools
Ross Carbide
Saber Diamond Tools
S & J Ind. Tool—Div.
Sanafrek Tool
Chas. G. Schmidt Co.
Stehle Tool—USA
Supreme Saw
SystiMatic Tool
Tel-Con Diamond
Tool-Pro Corp.
True-Cut Tool
U.S. Saw Corp.
Vits Mach. & Tool—USA
Wadkin—USA
L.A. Weaver Co.
Michael Weinig—USA
Western Cutterheads
H. Weigand Corp.
Wisconsin Knife Corp.
Yates-American Co.
York Saw

A special thanks to G.E. Specialty Products Division for years of work and research that have produced many of the unique materials in use today.

Professional Organizations

The American Furniture Manufacturers Association
The Architectural Woodworking Institute
Contract Manufacturers Association
Forest Products Research Society
Hardwood Manufacturers Assocation
Kitchen Cabinet Manufacturers Association
National Association of Store Fixture Manufacturers
National Dimension Manufacturers Association
National Forest Products Association—Canada
National Hardwood Lumber Association
National Kitchen and Bath Association
Wood Machinery Distributors Association
Wood Machinery Manufacturers of America
Wood Moulding & Millwork Producers Association
Wood Products Manufacturers Association
Woodworking Machinery Importers Association

A Woodworking Library

All woodworkers should note that one of the finest libraries of furniture in the U.S. is:

The Furniture Library
1009 N. Main St.
High Point, NC 27262
(919) 883-4011

Founded by curator Bernice Beinenstock, the library contains over 6,000 volumes and includes most significant works published in the field since 1640. Many have been duplicated and are offered for sale to the public. Among the library's rare book collection are the complete works of Chippendale, Sheraton, and Hepplewhite, and a complete set of *Diderot's Encyclopedias*. Request a catalog of published works offered for sale by writing or calling the number listed above.

Index